New Phenomena in Subnuclear Physics

―――――― *Part B* ――――――

THE SUBNUCLEAR SERIES

Series Editor: **ANTONINO ZICHICHI**
European Physical Society
Geneva, Switzerland

1. 1963   STRONG, ELECTROMAGNETIC, AND WEAK INTERACTIONS
2. 1964   SYMMETRIES IN ELEMENTARY PARTICLE PHYSICS
3. 1965   RECENT DEVELOPMENTS IN PARTICLE SYMMETRIES
4. 1966   STRONG AND WEAK INTERACTIONS
5. 1967   HADRONS AND THEIR INTERACTIONS
6. 1968   THEORY AND PHENOMENOLOGY IN PARTICLE PHYSICS
7. 1969   SUBNUCLEAR PHENOMENA
8. 1970   ELEMENTARY PROCESSES AT HIGH ENERGY
9. 1971   PROPERTIES OF THE FUNDAMENTAL INTERACTIONS
10. 1972   HIGHLIGHTS IN PARTICLE PHYSICS
11. 1973   LAWS OF HADRONIC MATTER
12. 1974   LEPTON AND HADRON STRUCTURE
13. 1975   NEW PHENOMENA IN SUBNUCLEAR PHYSICS

Volume 1 was published by W. A. Benjamin, Inc., New York; 2-8 and 11-12 by Academic Press, New York and London; 9-10 by Editrice Compositori, Bologna; 13 by Plenum Press, New York and London.

New Phenomena in Subnuclear Physics

―――― Part B ――――

Edited by
Antonino Zichichi
European Physical Society
Geneva, Switzerland

PLENUM PRESS • NEW YORK AND LONDON

Library of Congress Cataloging in Publication Data

International School of Subnuclear Physics, Erice, Italy, 1975.
New phenomena in subnuclear physics, part B.

(Subnuclear series)
Includes index.
1. Particles (Nuclear physics) — Congresses. I. Zichichi, Antonino. II. Sicily. III. Weizmann Institute of Science. IV. Title. V. Series.
QC793.I555 1975   539.7'21   77-6439
ISBN 0-306-38182-6 (part B)

Proceedings of the second half of the 1975 International School of
Subnuclear Physics (NATO-MPI-MRST Advanced Study Institute) held in
Erice, Sicily, July 11—August 1, 1975 and sponsored by the
Sicilian Government and the Weizmann Institute of Science

© 1977 Plenum Press, New York
A Division of Plenum Publishing Corporation
227 West 17th Street, New York, N.Y. 10011

All rights reserved

No part of this book may be reproduced, stored in a retrieval system, or transmitted,
in any form or by any means, electronic, mechanical, photocopying, microfilming,
recording, or otherwise, without written permission from the Publisher

Printed in the United States of America

Preface

In July 1975 a group of 122 physicists from 68 laboratories of 27 countries met in Erice to attend the 13th Course of the International School of Subnuclear Physics.

The countries represented at the School were: Australia, Austria, Belgium, Brazil, Canada, Chile, Denmark, France, Germany, Greece, India, Iran, Israel, Italy, Japan, Mexico, The Netherlands, Norway, Poland, Portugal, Spain, Sweden, Switzerland, Turkey, The United Kingdom, The United States of America and Yugoslavia.

The School was sponsored by the Italian Ministry of Public Education (MPI), the Italian Ministry of Scientific and Technological Research (MRST), the North Atlantic Treaty Organization (NATO), the Regional Sicilian Government (ERS) and the Weizmann Institute of Science.

The School was one of the most exciting, due to the impressive number of discoveries made not only in the field of the new particles by the MIT-BNL (reported by S. C. C. Ting) and by the SLAC-SPEAR (reported by M. Breidenbach) Groups, but also in the field of high energy neutrino interactions where Carlo Rubbia observes μ-pairs, together with bumps in the total energy of the hadronic system at $W_h \sim 4$ GeV and a discontinuity in the $<y>$ at $E_\nu \sim 50$ GeV plus a bump at $W_{min} \simeq 4$ GeV; all these phenomena being possibly connected.

To this remarkable amount of new and exciting results it has to be added the great discovery of DORIS (reported by B. Wiik) on the first example of a new particle P_c: the highlight of the Course. Needless to say that it was too easy this year to have discussions of great interest – the atmosphere of the School being such that, even in much more sober years of meagre discoveries, it was possible to have interesting discussions.

No doubt: a new era has been opened in particle physics; and this is the first volume of it.

At various stages of my work I have enjoyed the collaboration of many friends whose contributions have been extremely important for the School and are highly appreciated. I would like to thank most warmly: Dr. N. Craigie, Mrs. M. Denzler, Dr. A. Gabriele, Mrs. C. Giusti, Mrs. H. Kirk, Miss P. Savalli, Mrs. S. Vascotto, Mrs. K. Wakley, and Miss M. Zaini for the general scientific and administrative work, and Drs. R. K. Ellis and R. Petronzio for their work as Scientific Secretaries.

A final word of acknowledgment to all those who, in Erice, Bologna and Geneva, helped me on so many occasions and to whom I feel very much indebted.

A. Zichichi

Geneva, January 1976

Contents of Part A

Opening Ceremony

One Day All Men Will be Scientists 1
   A. Zichichi

Address by the Chairman of the CCSEM Scientific
       Advisory Committee 7
   I. I. Rabi

Address by H. E. the Minister of Science and
       Technology 11
   M. Pedini

Opening Lecture

Quarks: from Paradox to Myth 13
   K. G. Wilson

Theoretical Lectures

Charm Spectroscopy 33
   L. Maiani

Quarks and Strings on a Lattice 69
   K. G. Wilson

Quark Confinement Schemes in Field Theory 143
   S. D. Drell

Hadrodynamics with the Elusive Quarks 193
   G. Preparata

The MIT Bag 1975 241
   V. F. Weisskopf

Gauge Theory for Strong Interactions 261
    G. 't Hooft

Classical Lumps and Their Quantum Descendants 297
    S. Coleman

The Phase and the Modulus of the Scattering
            Amplitude 423
    A. Martin

Weak Non Leptonic Amplitudes in Unified Gauge Theories . . . 465
    G. Altarelli

Relationship Between Gauge Field Theories and
            Dual Resonance Models 493
    P. H. Frampton

The Small Oscillations and the Relativistic Bag 533
    C. Rebbi

Index xiii

Contents of Part B

Seminars on Specialized Topics

J Particles, and Search for More Long
       Lived Particles 559
  S. C. C. Ting

Scattering of Muons at 150 GeV 589
  R. Wilson

Properties of the ψ Resonances 609
  M. Breidenbach

Evidence for a New Resonance P_c and other Recent
       Results obtained at DORIS using DASP 635
  B. Wiik

Studies of Neutron Dissociation at FERMILAB Energies 663
  T. Ferbel

Acceleration of Heavy Ions to Relativistic Energies
       and their Use in Physics and Biomedicine 703
  M. G. White

Mini Black Holes 731
  S. Frautschi

Review Lectures

A Review of Recent Progress and Old Problems in
       Subnuclear Physics 741
  A. Zichichi

A Survey of the Theoretical Models proposed
       for the New Particles 803
  G. Morpurgo

New Particle Production by Neutrinos 865
  C. Rubbia

Recent Results from the CALTECH-FERMILAB
          Neutrino Experiment 897
  B. C. Barish

Review of Direct Lepton Production in
          Nucleon-Nucleon Collisions 929
  J. W. Cronin

Methods for Theoretical Understanding of Neutral
          Currents 967
  M. Gourdin

The Status of Non-Charmed Hadron Spectroscopy 1053
  R. H. Dalitz

Very High Energy Hadronic Interactions 1067
  A. N. Diddens

Results from Studies of High Energy Cosmic Radiation 1135
  C. B. A. McCusker

Highlights in Other Fields

Fermion Systems in Different Dimensions 1151
  L. N. Cooper

The Future of High Energy Physics in Europe

Possible Future Storage Rings at CERN: pp and ep 1173
  K. Johnsen

The Glorious Days of Physics

My Life as a Physicist's Wife 1183
  L. Fermi

Evening Seminar

Some Conclusions from CERN's History 1201
  L. Kowarski

CONTENTS OF PART B

Closing Ceremony 1213

Participants 1217

Index 1229

J PARTICLES, AND SEARCH FOR MORE LONG LIVED PARTICLES

Samuel C.C. Ting

Department of Physics and Laboratory for Nuclear Science,
Massachusetts Institute of Technology, Cambridge, MA, USA

INTRODUCTION

There have been many theoretical speculations [1] on the existence of long lived neutral particles with a mass larger than 10 GeV/c^2 which play the role in weak interactions that photons play in electromagnetic interactions. There is, however, no theoretical justification, and no predictions exist, for long lived particles in the mass region 1-10 GeV/c^2.

Even though there is no strong theoretical justification for the existence of long lived particles at low masses, there is no experimental indication that they should not exist. Until last year no high sensitivity experiment had been done in this mass region.

There are calculations based on parton models on the production of an e^+e^- continuum from pp interactions [2]. An early experiment at the AGS [3] studied the continuum of $\mu^+\mu^-$ from p + Uranium → $\mu^+\mu^-$ + X. This experiment gives approximately the size of $\mu^+\mu^-$ yield. In the last ten years there have been many experiments [4] at Brookhaven, at CERN I.S.R., at Fermi Lab, etc. to study the inclusive e (μ) production P+P → e(μ) + X. Again, these experiments gave no indication of a long lived particle.

My talk will consist of two parts:
1. The Discovery of the J Particle
2. The Origin of the J Particle

1. DISCOVERY OF THE J PARTICLE

The discovery of the J Particle [5,6,7] in proton-proton collisions by the MIT-BNL group at Brookhaven National Laboratory follows a decade of experiments associated with e^+e^- pair productions from hadron interactions at high energies. One learns three kinds of physics from the reaction

$$h + p \to e^+e^- + X.$$

i) Using a 7.5 GeV bremsstrahlung photon beam one can compare the e^+e^- yield with predictions of QED at large momentum transfer or small distances, $<10^{-14}$ cm [8].

ii) One can study the e^+e^- decay of photon-like particles with spin 1 and negative parity and charge conjugation, such as the ρ,[9] ϕ,[10] ω,[11] and ρ'[12] and measure the coupling strengths between photons and massive photon-like particles [13]. One can also study the production mechanisms of these photon-like particles produced by photons.

iii) Search for additional particles which decay to e^+e^- from $pp \to e^+e^- + X$ or $pp \to \mu^+\mu^- + X$.

1.1 Design Considerations

To perform a high sensitivity experiment, detecting narrow width particles over a wide mass region, we make the following four observations:

i) Since the e^+e^- comes from photon decays, the yield of e^+e^- is lower than hadron pairs ($\pi^+\pi^-$, K^+K^-, pp, K^+p, etc.) by a factor

$$\frac{\alpha^2}{m^4} F^2(m^2) \approx 10^{-6}.$$

The factor α^2 comes from the virtual photon decay, m^{-4} is the photon propagator and $F(m^2)$ the form factor of the target proton, where m is the invariant mass of the e^+e^- pair.

ii) Because of (i) one must design a detector which can stand a high flux of hadrons to obtain a sufficient yield of e^+e^- pairs.

iii) The detector must be able to reject hadron pairs by a factor $\sim 10^6 - 10^8$.

iv) In choosing the best kinematic region to detect the decay of new particles, one notes that at high energies, inclusive production of ρ, π and ω from p-p interactions can all be described in the c.m. system by a dependence of the form

$$\frac{d^3\sigma}{dp_\parallel^* dp_\perp^{*2}} = \frac{ae^{-bp_\perp^*}}{E^*}, \text{ independent of } p_\parallel^*,$$

where $p_\parallel^*$, $p_\perp^*$ and E^* have their usual meaning.

Thus the maximum yield will occur when the particle is produced at rest in the center of mass. If we look at the 90° decay of the e^+e^- pair, we note that they emerge at an angle $\theta = \arctan(\frac{1}{\gamma}) = 14.6°$ in the lab system for an incident proton energy of 28.5 GeV, independent of the mass of the decaying particle.

1.2 Experimental Set-Up

Figure 1 shows the plan and side views of the spectrometer and detectors. Bending is done vertically to decouple angle (θ) and momentum. The field of the magnets in their final location was measured with a three dimensional Hall probe at a total of 10^5 points C_B, C_O and C_e are gas threshold Cerenkov counters. C_B is filled with isobutane at 1 atm., C_O is filled with hydrogen at 1 atm. and C_e is filled with hydrogen at 0.8 atm. A_O, A, B and C are proportional wire chambers with 2 mm spacing and a total of 4,000 wires on each arm. Behind chambers A and B are situated two planes of hodoscopes, 8x8, for improved timing resolution.

Behind the C chamber there are two orthogonal banks of lead glass counters of three radiation lengths each, the first containing twelve elements, the second thirteen, followed by one horizontal bank of lead lucite shower counters, seven in number, each ten radiation lengths thick, to further reject hadrons from electrons and improve track identification.

The following are the unique features of this experiment:

i) To obtain a rejection against hadrons of 10^8 or better, the two gas Cerenkov counters in each arm, C_O and C_e

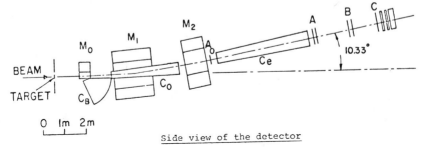

Side view of the detector

Fig. 1b

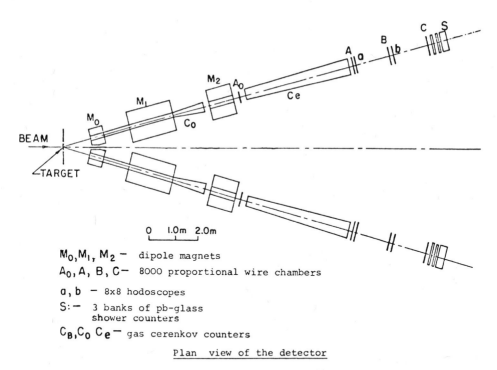

M_0, M_1, M_2 — dipole magnets
A_0, A, B, C — 8000 proportional wire chambers
a, b — 8×8 hodoscopes
S : — 3 banks of pb-glass shower counters
C_B, C_0, C_e — gas cerenkov counters

Plan view of the detector

Fig. 1a

(Fig. 2a, b) are filled with hydrogen and made with thin mylar windows to reduce knock-on and scintillation effects. The counters are painted black inside and are decoupled by the strong magnetic fields of M_1 and M_2, so that knock-on electrons produced in C_o do not enter C_e and only electrons along the beam trajectory will emit Cerenkov light which is focussed onto the photomultiplier tube. Special high gain, high efficiency phototubes of the type RCA C31000M were used so that the counters C_o and C_e can be operated at 100% efficiency with very low voltage. The counter C_o collects an average of 9 photoelectrons. To ensure the voltage was set on a single electron and not on e^+e^- pairs from π^o's, which would give ~18 photoelectrons, the counter C_o was filled with He and the location of the single photoelectron peak was found (Fig. 2c).

ii) To be able to handle a high intensity of 2×10^{12} protons per pulse with consequent single arm rates of ~20 MHz, there are eleven planes of proportional wires (2xA_0, 3xA, 3xB, and 3xC) rotated 20° with respect to each other as shown in Fig. 3a to reduce multitrack ambiguities. To ensure the chambers have a 100% uniform efficiency at low voltage (Fig. 3b) and a long life time in the highly radioactive environment, a special Argon-Methylal mixture at 2° C was used.

iii) To reduce multiple scattering and photon conversion, the material in the beam is reduced to a minimum. The front and rear windows of C_o are 125 μm thick respectively, both mirrors of C_o and C_e are made of 3 mm black lucite and hodoscopes are 1.6 mm thick.

The thickness of one piece of Beryllium target is 1.8 mm and the nine pieces are each separated by 7.5 cm so that the particles produced in one piece and accepted by the spectrometer do not pass through the next piece.

iv) To reduce photon and neutron contamination the location of all the hodoscopes and lead glass counters are such

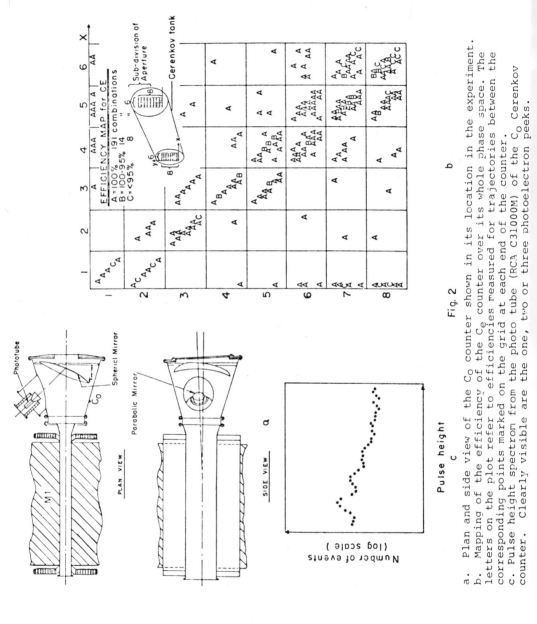

Fig. 2

a. Plan and side view of the C₀ counter shown in its location in the experiment.
b. Mapping of the efficiency of the Ce counter over its whole phase space. The letters on the plot refer to efficiencies measured for trajectories between the corresponding points marked on the grid at each end of the counter.
c. Pulse height spectrum from the photo tube (RCA C31000M) of the C₀ Cerenkov counter. Clearly visible are the one, two or three photoelectron peaks.

J PARTICLES, AND SEARCH FOR MORE LONG LIVED PARTICLES

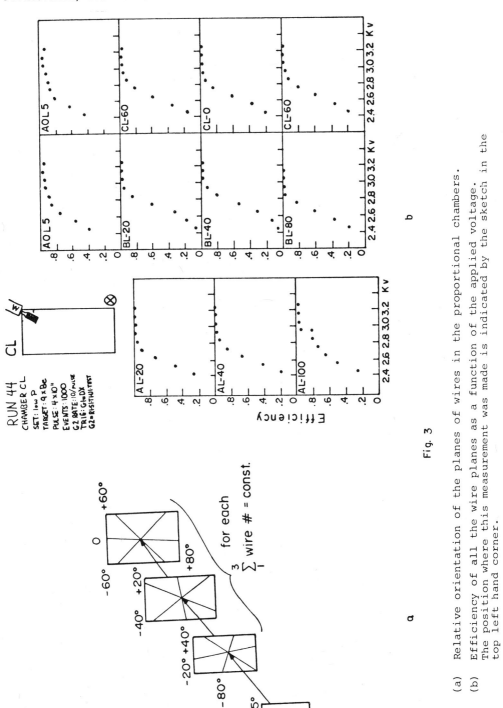

Fig. 3

(a) Relative orientation of the planes of wires in the proportional chambers.
(b) Efficiency of all the wire planes as a function of the applied voltage. The position where this measurement was made is indicated by the sketch in the top left hand corner.

that they do not see the target directly. To further shield the detectors from soft neutrons, 10,000 lbs. of Borax soap was placed on top of C_O between M_1 and M_2 and around the front part of C_e behind M_2.

v) To improve the rejection against $\pi^o \to \gamma\, e^+ e^-$, a very directional Cerenkov counter C_B was placed close to the target and below a specially constructed magnet M_O (Fig. 4a). This counter is painted black inside and is sensitive to electrons of 10 MeV/c and pions above 2.7 GeV/c. The coincidence between C_B and C_O, C_e, the shower counters and the hodoscopes indicates the detection of an $e^+ e^-$ pair from the process $\pi^o \to \gamma\, e^+ e^-$, and such events are rejected. A typical plot of the relative timing of this coincidence is shown in Fig. 4b. Conversely, one can trigger on C_B and provide a pure electron beam to calibrate C_O, C_e and the shower counters. This is a redundant check in setting the voltage of the C_O counters, since the coincidence between C_O and C_B will ensure that the counter is efficient for a single electron and not a zero degree pair.

vi) The spectrometer has a very large mass acceptance of 2 GeV/c^2 and enables us to study the mass region 1.5 - 5.5 GeV/c^2 in three overlapping settings. For a point target the acceptance in θ is $\pm 1°$, in ϕ it is $\pm 2°$ and in momentum it varies from $0.6 \times p_o$ to $1.8 \times p_o$ (where p_o is the principal axis momentum), all in the lab. system.

The following table summarizes the important properties of the MIT-BNL experiment:

Unique Features

1. Can stand incident flux of 2×10^{12} protons/pulse or 20 MC single rates.
  $\sigma\, B \longrightarrow 10^{-36}\, cm^2$.
2. Sort out 8 tracks per arm.
3. Rejection $\dfrac{e^- e^+}{h^- h^+} \ll 10^{-8}$.

J PARTICLES, AND SEARCH FOR MORE LONG LIVED PARTICLES

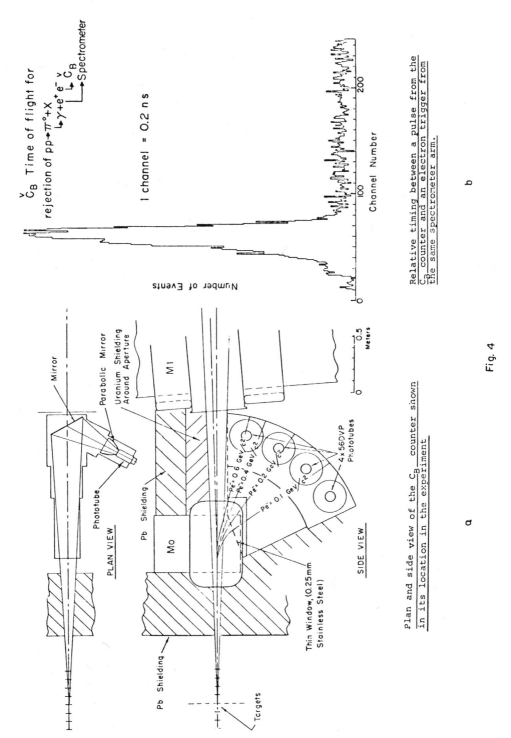

Fig. 4

a. Plan and side view of the C_B counter shown in its location in the experiment

b. Relative timing between a pulse from the C_B counter and an electron trigger from the same spectrometer arm.

    4. Mass resolution ± 5 MeV.
    5. Mass acceptance 2 BeV.

1.3 First Results

The first data from August 1974 are shown in Fig. 5a. There is a clear, sharp enhancement at a mass of 3.1 GeV/c^2.

To ensure that the observed peak is a real particle, from August to November experimental checks were made on the data. I list six examples:

i) The magnet currents were decreased by 10%, the peak remained fixed at 3.1 GeV/c^2 (see Fig. 5a).

ii) To check second order effects on the target, the target thickness was increased by a factor of two. The yield increased by a factor of two, not by four.

iii) To check for pile up in the lead glass and shower counters, different runs were made with different voltage settings on the counters. No effect was observed on the yield of J.

iv) To insure that the peak is not due to scattering from the sides of the magnets, cuts were made in the data to reduce the effective aperture. No significant reduction in the J yield was found.

v) To check the read-out system of the chambers and the triggering system of the hodoscopes, runs were made with a few planes of chambers deleted and with sections of the hodosopces omitted from the trigger. No effect was observed on the J yield.

vi) Runs with different beam intensity were made and the yield did not change.

These and many other checks convinced us that we have observed a real massive particle $J \rightarrow e^+e^-$.

Partial analysis of the width of the J particle shown in Fig. 5b indicates it has a width smaller than 5 MeV/c^2.

If we assume a production mechanism for J to be

$$\frac{d^3\sigma}{dp_\perp^{*2} dp_\parallel^*} = \frac{e^{-6p_\perp^*}}{E^*} \text{ , independent of } p_\parallel^*$$

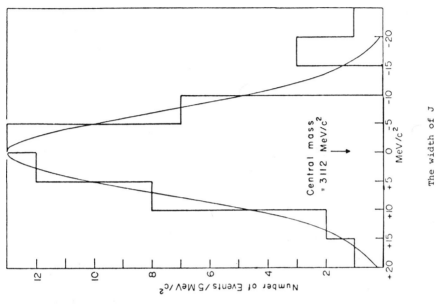

Mass spectrum for events in the mass range 2.5 < m_{ee} < 3.5 GeV/c^2. The shaded events correspond to those taken at the normal momentum setting, while the unshaded ones correspond to a momentum setting 10% below normal. The acceptance is a smooth function of m.

The width of J

Fig. 5

and an isotropic decay in the rest system of the J, we obtain a J → e^+e^- yield of 10^{-34} cm^2/nucleon at 28.5 GeV.

Fig. 6 shows the yield of e^+e^- in the region 3.2 to 4.0 GeV/c^2, normalized to Fig. 5a. The acceptance in this region is a smooth function and varies at most by a factor of two. The observed events are consistent with purely random coincidences. To a level of 1% of the J yield, with a confidence level of 95%, no heavier J particles were found. We note that this upper limit is independent of any production mechanism of the J, we obtain an upper limit of 10^{-36} cm^2/nucleon for the production of heavier J's with a 95% confidence level.

2. THE ORIGIN OF THE J PARTICLE

There have been hundreds of theoretical papers appearing in journals on the new particles.

The model of Glashow[14] views the J particle as a bound (Ortho) state of Charm (C) and Anti-Charm. In this model there would be a (Para) state J → $\bar{p}p$ near 3.1 GeV. Furthermore, the charmed particle C → Kπ should be observed in the mass region 1.5 - 2.5 GeV.

The model of Yang and Wu[14], which views the J particle production from p p reaction to be via p + p → J θ + X to conserve the possible additional quantum numbers of the J particle, would predict a long lived θ → π^-p particle near the mass of 2 GeV.

Independent of theories, one can ask two important experimental questions.

How many other narrow resonances exist?
To answer these questions the MIT group has just completed a systematic search of

$$p + p \rightarrow \quad \pi^-p + X \quad \quad A$$
$$\pi^+\pi^- + X$$
$$\bar{p}p + X$$
$$K^-p + X \quad \quad B$$
$$K^+\pi^- + X$$

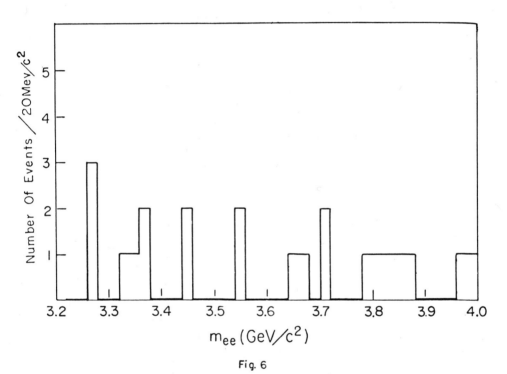

Fig. 6

Mass spectrum for events in the mass range $3.2 < m_{ee} < 4.0$ GeV/c^2, normalized to fig. 5a.

$$K^+K^- + X$$
$$K^-\pi^+ + X \quad\quad C$$
$$K^{\pm}\bar{p} + X$$
$$\pi^+p^- + X$$

in the pair mass region 1.2 - 5.0 BeV with a mass resolution of 5 MeV.

The experimental set up for this experiment is very similar to the original e^+e^- experiment. However since there are much more hadrons than electrons the random accidentals are more serious. To reduce the accidentals to minimum a new target system was put in. This consists of 5 pieces of 4 mm x 4 mm x 4 mm Be target separated from each other by six inches. The targets are supported by thin piano wires. This arrangement enables one to locate the point of intersection between the two trajectories. Comparing it with the target location enables us to reduce the random accidentals (Fig. 7a). To further reduce the accidentals additional scintillation counters were installed to tighten the two arm coincidence to 0.9 ns (Fig. 7b). Two high pressure (300 psi) Cerenkov counters were installed replacing the shower counters to identify K's. The counters C_e (Fig. 1) were filled with 1 atm. isobutane to identify π's. The Cerenkov counters set the mass acceptance to $\approx$1 BeV. In this way all 9 combinations were measured simultaneously. To avoid systematic errors, 6 overlapping magnet settings were made for the measurement.

Figures 8-16 show the results of the 9 reactions. Without acceptance corrections the yield increases with mass due to an increase in acceptance and then decreases due to the decrease in production cross sections. As seen there are no sharp narrow resonances in any of the 9 reactions. There may be of course very wide "ordinary" resonances with widths of 300 MeV or more. A search for these depends on exact calculations of acceptance and has not yet been made. To obtain a feeling of the sensitivities of the measurement,

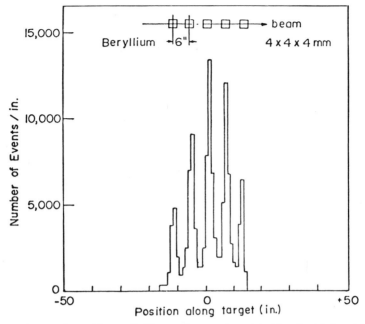

a. Reconstruction of the pair vertex at the target; using information from the proportional chambers. The five pieces of beryllium are seen clearly.

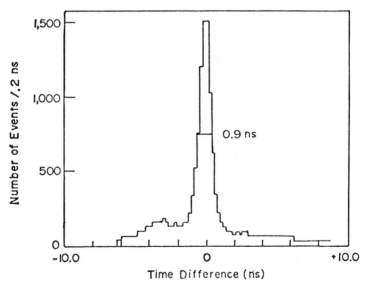

b. Time difference between additional scintillation counters in the left and right arms. The resolution obtained is 0.9ns and little background is present.

Fig. 7

we take the production of the 9 reactions to be the same as the assumed J production mechanisms. From this we obtain the following table:

Sensitivity (cm^2)
For Narrow Resonances

h^+X^- \ m	2.25 GeV	3.1 GeV	3.7 GeV
$\pi^+\pi^-$	8 × 10^{-33}	5 × 10^{-34}	3 × 10^{-35}
π^+K^-	1 × 10^{-33}	4 × 10^{-35}	1 × 10^{-35}
$\pi^+\bar{p}$	2 × 10^{-33}	4 × 10^{-35}	7 × 10^{-36}
$K^+\pi^-$	4 × 10^{-33}	8 × 10^{-35}	4 × 10^{-35}
K^+K^-	1 × 10^{-33}	5 × 10^{-35}	1 × 10^{-35}
$K^+\bar{p}$	2 × 10^{-33}	4 × 10^{-35}	8 × 10^{-36}
$p\pi^-$	4 × 10^{-32}	4 × 10^{-33}	5 × 10^{-34}
$p K^-$	7 × 10^{-33}	4 × 10^{-34}	3 × 10^{-35}
$p \bar{p}$	-	4 × 10^{-34}	2 × 10^{-35}

Whereas the spectra do not show any sharp resonance states the cross sections $\frac{d\sigma}{dm}$ vs m for groups A (π^-p), B ($\pi^+\pi^-$, $\bar{p}p$, $K^+\pi^-$, K^-p) and C (K^+K^-, $K^+\bar{p}$, $\pi^+\bar{p}$, $K^-\pi^+$) do exhibit some simple degeneracies above the mass of J. The cross section for each group decreases with m as $\approx e^{-4m}$ and differs from each other by an order of magnitude.

To search for multibody final states additional reactions;

$$p + Be \rightarrow \pi^-\pi^- + ...$$
$$K^-K^- + ...$$
$$\bar{p}\,\bar{p} + ...$$
$$K^-\bar{p} + ...$$
$$\pi^-\bar{p} + ... \quad \text{were measured.}$$

When one measures the two particle spectrum to search for multibody decays like C → $K^-\pi^+\pi^-\pi^+$, etc. one would expect a discontinuity in the $K^-\pi^+$ spectrum near the mass of C:

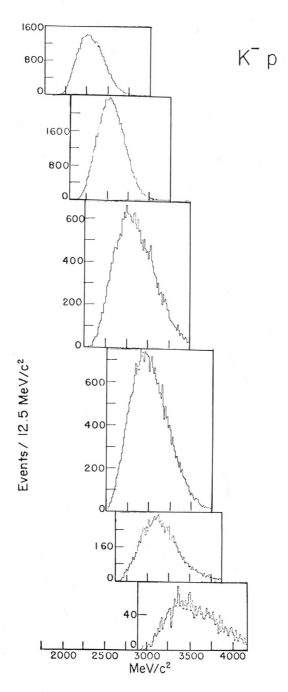

Fig. 8. K⁻p mass spectra for five mass ranges from MIT-BNL.

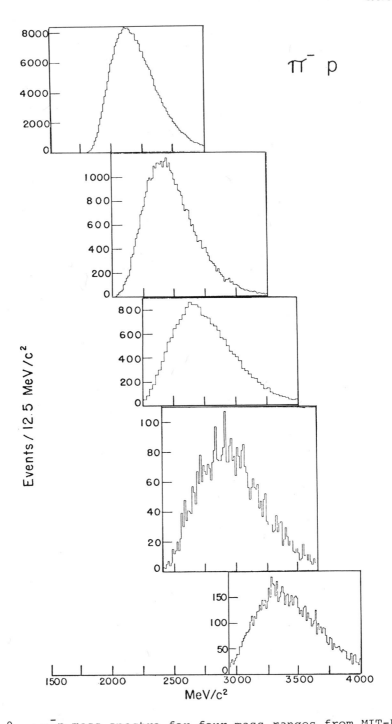

Fig. 9. $\pi^- p$ mass spectra for four mass ranges from MIT-BNL.

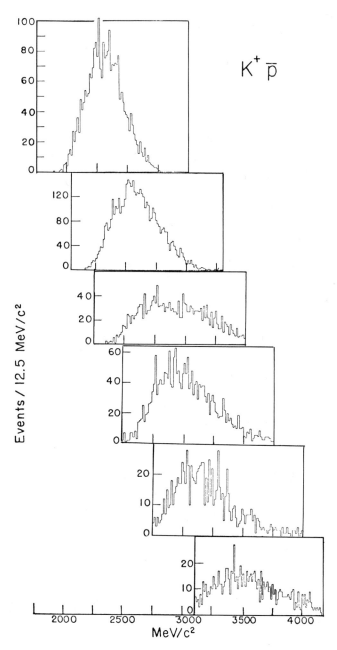

Fig. 10. $\bar{p}$ K^+ mass spectra for five mass ranges from MIT-BNL.

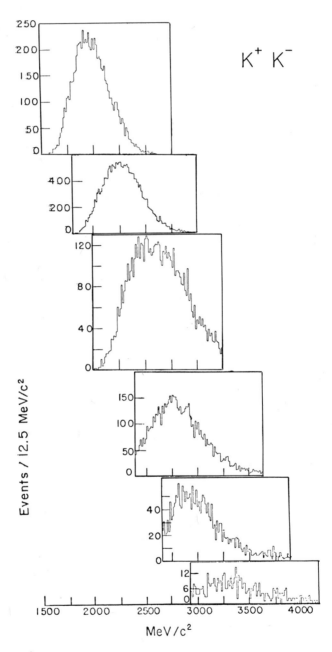

Fig. 11. K^-K^+ mass spectra for five mass ranges from MIT-BNL.

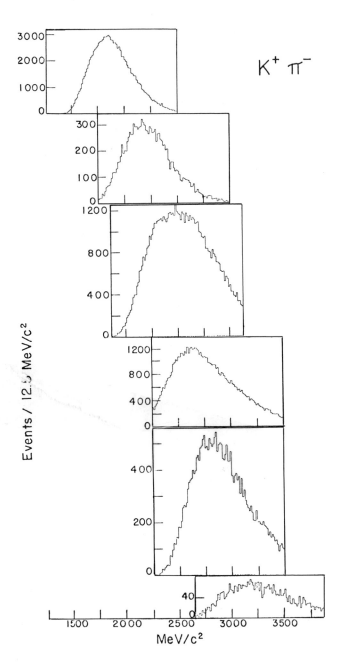

Fig. 12. $\pi^- K^+$ mass spectra for five mass ranges from MIT-BNL.

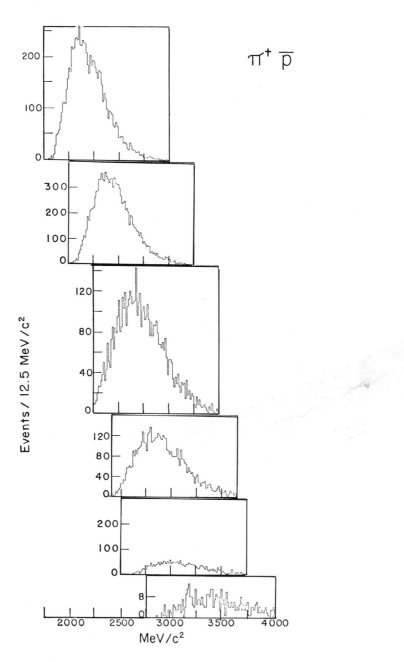

Fig. 13. $\bar{p}\pi^+$ mass spectra for five mass ranges from MIT-BNL.

J PARTICLES, AND SEARCH FOR MORE LONG LIVED PARTICLES

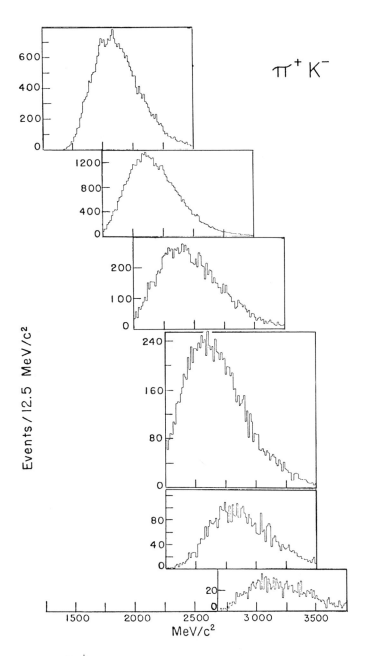

Fig. 14. $K^-\pi^+$ mass spectra for five mass ranges from MIT-BNL.

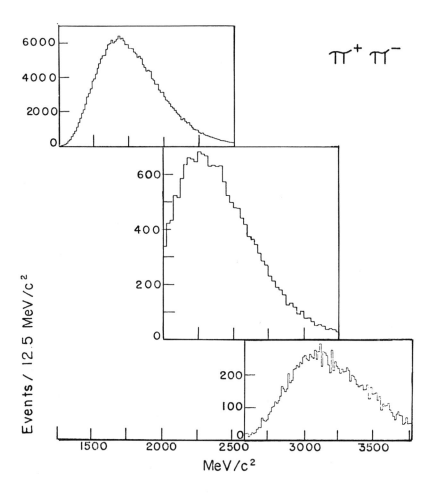

Fig. 15. $\pi^-\pi^+$ mass spectra for two mass ranges from MIT-BNL.

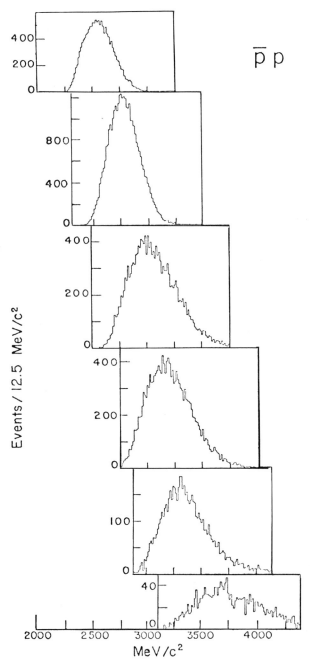

Fig. 16. p̄ p mass spectra for five mass ranges from MIT-BNL.

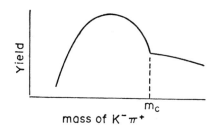

mass of $K^-\pi^+$

ACKNOWLEDGEMENT

I wish to thank Professor A. Zichichi for his hospitality during the Summer School.

* * *

REFERENCES

1) T.D. Lee, Phys. Rev. Lett. <u>26</u>, 801 (1971);
S. Weinberg, Phys. Rev. Lett. <u>19</u>, 1264 (1967); Phys. Rev. Lett. <u>27</u>, 1688 (1971); Phys. Rev. Lett. <u>D5</u>, 1412 (1972) and Phys. Rev. <u>D5</u>, 1962 (1972);
A. Salam, in Elementary Particle Theory, edited by N. Svartholm (Almquist and Farlag, Stockholm, 1968).

2) S.D. Drell and T.M. Yan, Phys. Rev. Lett. <u>25</u>, 316 (1970).

3) J.H. Christenson et al., Phys. Rev. Lett. <u>25</u>, 1523 (1970).

4) R. Burns et al., Phys. Rev. Lett. <u>15</u>, 830 (1965);
J.P. Boymond et al., Phys. Rev. Lett. <u>33</u>, 112 (1974);
J.A. Appel et al., Phys. Rev. Lett. <u>33</u>, 722 (1974);
F.W. Buesser et al., Phys. Rev. Lett. <u>53B</u>, 212 (1974);
L.B. Leipuner et al., Phys. Rev. Lett. <u>34</u>, 103 (1975).

5) J.J. Aubert et al., Phys. Rev. Lett. <u>33</u>, 1404 (1974).

6) J.E. Augustin et al., Phys. Rev. Lett. <u>33</u>, 1406 (1974).

7) C. Bacci et al., Phys. Rev. Lett. <u>33</u>, 1408 (1974).

8) J.G. Asbury et al., Phys. Rev. Lett. <u>18</u>, 65 (1967);
H. Alvensleben et al., Phys. Rev. Lett. <u>21</u>, 1501 (1968).

9) J.G. Asbury et al., Phys. Rev. Lett. <u>19</u>, 869 (1967).

10) U. Becker et al., Phys. Rev. Lett. <u>21</u>, 1504 (1968).

11) H. Alvensleben et al., Phys. Rev. Lett. <u>25</u>, 1373 (1970);
P. Biggs et al., Phys. Rev. Lett. <u>24</u>, 1197 (1970).

12) G. Barbarino et al., Lett. al Nuovo Cimento 3, 693 (1972).

13) J.G. Asbury et al., Phys. Rev. Lett. 19, 865 (1967);
H. Alvensleben et al., Phys. Rev. Lett. 24, 786 (1970);
H. Alvensleben et al., Phys. Rev. Lett. 28, 66 (1972);
K. Gottfried, 1971 International Symposium on Electron/Photon Interactions at High Energies, Cornell (1971);
see references listed therein.

14) H.T. Nieh, T.T. Wu, and C.N. Yang, Phys. Rev. Lett. 34, 49 (1975);
A. de Rujula and S.L. Glashow, Phys. Rev. Lett. 34, 46 (1975);
C.G. Callam, R.L. Kingsley, S.B. Treiman, F. Wilczek and A. Zee, Phys. Rev. Lett. 34, 52 (1975);
J.J. Sakurai, Phys. Rev. Lett. 34, 56 (1975).

DISCUSSION

CHAIRMAN: S.C.C. Ting

Scientific Secretaries: F. Dydak and R. Rückl

DISCUSSION

- *DYDAK:*

Can the degeneracies in the double inclusive cross-section $d\sigma/dm$ versus pair mass be understood in terms of the inclusive single particle production yields?

- *CRONIN:*

Taking single particle production, and assuming the same correlation between the particles, one gets roughly the observed effect. With regard to the like-charge data, one would expect the same cross-section as for the opposite-charge data. I would bet that those like-charge data do not agree.

- *TING:*

The data for the $\pi^+\pi^-$, $\pi^+\pi^+$, and $\pi^-\pi^-$ cross-sections are fairly close, but they are not exactly the same.

- *FERBEL:*

The correlations of the particle production have to be taken into account.

- *CRONIN:*

For a relative comparison of the cross-sections they should not play a substantial role.

- *MILLER:*

Is $x = 0$ and $p_T = 0$ a saddle point of the J-production cross-section?

- *TING:*

We have measured the p_T dependence of the cross-section to be $\exp(-1.6 \times p_T^2)$. The x-dependence has not yet been done.

- *DRELL:*

Do you see any structure at 4.1 GeV/c^2?

- *TING:*

We have no evidence for any structure besides the 3.1 GeV/c^2 particle. However, in the region 4 to 5 GeV/c^2, so far we have only a few events.

SCATTERING OF MUONS AT 150 GEV

Richard Wilson

Harvard University, Cambridge, MA, USA

The muon and electron are, as far as we now know, indistinguishable except for their masses and a special quantum number. The mass makes the muon unstable, and this has led Feynman to ask his famous question--why is the muon? it seems to have no purpose in the scheme of things.

Many of the experiments with muons, therefore, have been directed to trying to find another difference. These have mostly been in the realm of limits of quantum electrodynamics, which might be different for muons and electrons. In this class we can include, for example, the g-2 experiments. More usual are the comparison of electron and muon scattering, both elastic and inelastic. The latest muon experiments of this type were done at A.G.S. with a muon beam of 7 GeV and the scattering at 4 momentum transfers (q^2) up to $4(GeV/c)^2$ was compared with SLAC data to about 5%. Alternatively, we can study time-like momentum transfers with electron-positron annihilation, and again muon and electron production agree with predictions of Q.E.D. to about 5% at momentum transfers which at SPEAR go beyond $q^2 = 30(GeV/c)^2$. The electron and muon both seem to be point particles with spin 1/2.

This talk, however, will be concerned with the use of muons as probes of nucleon structure, in similar ways to our use of electrons. One of the first uses of muons, by Fitch and Rainwater, was just such a use; muons, because of their heavy mass, can displace electrons in muonic atoms; their Bohr orbits enter the nucleis and a measurement of their energy levels allows a probe of nuclear structure.

The results confirm and agree with electron scattering results to a percent, although a few discrepancies exist. At multi-GeV energies there are more powerful reasons for using muons. The highest energy electron accelerator is SLAC at 20 GeV, and the cost rises linearly. The highest energy proton accelerator,

FNAL, has produced 400 GeV with not much more expense (allowing for inflation). Therefore, secondary beams of muons or electrons can be made at energies which exceed those available at SLAC. This is illustrated in figure 1 which shows the accessible range for a 150 GeV secondary beam.

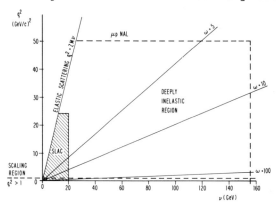

Figure 1

μp NAL-DOMAIN OF MEASUREMENT

When a primary beam can be used, clearly electrons are the best because they boil off a hot filament whereas muons do not. But once secondary beams have to be used, the possible intensities are comparable. In each case π mesons are produced and the charged mesons are allowed to decay (to produce muons) or the neutral mesons to produce γ rays which convert and produce electrons. The decay of the pion must take place over many meters--and both the π beam and the muon beam must be held together over that distance. The phase space in the decay makes this a large aperture and expensive channel. Muons are separated from the pions by demanding that they penetrate some 20 meters of polyethelene absorber.

On the other hand, the electrons are less numerous--at high energies at least--since the γ rays must convert in a radiator and only a few percent do so without degradation--but they can be well focused. The proposed method of separation from π mesons is to focus and bend them; the synchrotron radiation will cause the electron beam to be displaced.

When it comes to doing experiments, there are advantages

and disadvantages of each. I will describe the Fermilab experiments with muons and display the advantage of muons.

The first muon beam designed according to the principles roughly discussed was by Tim Toohig for the 200 GeV summer study at Berkeley nine years ago. It was expensive--at Fermilab the stringent budget forced economies. It was decided that a muon beam had to be a parasite off a neutrino beam using the same decay pipe. Although an intensity of $10^8 - 10^9$ muons/pulse is available in a long channel, the cost of the magnets was too great and a cheaper beam was made, intended to be 10^7/pulse, but actually 10^6/pulse as shown in figure 2.

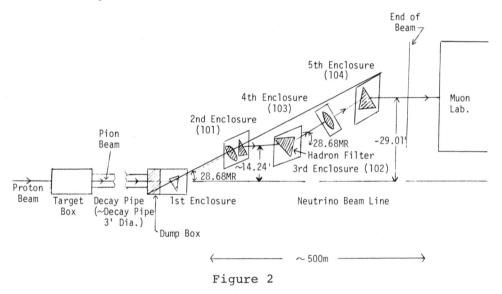

Figure 2

There are three main problems:
    i)   it was meant to run simultaneously with spark chamber neutrino experiments, because <u>both</u> wanted a long spill.
    ii)   the apertures were limited to save money and the decay length was shortened for the same reason--these reduce intensity.
   iii) the proton beam focus is not quite as good as hoped and the pion production not as big as hoped and these reduce intensity to 10^6/pulse.

iv) muons not in the beam still tend to stay close to it and become a halo; they can confuse track finding for small angle scattered events.

v) in retrospect, we should have broadened the energy acceptance of the beam (from 1% to 10%) to increase intensity; this could have been done as shown by a straight line from the first enclosure to the fifth enclosure in figure 2.

We expect all these to be overcome, with some expense, at the CERN SPS and a beam of 10^8 muons/pulse at 200 GeV and 10^7 muons/pulse at 300 GeV to be obtained. Meanwhile, we have some physics results.

Muons do not radiate much in matter and therefore penetrate absorbers. This large range of muons enables thick targets to be used. A "small" group at Fermilab have done a "quick" experiment on scattering from 600 gm/cm^2 of iron. The group presented results at Palermo,[1] available to me only since Friday, and I have been graciously permitted to discuss them here. This is a dangerous task because I fear that I may not do justice to their exciting results. Figure 3 shows their apparatus.

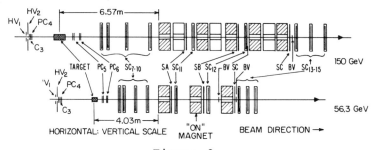

Figure 3

Apparatus for the scaling test. Shaded magnets are on, others are degaussed. The spark-chamber modules have four planes each and are labeled SC_i. Multiwire proportional chambers of two planes each are shown as PC; upstream PC and scintillation counters are not shown.

The muon spectrometer is magnetized iron with spark chambers interspersed. Multiple scattering limits the resolution to about 8% rms.

They have run the apparatus with incident energies of 56 GeV and 150 GeV and adjusted positions so that the apparatus scales as we shall see.

The aim of this scattering experiment is to study the scaling in scattering of leptons by hadrons.

You will remember that the differential cross-section can be expressed in the form

$$\frac{d^2\sigma}{dq^2 d\nu} = \frac{4\pi\alpha^2}{q^4} \frac{E'}{E} W_2(q^2, \nu) \left[1 - \frac{q^2}{4EE'} + \frac{2}{1+R} \left(\frac{q^2+\nu^2}{4EE'} \right) \right] \quad (1)$$

$$R = \frac{\sigma_o(q^2, \nu)}{\sigma_T(q^2, \nu)}$$

where $W_1(q^2, \nu)$ and W_2 are the invariant structure functions discussed by Bjorken. Bjorken[2] suggested W_1 and W_2 are functions of $\omega = 1/x = 2M\nu/q^2$ only once all energies are high enough $\nu \gg$ nuclear mass and excitations $q^2 \gg$ masses of exchanged particles. As we all know, this was found to be approximately true at SLAC for $q^2 > 1(GeV/c)^2$ and $\omega > 4$.[3] The purpose is to extend these experiments to higher W and q^2; this gives a better test of scaling since the conditions of scaling are more easily satisfied.

The experiment detects scattered muons within a large angular range, and large energy range in contrast to the work of reference 3. Therefore, data is not taken at one or two well-defined points. It is therefore convenient to express the data as ratios to an assumed cross-section. The assumed is shown in figure 4 from data in reference 3 using a Monte Carlo calculation.

The results are shown in figure 5. There is a slightly confusing point here; these Monte Carlo calculations assume scaling in a variable $\omega' = \omega + M^2/q^2$ not in ω, because in

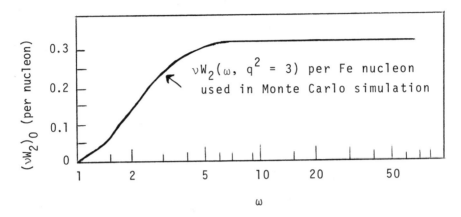

Figure 4

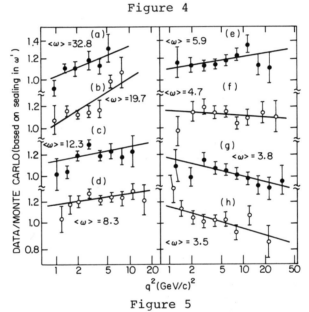

Figure 5

reference 3, this was found to be a better scaling variable. Note that at FNAL energies, ω is greater, and ω' → ω.

In figure 5, I first draw your attention to the data for $\langle\omega\rangle \approx 4$. In this region, the SLAC data showed scaling over the largest range, $1 < q^2 < 10 (GeV/c)^2$. Note that this is still found. The straight lines are fits to the data, but at ω ≈ 4, they could obviously be horizontal.

But at higher ω, deviations are marked at high q^2, with cross-sections rising 30%, and at lower ω, deviations are in the opposite direction. In reference 1, data was also taken at two separate incident energies, 56 and 150 GeV, which enables a check on scaling to be performed independent of some possible systematic errors.

The cross section can be parametized:

$$\frac{d^2\sigma}{dx\,dy} = \frac{4\pi\alpha^2}{2ME}\,\frac{\nu W_2(q^2, \nu)}{x^2 y^2}\left[1 - y + \frac{y^2}{2(1+R)}\right] \quad (2)$$

where $x = \frac{1}{\omega} = \frac{q^2}{2M\nu}$, $y = \frac{\nu}{E}$ and $R = \sigma_S/\sigma_T$ (in Hand's notation) then scaling implies that the ratio

$$r = \left[E\,\frac{d^2\sigma}{dx\,dy}\right]_{E=150} \Big/ \left[E\,\frac{d^2\sigma}{dx\,dy}\right]_{E=56} \quad (3)$$

be a constant. Figure 6 shows that it is not, but instead can be fitted by a curve $r = (\omega/\omega_o)^n$ where $n = 0.096 \pm 0.028$, $\omega_o = 6.1$. Many parametizations of this data, all leading to the same general conclusion, can be found in reference 1. One speculation is that at high energies (high ν, high ω) another degree of freedom enters and a new scaling regime might be established. In future extensions of this work, higher beam currents and energies will be used, and greater precision can be expected. I note some problems with this experiment:

1) the energy resolution, mainly due to multiple scattering, is large. The ratio experiment, in particular, avoids problems due to this cause.
2) the target is iron, not a nucleon. Corrections must be made for Fermi motion of the nucleons (small at high ω) and possible shadowing effects of the virtual photons do not yet display agreement between theory and experiment; there may be a problem here.
3) the radiative correction (wide angle bremsstrahlung) is large for low q^2, high ν (low E') but can be calculated well.
4) the apparatus detection efficiency is large because

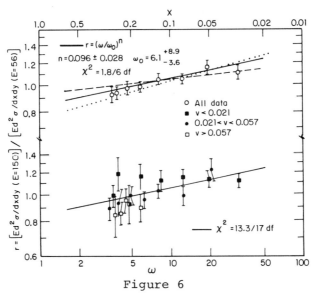

Figure 6

a single muon is detected behind the magnet, and this can be picked up independent of a hadron shower. μe events in the iron magnet can lead to a shower which confuses the track finder. The loss of efficiency can be evaluated and is about 8%.

I now turn to another muon experiment with another group of authors. Only preliminary results are yet available.[4]

The apparatus is shown in figure 7. A liquid hydrogen

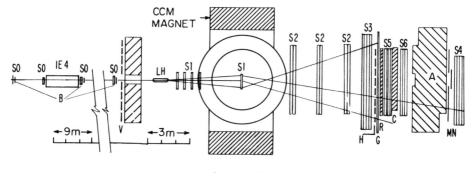

Figure 7

Layout of the muon scattering spectrometer. S0, S1, multi-wire proportional chambers; S2, S3, S4, S5, S6 multiwire spark chambers. B, H, G, M, N, V counter hodoscopes; 1E4, CCM magnets; R, C, A, absorbers.

9 gm/cm^2 or liquid deuterium 20 gm/cm^2 is used.

Scattered muons and produced hadrons are detected by multiwire proportional chambers downstream, are bent in a large magnet (Δp_T = 3 GeV/c) and detected in an array of spark chambers; muons are identified by their penetration of an absorber.

There are several disadvantages compared with the experiment just discussed:

1) The target thickness is 100 times less (to allow hadrons to escape) and therefore the counting rate is 100 times less.
2) The muons must be identified among all the hadrons and often close to a region where beam muons pass.
3) The apparatus is larger and we can be confused by "halo" muons passing within the sensitive time of the spark chambers.

Figures 8a and 8b illustrate how the computer finds tracks. Because of item 2, we have trouble in presenting a precise inclusive muon cross-section at this time. We can, however, examine the hadron spectra whenever a muon scatter is seen--the reaction $\mu + p \to \mu + h^{\pm} + x$.

The discovery of precocious "scaling" in inelastic electron scattering from protons[5] has raised many interesting questions about the electromagnetic structure of the proton and about strong interactions in general.

Muon-proton scattering is dominated by the exchange of a single virtual photon and the kinematics can be characterized by q^2, the magnitude of the four momentum transfer squared, and by ν, the energy transferred by the photon. The differential muon scattering cross

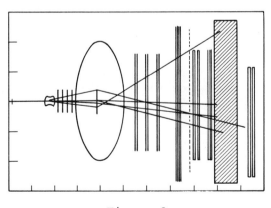

Figure 8a

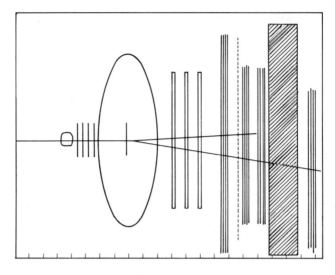

Figure 8b

section $d^2\sigma/dq^2 d\nu$ can be expressed in terms of a virtual photon flux $\Gamma(q^2, \nu)$ and the virtual photon total cross section $d^2\sigma/dq^2 d\nu = \Gamma(q^2, \nu)\sigma(q^2, s)$.

We first identify all events where a muon is scattered from a proton and measure the parameters q^2, ν. Then we measure the hadrons for each of these events. The fraction of hadrons, corrected for acceptance, is related directly to the normalized invariant hadron differential cross section for an incident virtual photon which can be written as:

$$\frac{1}{\sigma(q^2, s)} E \frac{d^3\sigma(q^2, s)}{dp^3} = \frac{1}{\sigma(q^2, s)} \frac{1}{\pi} \frac{E^*}{(p_{max}^{*2} - p_T^2)^{1/2}} \frac{d^2\sigma(q^2, s)}{dp_T^2 \, dx'} \quad (4)$$

E and p are the laboratory values of the hadron's energy and momentum, E^* the hadron energy in the center of mass.

The invariant structure function is defined by:

$$F(x') = \frac{1}{\sigma(q^2, s)} \frac{1}{\pi} \int_0^\infty dp_T^2 \frac{E^*}{(p_{max}^{*2} - p_T^2)^{1/2}} \frac{d^2\sigma(q^2, s)}{dp_T^2 \, dx'} \quad (5)$$

The transverse momentum distributions are given as values of:

$$\frac{1}{\sigma(q^2, s)} \frac{d\sigma(q^2, s)}{dp_T^2} = \frac{1}{\sigma(q^2, s)} \frac{1}{\pi} \int_{x_1'}^{x_2'} dx' \frac{d^2\sigma(q^2, s)}{dp_T^2 \, dx'} \quad (6)$$

This distribution has been fitted with the form $\exp(-bp_T^2)$ for the interval $0 < p_T^2 < 0.5 (GeV/c)^2$.

Events where a muon has scattered are identified and reconstructed. Then the hadrons are defined by tracks which do not penetrate the absorber.

Corrections are made to the yield of muons and hadrons for the effects of acceptance, radiative corrections, hadron interaction in the target, and reconstruction efficiency. We discuss these in turn and principal radiative corrections.

Figure 9 shows corrected transverse momentum distributions for positive and negative hadrons. For $p_T^2 > 0.4 (GeV/c)^2$ these distributions are consistent with the high transverse momentum behavior observed at the ISR[6] and Fermilab.[7] At low p_T^2 the form $\exp(-bp_T^2)$ is apparent. In figure 10 we show the slopes b of $(1/\sigma)(d\sigma/dp_T^2)$ in the range $0 < p_T^2 < 0.5$.

There is no apparent variation of b with q^2 or charge. But for $0 < x' < 0.5$ the slopes at smaller s are consistently smaller than those at larger s. The measurements of b are

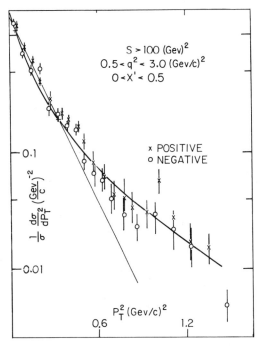

Figure 9

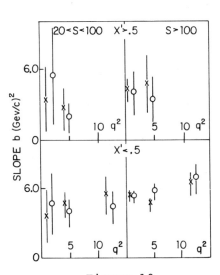

Figure 10

consistent with those from lepto-production at lower values of s and q^2,[8] but we note that our p_T^2 distributions are more extensive. They are also typical of hadron interactions at similar s. At large p_T, as figure 9 shows, the data fit exp $(-5p_T)$. We defer discussion of this to another paper.

In figure 11, we show $F(x')$ as a function of x' for various regions of q^2 and s for both charges. The straight lines are the function 0.35 exp $(-3.25 x')$ and are inserted to indicate the extent that $F(x')$ is independent of q^2 and s. For $x' > 0.5$ these differences are dependent upon charge and q^2 and s. To emphasize this, we show in figure 12 the charge rates for each region and for $0 < x' < 0.5$ and $0.5 < x' < 1$.

We can also identify two pion states and in particular ρ mesons. I show in figure 13 an energy balance for neutral hadron pairs $E^\mu_{incident} - E^\mu_{scatt} - E^\mu_{hadrons}$. A peak close to 0 is found, with a radiative tail, corresponding to $\mu + p \rightarrow \mu + \rho^0 + p$. In this sample are a few e^+e^- pairs masquerading

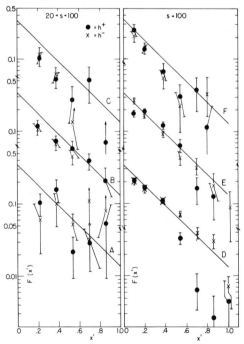

Figure 11

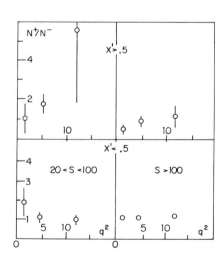

Figure 12

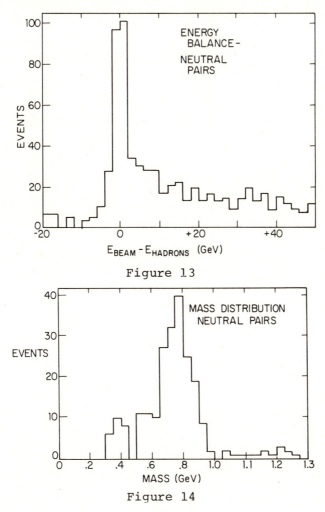

Figure 13

Figure 14

as ρ^0 mesons; these are excluded because they have a small opening.

The mass plot of the remaining neutral pairs is shown in figure 14, and shows a distinct ρ^0 peak at the ρ mass. The number of these ρ^0 mesons can be plotted as a function of the virtual photon 4-momentum q^2, and should be proportional to the number of virtual photons (Γ_T in Hand's notation) and a rho propagator ($m\rho^2/m\rho^2 + q^2$). This is shown in figure 15. Others, in particular Berkelman and collaborators at Cornell, already found this behavior. We can deduce other features about the rho production.

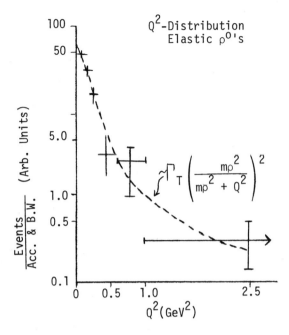

Figure 15

Figure 16 shows the distribution of the transverse momentum of the recoiling muon. We see the characteristic exp (-8t) found at lower energies. We do not see a reduction in the slope at high q^2 for which T.T. Wu and others have urged us to search.

In the scattering of a lepton the virtual photons become transversely polarized. (Actually since the muons have spins aligned along the direction of motion, the polarization is elliptical--but this is irrelevant for this purpose.) In figure 17 we plot the rho mesons as a

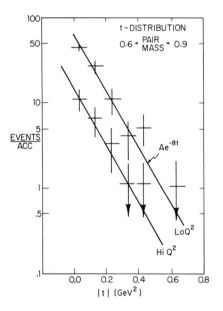

Figure 16

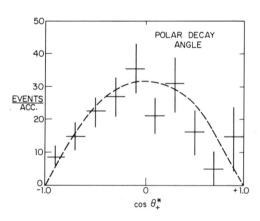

Figure 17

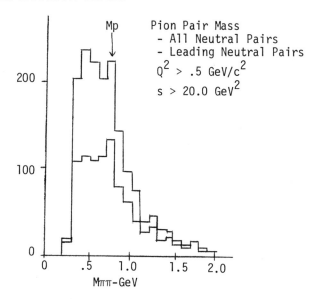

Figure 18

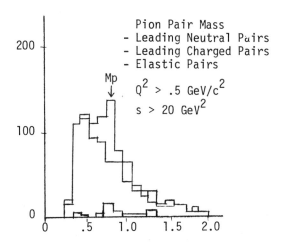

Figure 19

function of the polar angle of the decay. The $\sin^2 \theta$ plot shows that we have transversely polarized rhos and scaler virtual photons play little part.

It is harder to search for rho mesons which <u>do not</u> balance energy. But in figures 18 and 19 we show how we begin to find them. Figure 18 shows the mass plot for all neutral pairs and the <u>leading</u> (highest energy) neutral pairs. By taking the leading pairs a rho peak begins to stand out. In figure 19 we compare the leading neutral and the charged pairs. By subtraction we find a small rho peak. This whets our appetite for the future. I personally speculate that when we add up

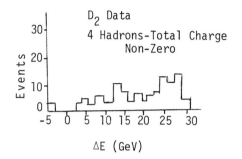

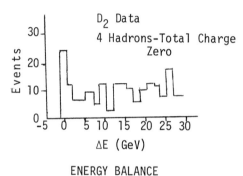

ENERGY BALANCE

Figure 20

all rho mesons, they will appear in 15% of the muon events independent of q^2 in contrast to the behavior of the diffractive rho production shown in figure 15.

We have begun to search for rho prime mesons diffractively produced with a mass of 1600 MeV decaying into 4 pions. In contrast with the previous data from hydrogen, this data, purely for reasons of convenience, is from deuterium. In figure 20 we show that there are neutral 4 pion events produced with an energy balance of 0, whereas no charge 4 pion events have a 0 energy balance. Figure 21 shows the mass plot of these events, which is unfortunately very broad and hard to separate from phase space.

If these events are indeed rho prime mesons, they should decay into a rho and an epsilon: $\rho' \to \rho^0 \epsilon \to (\pi^+\pi^-)(\pi^+\pi^-)$.

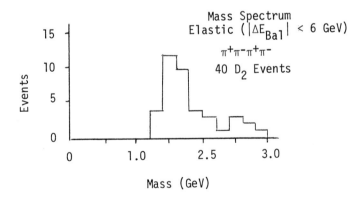

Figure 21

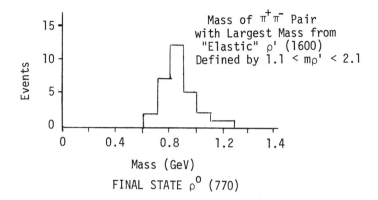

Figure 22

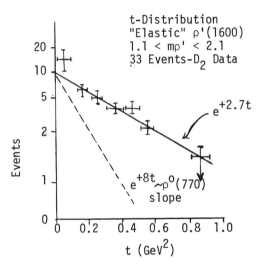

Figure 23

Figure 22 shows the mass plot for the leading $\pi^+\pi^-$ pair which shows a rho peak inconsistent with phase space. The number of rho prime events is 10% of the rho events which is consistent with SLAC photoproduction data. On the assumption that these are rho prime events, we can study their characteristics.

Figure 23 shows the distribution of transverse momentum of the recoil proton--exp(+2.7t) which is very different from the exp(+5t) of a single proton or exp(+8t) of the rho meson. The slight peak, statistically insignificant as yet. near t = 0, may demonstrate coherent production of the rho prime on deuterium. As noted before, the rho production varies as the rho propagator: $\Gamma_t \, (m_\rho^2/m_\rho^2 + q^2)$. We might expect rho prime mesons to have a different propagator and the production to vary as: $\Gamma_t \, (m_{\rho'}^2/m_{\rho'}^2 + q^2)^2$.

Because of the larger mass of the rho prime, we therefore expect relatively <u>more</u> rho prime mesons at high q^2. By calculation, we expect:

$$(N_{\rho'}/N_{\rho}) \ q^2 > 1/(N_{\rho'}/N_{\rho}) \ q^2 \quad 1 = 2 \qquad (9)$$

We find 2^{+3}_{-1} which so far tells us nothing.

Although I have told you it is hard, in general, to give a cross section for inclusive muon events, it is possible near $q^2 = 0$ and high ν where the low energy of the scattered muon allows it to be swept by the magnet away from the beam.

In figure 24 I show the cross-section expressed as a cross section for transverse photons:

$$\frac{d^2\sigma}{dq^2 d\nu} = \Gamma_t \sigma_T(q^2, \nu) \quad \text{where} \quad \sigma_T(\sigma, \nu) = \sigma_\gamma(k) \quad \text{and} \quad k = \nu \qquad (10)$$

By extrapolating σ_T to $q^2 = 0$, we find $\sigma_\gamma(k)$ (figure 25), the total cross-section for protons on nucleons. The extraction of σ_γ this way involves a large radiative correction. I remind you that at SLAC radiative corrections limited the accuracy of such measurements. However, in our case, the radiative correction is reduced by $[\log(q^2/m_\mu^2)]/[\log(q^2/m_e^2)] \approx 4$.

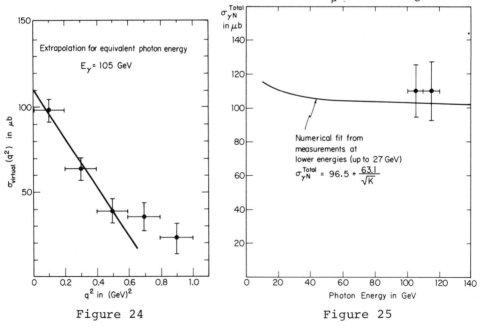

Figure 24 Figure 25

Presently, our accuracy is 15% as shown and σ_γ nucleon agrees with the vector dominance model predictions based on lower energy data.

In conclusion, muon scattering has shown that scale invariance is, at least, not as simple as SLAC data suggested. The hadron data has the potential of being more detailed than any data before, but so far shows no surprises at this higher energy. In particular, there are not as yet enough statistics to see whether there is any dramatic change at $\omega \approx 40$ and $q^2 > \omega$ as suggested by the scaling deviations.

I would like to thank the whole group of reference 1 for the privilege of allowing me to discuss their beautiful data, and my hard working colleagues of reference 5 for our stimulating collaboration with the hadron data.

REFERENCES

1. Y. Watanabe, L.H. Hand, S. Herb, A. Russell (Cornell); C. Chang, K.W. Chen, D.J. Fox, A. Kotlowski, P.F. Kunz (Michigan State); and S.C. Loken, M. Strovink (Berkeley); preliminary results, Phys. Rev. Letts. 33, 1504 (1974); reports CLNS 302, 308; MSU/CSL 22, 23; LBL 3870, 3886 to be published in Phys. Rev. Letts.
2. J.D. Bjorken, Phys. Rev. 179, 1547 (1969).
3. E.D. Bloom, et al, Phys. Rev. Letts. 23, 930 (1969); M. Briedenbach, et al, Phys. Rev. Letts. 23, 935 (1969). Latest review, F.J. Gilman, Proceedings of XVII International Conference on High Energy Physics, London, 1974.
4. W.A. Loomis, F.M. Pipkin, L.J. Verhey, Richard Wilson (Harvard); W.R. Francis, T.B.W. Kirk, R.G. Hicks (Illinois); H.L. Anderson, L.W. Mo, S.C. Wright, R.M. Fine, R.H. Heisterberg, H.S. Matis, L.C. Myrianthopolous (Fermi Institute); and N.E. Booth, T.W. Quirk, A. Skuja, W.S.C. Williams, V.K. Bharadwaj, G.I. Kirkbride (Oxford); reported at the Washington 1975 meeting of the A.P.S. and shortly to be published.
5. J.I. Friedman and H.W. Kendall, Ann. Rev. Nucl. Sci, 22 203 (1972).
6. R. Slasky, Physics Reports 11C, 101 (1974).
7. J.W. Cronin, et al, Phys. Rev. Letts. 31, 1426 (1973).
8. J.T. Dakin, et al, Phys. Rev. 10D, 1401 (1974).

PROPERTIES OF THE ψ RESONANCES*

M. Breidenbach
Stanford Linear Accelerator Center
Stanford University, Stanford, California, USA

ABSTRACT

Sharp peaks are seen in the electron-positron annihilation cross section corresponding to the ψ at 3.095 GeV and the ψ' at 3.684 GeV. Cross sections for the ψ and ψ' decay into hadrons, e^+e^- pairs, and $\mu^+\mu^-$ pairs are used to deduce the widths and quantum numbers of the ψ and ψ'. Studies of the decay modes are used to determine the G-parity of the ψ and the existence of a two-pion cascade decay from the ψ' to the ψ. No other narrow resonances have been found with masses between 3.2 and 7.6 GeV.

In early November of 1974 the total annihilation cross section for electrons and positrons into hadrons at high energy seemed essentially featureless. After the structure of the vector mesons the cross section decreased slowly with increasing energy. Within a month, this relatively simple picture had become somewhat chaotic due to the discovery of two very narrow resonances, the ψ at a mass of 3.1 GeV[1] and the ψ' at a mass of 3.7 GeV.[2] The most surprising features of these particles are their extremely narrow widths (or long lifetimes) for so massive states and the magnitudes of the peak cross sections. At the ψ, the experimentally observed cross section increases by a factor of more than 100 from that observed at a beam energy only 2.5 MeV lower. It seems that some basically new physics is manifesting itself, such as a new quantum number or selection rule. We will describe some of the properties of the ψ's as determined at SPEAR[3] by the SLAC-LBL collaboration.[4]

The data to be described were acquired with the SLAC-LBL magnetic detector, which is shown in Fig. 1 and described in more detail in Appendix I. Basically, it is a cylindrical detector 3 meters in diameter and 3 meters long, and has a uniform axial magnetic field of 4 kilogauss. The detector encompasses the full azimuthal coordinate and includes a solid angle of $0.65 \times 4\pi$.

---

* Work supported by the U.S. Energy Research and Development Administration.

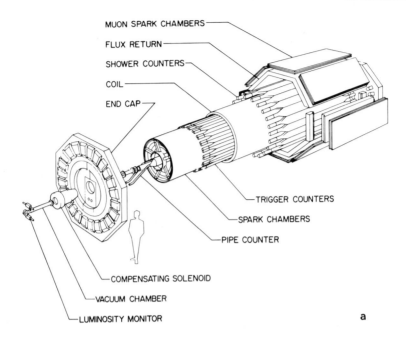

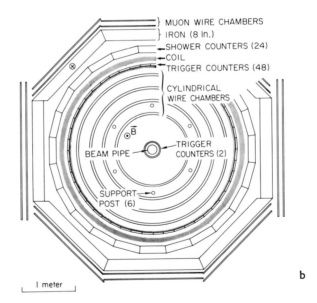

Fig. 1 The SLAC-LBL Magnetic Detector.
a) Telescoped view.
b) End view.

PROPERTIES OF THE ψ RESONANCES

Charged particles with momenta greater than 100 MeV/c are detected, and two particles with transverse momenta greater than about 200 MeV/c are required for a trigger. Yields are normalized by measurements of Bhabha scattering at small angles (25 mrad). Backgrounds at the resonances are negligible (< 0.1%).

MEASURED CROSS SECTIONS AND DECAY WIDTHS

Measurements of the cross sections for the resonant decays into hadrons, e pairs, and μ pairs allow, with assumptions about the shape and quantum numbers of the resonant state, the calculation of the true resonant width and the partial widths for e pairs and μ pairs. The assignments of the quantum numbers will be deduced from interference effects between the resonant amplitude and the direct QED amplitude in the μ-pair channel.

Figure 2a shows the total cross section for hadron production σ_T versus center-of-mass energy $E_{c.m.} = 2 E_{beam}$ for the ψ.[5] The observed full width at half maximum (FWHM) is 2.6 MeV, which is the width expected from the convolution of a much narrower resonance shape, the inherent spectral resolution of the storage ring, and radiative corrections in the production of a virtual photon. Figures 2b and 2c show the cross sections for the final states $\mu^+\mu^-$ and e^+e^-,

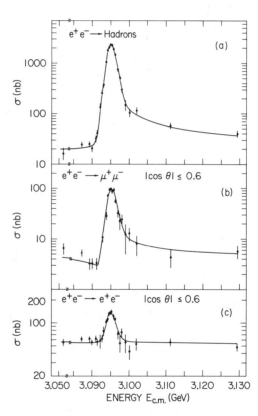

Fig. 2 Cross sections for the indicated modes in the vicinity of the ψ. Only the hadronic cross sections are corrected for detector acceptance. The curves show the fitted cross sections.

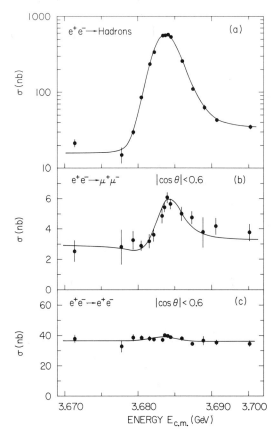

Fig. 3 Cross sections for the indicated modes in the vicinity of the ψ'. Only the hadronic cross sections are corrected for detector acceptance. The curves show the fitted cross sections.

respectively.[6] The cross sections are integrated over the angular region $|\cos \theta| < 0.6$, where θ is the angle between the outgoing positive lepton and the positron beam. The corresponding cross sections for the ψ'[7] are shown in Figure 3. The observed FWHM is about 4.6 MeV, which is also compatible with a much smaller true width. The peak cross section is about 550 nb, approximately one-quarter that of the ψ. The integrated resonance strengths, with radiative corrections, are:

ψ: $\Sigma_h = \int \sigma_h^\psi (E_{c.m.}) \, dE_{c.m.} = 10,400 \pm 1500$ nb-MeV

ψ': $\Sigma_h = \int \sigma_h^{\psi'} (E_{c.m.}) \, dE_{c.m.} = 3700 \pm 600$ nb-MeV.

The extraction of the leptonic cross sections for the ψ' is complicated by the large branching ratio for ψ' to ψ decay, followed by the subsequent leptonic decay of the ψ. The decay of the ψ' into leptons may be separated from the cascade decay by reconstructing the invariant mass of the lepton pair. This can be done if the detector has sufficient momentum resolution. Such a separation is shown in Figure 4. The e^+e^- pairs from the ψ may be seen above the radiative tail of the ψ' in Fig. 4a. The $\mu^+\mu^-$ pair separation is shown in Fig. 4b, and is very clean. The difference between the electron pairs and the muon pairs is that the resonant electron cross section sits on a QED background of about 37 nb[8]) (electron pairs are produced from spacelike scattering processes as well as the annihilation of the e^+ and e^-), but the $\mu^+\mu^-$ pair cross section is on a background of only 3 nb. (The $\mu^+\mu^-$ pairs come only from annihilation.) Assuming μ-e universality, the resonant μ pair and e pair cross sections are equal.

These data may be used to calculate the true width Γ, as well as the partial widths for decays into hadrons, Γ_h; e^+e^- pairs, Γ_e; and $\mu^+\mu^-$ pairs, Γ_μ for each of the resonances. If we assume that the resonance has a Breit-Wigner shape, then the resonant cross section σ_f for the decay to any set of final states f is

$$\sigma_f = \frac{(2J+1)\pi}{m^2} \frac{\Gamma_e \Gamma_f}{(E_{c.m.} - m)^2 + \Gamma^2/4},$$

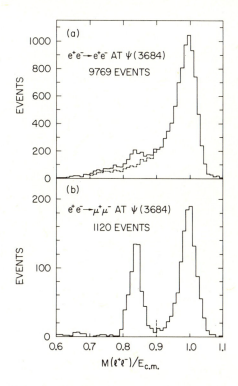

Fig. 4 Invariant mass of lepton pairs from the ψ'. The dotted lines indicate the cuts that were made.
a) $e^+e^- \to e^+e^-$.
b) $e^+e^- \to \mu^+\mu^-$.

where J is the spin of the resonance, m is the resonance mass, and Γ_f is the width to the set of final states. Γ is the total width for the decay to all final states. If this expression is integrated over the center-of-mass energy, it may be compared to the integrated experimental cross sections (after appropriate radiative corrections)[9]) without explicit dependence on the storage ring energy resolution. Thus:

$$\Sigma_f = \int \sigma_f \, dE_{c.m.} = \frac{(2J+1) \, 2\pi^2}{m^2} \frac{\Gamma_e \Gamma_f}{\Gamma}.$$

Then, assuming J = 1 (which will be justified shortly) and that $\Gamma = \Gamma_h + \Gamma_e + \Gamma_\mu$, i.e., there are no neutral modes, $\Gamma_e = \frac{m^2}{6\pi^2} \Sigma_{all}$, and $\Gamma = \Gamma_e \frac{\Sigma_{all}}{\Sigma_{ee}}$.

Since Σ_{ee} is small and sitting on a (relatively) large background, it is advantageous to assume μ-e universality and substitute $\Sigma_{\mu\mu}$ for Σ_{ee} in the expression for Γ, particularly at the ψ'. Alternatively, one may simultaneously fit the three cross sections to determine the mass and the widths. (Each resonance is done separately.) The fit folds the Gaussian energy resolution of the ring with radiative effects and a Breit-Wigner cross section. Also included are the effects of interference between the resonant channel $e^+e^- \to \psi \to f$ and the direct process $e^+e^- \to f$. The results of the two techniques agree, and the results are shown in Table I. The errors are mainly due to systematic uncertainties, such as uncertainty in the hadron detection efficiency, and are listed in the tables.

TABLE I

Properties of the ψ-particles as obtained from fit to cross sections σ_{HAD}, $\sigma_{\mu\mu}$, and σ_{ee}

	$\psi(3095)$	$\psi(3684)$
Mass	3.095 ± 0.004 GeV	3.684 ± 0.005 GeV
J^{PC}	1^{--}	1^{--}
$\Gamma_e = \Gamma_\mu$	4.8 ± 0.6 keV	2.2 ± 0.3 keV
Γ_H	59 ± 14 keV	220 ± 56 keV
Γ	69 ± 15 keV	225 ± 56 keV
Γ_e/Γ	0.069 ± 0.009	0.0097 ± 0.0016
Γ_H/Γ	0.86 ± 0.02	0.981 ± 0.003
Γ_μ/Γ_e	1.00 ± 0.05	0.89 ± 0.16

Errors accounted for: (a) statistical, (b) 15% uncertainty on hadron efficiency, (c) 100 keV setting error in $E_{c.m.}$, (d) 2% point-to-point errors, uncorrelated, (e) 3% luminosity normalization.

The full width of the ψ is about 70 keV, and that of the ψ' is about 225 keV. All of the widths in the table except the last row were calculated assuming $\Gamma_e = \Gamma_\mu$. This hypothesis of μ-e universality is tested in the last row, and good agreement with universality is found.

SPIN AND PARITY

If the resonances are eigenstates of P and C, there should be no asymmetry in the angular distribution of the leptonic decays. The front-back asymmetry for μ pairs as a function of energy is shown for the ψ and ψ' in Fig. 5. The

PROPERTIES OF THE ψ RESONANCES

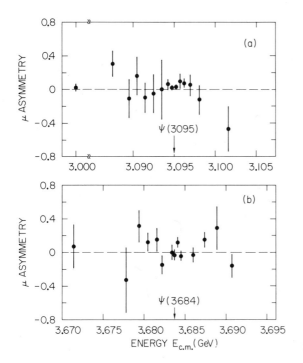

Fig. 5 The front-back asymmetry for $\mu^+\mu^-$ pairs
  a) near the ψ
  b) near the ψ'.

asymmetries are consistent with zero, indicating no significant violation of parity or charge conjugation, and that the resonances are not mixtures of states of opposite parity.

Since the ψ's are copiously produced in e^+e^- annihilation, one might expect them to couple directly to the annihilation photon and have the same quantum numbers, $J^{PC} = 1^{--}$. This assignment of J^{PC} would not be necessary if a ψ were produced by a direct interaction of the electron and positron.

If the ψ's couple directly to the photon, one would expect interference between the muonic decays of the ψ's,

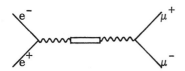
$$A = \left(\frac{(2J+1)\pi}{E_{c.m.}^2}\right)^{\frac{1}{2}} \frac{\Gamma_e}{m - E_{c.m.} - i\Gamma/2}$$

and the direct QED production of muon pairs

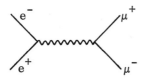
$$A = \left(\frac{3\pi}{E_{c.m.}^2}\right)^{\frac{1}{2}} \left(\frac{-2\alpha}{3}\right).$$

These two amplitudes would interfere, yielding a cross section proportional to

$$\left| \frac{-2\alpha}{3} + \frac{\Gamma_e}{m - E_{c.m.} - i\Gamma/2} \right|^2 .$$

The addition is shown graphically in Figure 6, and it can be seen that there will be destructive interference at energies slightly below the peak.

In practice, the ratio of the μ-pair to e-pair cross sections is compared with predictions in order to eliminate systematic normalization errors. Such comparisons are shown in Fig. 7. The data are fully compatible with the hypothesis of interference, and disagree with the hypothesis of no interference (e.g., J = 0) by 2.7 standard deviations for the ψ and 4.9 standard deviations for the ψ'. Because the detector does not cover the full solid angle, the observation of interference does not necessarily mean J = 1. (But because the detector is symmetric in space, and with respect to charge, and sums over spins, the observation of interference does imply P = C = -1.)

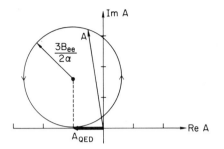

Fig. 6 Addition of interfering resonant and direct QED amplitudes.

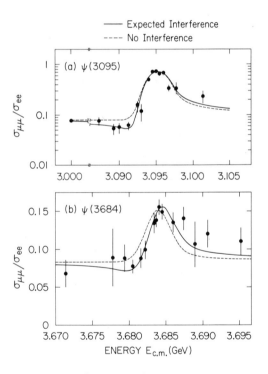

Fig. 7 The ratio of μ pair to e pair cross sections
a) near the ψ
b) near the ψ'.

PROPERTIES OF THE ψ RESONANCES

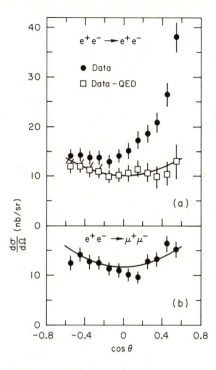

Fig. 8 a) The angular distribution of electron pairs at the ψ, $3.0944 < E_{c.m.} < 3.0956$.
b) The angular distribution of muon pairs in the same energy range.

However, spins 2 and 3 cause constructive interference (for the detector geometry) and generate predictions even further from the data than the $J = 0$ assumption.

Figure 8 shows the angular distribution for the leptonic decays of the ψ. For the e^+e^- pairs, the open boxes show the data with the QED contribution removed. The curves represent $1 + \cos^2\theta$. Figure 9 shows the angular distributions for the ψ'. The dashed curve is the cross section expected from QED, while the solid curves include the resonant contribution calculated from the widths discussed earlier. The angular distribution for both particles confirms the $J = 1$ assignment from the interference results.

ψ DECAYS

The hadronic decays of the ψ may be either direct decays

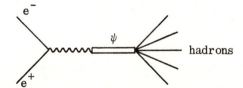

or "2nd order electromagnetic" decays

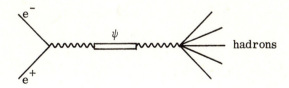

The G-parity of the ψ may be obtained by observing whether the direct decay of the ψ is into states containing even or odd numbers of pions. The photon in the "2nd order electromagnetic" decay does not have definite isospin and can decay into either even or odd numbers of pions.

The resonant muon pairs presumably come from a 2nd order decay

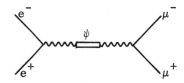

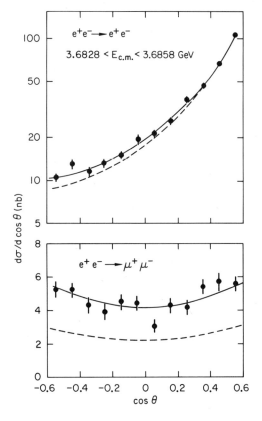

Fig. 9 a) The angular distribution of electron pairs at the ψ'.
b) The angular distribution of muon pairs at the ψ'.

Away from the resonance, let us define $R_{off} = \dfrac{\sigma_h^{off}}{\sigma_\mu^{off}}$, where σ_h is the cross section for a particular hadronic channel and σ_μ is the $\mu^+\mu^-$ cross section. Similarly, define R_{on} as the corresponding ratio on the resonance. If all of the hadronic decays were 2nd order electromagnetic, then $R_{on} = R_{off}$. If there is a direct decay, $R_{on} > R_{off}$. In order to cancel most of the systematic errors, we will examine the ratio $\alpha = R_{on}/R_{off}$, so that $\alpha = 1$ for the second order process and $\alpha > 1$ for direct decays.

The first step is to isolate the different multipion final states. Since the time-of-flight particle identification system does not isolate pions with momenta greater than about 650 MeV/c, the identification of exclusive states depends on energy and momentum balance for states with all particles observed,

PROPERTIES OF THE ψ RESONANCES

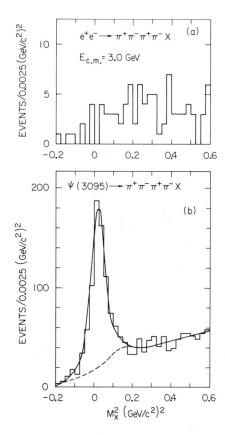

Fig. 10 The missing mass squared recoiling against four charged pions
a) at $E_{c.m.} = 3.0$ GeV
b) at the ψ.

can be isolated in a similar manner.

In Figure 12, α is shown for final states containing from 3 to 7 pions. The states containing 4 and 6 pions have $\alpha = 1$, the value expected from the "2nd order electromagnetic" decays, while

and on the existence of a missing mass peak for events with one missing neutral. Figure 10 shows the missing mass recoiling against four charged pions. At the resonance, a large peak is seen,[10] corresponding to $\psi \to \pi^+\pi^-\pi^+\pi^-\pi^0$. Off the resonance at $E_{c.m.} = 3.0$ GeV, no peak is discernable and only an upper limit on the five-pion yield may be determined. Figure 11 shows the total energy in four-prong events at the ψ, assuming that each of the prongs is a π. The peak at 3.1 GeV is due to $\psi \to \pi^+\pi^-\pi^+\pi^-$, and the peak at 2.8 GeV is due to $\psi \to \pi^+\pi^-K^+K^-$. The 4π cross section at $E_{c.m.} = 3.0$ GeV

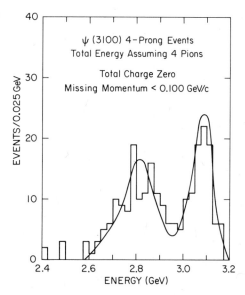

Fig. 11 Total observed energy in four-prong events at the ψ. The curve is a Monte Carlo fit to the data.

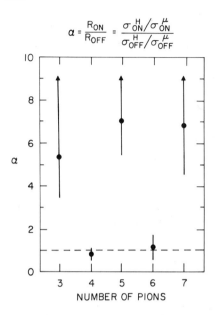
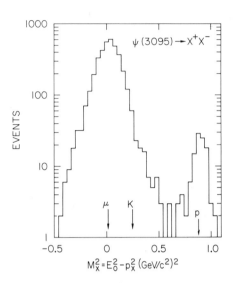

Fig. 12 Comparison of α for different multipion final states at the ψ (ON) and at 3.0 GeV (OFF).

Fig. 13 Mass squared for collinear 2-prong events from the ψ. e^+e^- pairs have been removed by shower-counter pulse-height cuts.

the states containing odd numbers of pions have α greater than one. We conclude that the direct decay of the ψ is to states of odd numbers of pions and thus the G-parity of the ψ is odd. For final states having only pions, $G = C(-1)^I$, so the isospin of the ψ must be even. Also, the value of $\alpha = 1$ for the even number of pion states indicates that the ψ does not have direct leptonic decays.

Figure 13 shows the invariant mass distribution for the ψ decaying into a pair of charged particles. A peak corresponding to the $\bar{p}p$ decay is clearly visible. The absence of an observable peak off the resonance indicates that the $\bar{p}p$ mode is a direct decay of the ψ. Since the isospin of a $\bar{p}p$ state can be only 0 or 1, and since 1 is excluded by the odd G-parity, the isospin of the ψ must be 0.

Another technique to determine the isospin of the ψ is through the decay $\psi \rightarrow \pi^+\pi^-\pi^0$. Figure 14 shows the Dalitz plot for this decay, indicating that the dominant mode is $\rho\pi$. The horizontal, vertical, and diagonal bands correspond to the $\rho^-\pi^+$, $\rho^+\pi^-$, and $\rho^0\pi^0$ modes, respectively. Since these three modes occur with equal frequency, the $I = 0$ assignment is selected over $I = 2$, which

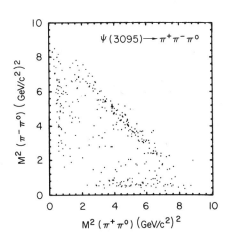

Fig. 14 Dalitz plot for the $\psi \to \pi^+\pi^-\pi^0$ decay.

would require $\rho^0\pi^0$, $\rho^+\pi^-$, and $\rho^-\pi^+$ to be in the ratio 4:1:1.

The study of the ψ decay modes is still incomplete. A "progress report" is shown in Table II. These decay modes offer a clue to the SU_3 nature of the ψ,[11] i.e., whether the ψ is an SU_3 singlet, the eighth component of an octet, or a mixture of the two, such as the ω or ϕ. For an SU_3 singlet, all decays into pairs of pseudoscalar mesons are forbidden, e.g., $\psi \to \pi^+\pi^-$ or $\psi \to K^+K^-$. If the ψ were the eighth component of the octet, or a mixture, the $\psi \to \pi^+\pi^-$ decay is still forbidden, but the decays $\psi \to K^+K^-$ and $\psi \to K_S K_L$ occur with equal rates. For second order electromagnetic decays, exact SU_3 forbids the $\psi \to \gamma \to K_S K_L$ decay. Thus, the absence of the $\psi \to K_S K_L$ decay would support the singlet hypothesis. An upper limit of 0.02% for this decay mode has been set.

The decay mode $\psi \to \Lambda\bar{\Lambda}$ has also been seen, with a branching ratio of 0.16 ± 0.08%. The approximate equality of the $\psi \to \Lambda\bar{\Lambda}$ branching ratio with the $\psi \to p\bar{p}$ branching ratio also supports the singlet hypothesis. However, the ratio of KK* to $\pi\rho$ decays is in disagreement with the singlet predictions.

ψ' HADRONIC DECAYS

The ψ' decays predominantly to the ψ with the emission of two pions.[12] While this decay mode has been studied, almost nothing is known about other decay modes, except that they do not seem to be simple.

The mode $\psi' \to \psi$ + anything is studied by observing the leptonic decay of the ψ. The mode $\psi' \to \psi + X$, $\psi \to \mu^+\mu^-$ is shown distinctly in Figure 4b, which shows the invariant mass distribution of μ pairs from ψ decays. The high mass peak is from the QED and resonant production of muon pairs, and the well separated

TABLE II

Decay Modes of the $\psi(3095)$

Mode	Branching Ratio (%)	No. of Events Observed	Comments
e^+e^-	6.9 ± 0.9	ca 2000	
$\mu^+\mu^-$	6.9 ± 0.9	ca 2000	
$\rho\pi$	1.3 ± 0.3	153 ± 13	$>70\%$ of $\pi^+\pi^-\pi^0$
$2\pi^+ 2\pi^-$	0.4 ± 0.1	76 ± 9	
$2\pi^+ 2\pi^- \pi^0$	4.0 ± 1.0	675 ± 40	$20\% \,\omega\pi^+\pi^-$ $30\% \,\rho\pi\pi\pi$
$3\pi^+ 3\pi^-$	0.4 ± 0.2	32 ± 7	
$3\pi^+ 3\pi^- \pi^0$	2.9 ± 0.7	181 ± 26	
$4\pi^+ 4\pi^- \pi^0$	0.9 ± 0.3	13 ± 4	
$\pi^+\pi^- K^+K^-$	0.4 ± 0.2	83 ± 18	not including $K^*(892) K^*(1420)$
$2\pi^+ 2\pi^- K^+K^-$	0.3 ± 0.1		
$K_S K_L$	<0.02	≤ 1	90% C.L.
$K^0 K^{0*}(892)$	0.24 ± 0.05	57 ± 12	
$K^\pm K^{\mp*}(892)$	0.31 ± 0.07	87 ± 19	
$K^0 K^{0*}(1420)$	<0.19	≤ 3	90% C.L.
$K^\pm K^{\mp*}(1420)$	<0.19	≤ 3	90% C.L.
$K^{*0}(892)\bar{K}^{*0}(892)$	<0.06	≤ 3	90% C.L.
$K^{*0}(1420)\bar{K}^{*0}(1420)$	<0.18	≤ 3	90% C.L.
$K^{*0}(892)K^{*0}(1420)$	0.37 ± 0.10	30 ± 7	
$p\bar{p}$	0.21 ± 0.04	105 ± 11	assuming $f(\theta) \sim 1+\cos^2\theta$
$\Lambda\bar{\Lambda}$	0.16 ± 0.08	19 ± 5	
$p\bar{p}\pi^0$, $\bar{n}p\pi^-$, $\bar{p}n\pi^+$	0.37 ± 0.19	87 ± 30	

lower peak is due to the ψ decay. Using the previously measured branching ratio for $\psi \to \mu^+\mu^-$, we find:

$$\frac{\Gamma(\psi' \to \psi + \text{anything})}{\Gamma(\psi' \to \text{all})} = 0.57 \pm 0.08$$

The ψ may also be identified by calculating the mass recoiling against a pion pair, as in the decay $\psi' \to \psi + \pi^+ + \pi^-$. The mass recoiling against all pairs of charged particles is shown in Fig. 15a, revealing a sharp spike at a mass of 3.1 GeV, with a width equal to that expected from the detector momentum resolution. Folding the calculated detection efficiencies with this yield leads to:

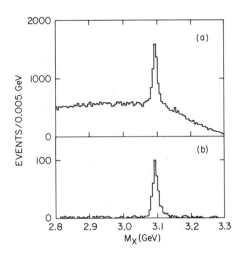

Fig. 15 a) Invariant mass M_x recoiling against all pairs of oppositely charged particles.
b) Same as (a) but for those events in which the observed charged particles satisfy, within errors, conservation of total momentum and energy.

$$\frac{\Gamma(\psi' \to \psi + \pi^+ + \pi^-)}{\Gamma(\psi' \to \text{all})} = 0.32 \pm 0.04$$

These two branching ratios may be combined to find the fraction of $\psi' \to \psi$ cascades with a charged pion pair:

$$\frac{\Gamma(\psi' \to \psi + \pi^+ + \pi^-)}{\Gamma(\psi' \to \psi + \text{anything})} = 0.56 \pm 0.03$$

If the "anything" consists of only the charged pions and π^0 pairs, then the isospin of the pion system determines the above ratio. Values of approximately[13] 2/3, 1, and 1/3 correspond to isospin states of 0, 1, and 2, respectively. The isospin assignment 0 is preferred. The apparent absence of a single π^0 cascade decay and the large fraction of cascade decays with two pions indicate that the ψ' and ψ have the same G-parity. If the $I = 0$ assignment is correct, then the difference between the branching ratio of 0.56 and the predicted values may indicate that

$$\frac{\Gamma(\psi' \to \psi + \text{not } \pi\pi)}{\Gamma(\psi' \to \psi + \text{anything})} \leq 10\% \ .$$

Figure 15b shows events where all the final particles in the decay $\psi' \to \psi + \pi^+\pi^-$, $\hookrightarrow \mu^+\mu^-$ are observed. There is almost no background, and these data have been used to study the angular distributions. The data appear to indicate that the dipion is in an S-wave state with respect to the leptons.

A small signal seems to be seen for the decay $\psi' \to \psi + \eta$ with a branching ratio of $4 \pm 2\%$. Upper limits on the radiative decay of the ψ' to an intermediate state are approximately 7%. Such states might be the $3P_{0,1,2}$ states expected in the charmonium models of the ψ's.[14]

Approximately forty percent of the ψ' decays remain unaccounted for. It is interesting to remove the cascade decays from the ψ' sample and compare it with the ψ. This has been done for the class of 4-prong, $Q = 0$ events in Fig. 16. Each scatter plot displays the missing momentum against the total charged energy observed in each event, assuming pion kinematics. Figure 16a shows the ψ decays. The concentrations against the energy axis are the four body decays: $\pi^+\pi^-\pi^+\pi^-$ at $E = 3.1$ GeV, $\pi^+\pi^-K^+K^-$ around 2.8 GeV, and various combinations of π's, K's, and nucleons at lower energies. The strong diagonal concentration corresponds to $\pi^+\pi^-\pi^+\pi^-\pi^0$ decays. (b) shows the full sample of ψ' decays, and (c) shows the pion cascade decay. (d) shows (b) with the cascade contribution removed. Very little structure is left, but these decays are still being actively investigated. Upper limits on the decays $\rho^0\pi^0$ and $\pi^+\pi^-\pi^+\pi^-\pi^0$ have been set. It appears that the decays into odd numbers of pions are suppressed in comparison with the ψ. Perhaps there is some new principle or dynamics that ensures two neutrals in all the final states. A summary of the ψ' decay modes is listed in Table III.

OTHER STRUCTURES

Measurements of the total cross section σ_h now extend to an $E_{c.m.}$ of 7.4 GeV, and are shown in Figure 17. The principal structure is the peak at about 4.1 GeV.[15] The width is about 250-300 MeV, and it has a peak cross section of about 32 nb sitting on a background level of about 18 nb. If this structure is a resonance, its width and peak cross section are very different from the ψ's. However, the integrated cross section is about 5500 nb-MeV, a value

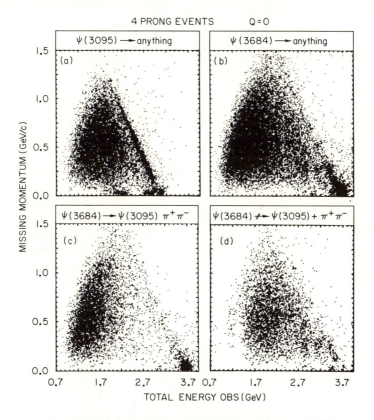

Fig. 16 Comparison of 4-prong events at the ψ and ψ'.
Missing momentum versus total charged energy,
assuming pion kinematics.

a) All 4-prongs at the ψ.
b) All 4-prongs at the ψ'.
c) All 4-prong $\psi' \to \psi + \pi^+\pi^-$ cascade decays.
d) Non-cascade ψ' decays, i.e., (c) removed from (b).

comparable with the ψ's. The study of exclusive decay modes from the 4.1 GeV structure is just beginning. The problem is complicated by the much smaller data sample here compared with the ψ's, because the peak cross section at 4.1 GeV is a relatively civilized 32 nb.

Other sharp structure has been sought.[16] Shortly after the discovery of the ψ, SPEAR and the magnetic detector were modified to allow scanning

TABLE III

Decay Modes of the $\psi(3684)$

Mode	Branching Ratio (%)	Comments
e^+e^-	0.97 ± 0.16	μ-e universality assumed
$\mu^+\mu^-$	0.97 ± 0.16	
$\psi(3095)$ anything	57 ± 8	
$\psi(3095)\, \pi^+\pi^-$	32 ± 4	these decays are included in the fraction for ψ + anything via an intermediate state
$\psi(3095)\, \eta$	4 ± 2	
$\psi(3095)\, \gamma\gamma$	$< 6.6^*$	
$\rho^0 \pi^0$	$< 0.1^*$	
$2\pi^+ 2\pi^- \pi^0$	$< 0.7^*$	* 90% confidence limit based on a preliminary analysis
$p\bar{p}$	$< 0.03^*$	

measurements of the cross section in steps of c.m. energy of approximately 1.9 MeV. The time spent at each step was sufficient for the production of 1-2 hadronic events; this was typically a minute. The data were subject to the "offline" analysis in real time and cross sections were determined after each step.

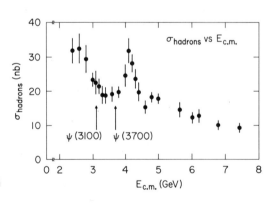

Fig. 17 Total hadron cross section vs. $E_{c.m.}$.

The ψ' was discovered by this system in its first few hours of use. Since then, the region between 3.2 GeV and 5.9 GeV has been scanned, as shown in Fig. 18. The scan has since been extended to 7.6 GeV. The sensitivity of the scan was such that a narrow resonance with an integrated strength between 5 and 10% of that of the ψ would have been seen. A set

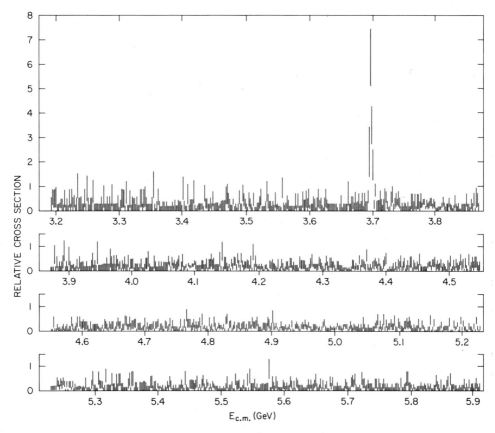

Fig. 18 Relative cross sections for e^+e^- into hadrons from the fine mesh narrow resonance search.

of upper limits for narrow resonances is shown in Table IV.

SUMMARY AND QUESTIONS

The ψ and ψ' seem to couple directly to photons, and have quantum numbers $J^{PC} = 1^{--}$. Both have odd G-parity and isospin zero. Photoproduction of the ψ at SLAC[17] and FNAL[18] implies a ψ-nucleon cross section of about 1 mb. These results seem consistent with the idea that the ψ's are hadrons.

On the other hand, one must contend with the narrow widths of the ψ's, about 70 keV for the ψ and 225 keV for the ψ'. At this time, no thoroughly satisfying theoretical explanation has emerged for the existence and properties of

TABLE IV

Results of the search for narrow resonances. Upper limits (90% confidence level) for the radiatively corrected integrated cross section of a possible narrow resonance. The width of this resonance is assumed to be small compared to the mass resolution.

Mass Range (GeV)	Limit on $\int \sigma_H \, dE_{cm}$ (nb MeV)
3.20 → 3.50	970
3.50 → 3.69	780
3.72 → 4.00	850
4.00 → 4.40	620
4.40 → 4.90	580
4.90 → 5.40	780
5.40 → 5.90	800
5.90 → 7.60	450

the ψ's. The leading contenders appear to be color and charm.[19] The color theories forbid strong decays of the ψ's because of a new nonadditive quantum number, so that electromagnetic decays with the emission of a photon would be expected. No evidence for such decays is seen. For example, in the four pion plus neutral decay of the ψ, there is a strong $\omega\pi\pi$ signal, indicating that the neutral is a π^0 and not a γ. The charm theories are based on the existence of a new additive quantum number, and rely on Zweig's[19] rule to suppress strong decays. The charm schemes imply the existence of many other states. A search[20] for some of these charmed states was performed at SPEAR and no evidence for them was found.

Another open question is to the meaning of the structure at 4.1 GeV. Is it another resonance related to the ψ and ψ' but having decay modes kinematically inaccessible to the ψ's? Are the resonances related to the step in $R = \sigma_h/\sigma_\mu$ around 4 GeV? Many questions have been raised by these data in the past few months – perhaps the next few months will see answers.

REFERENCES AND FOOTNOTES

1) J.-E. Augustin et al., Phys. Rev. Letters 33, 1406 (1974).
   J. J. Aubert et al., Phys. Rev. Letters 33, 1404 (1974).
   C. Bacci et al., Phys. Rev. Letters 33, 1408 (1974).

2) G. S. Abrams et al., Phys. Rev. Letters 33, 1453 (1974).

3) SPEAR Storage Ring Group, Proc. IXth Int. Conf. on High Energy Accelerators, Stanford Linear Accelerator Center, 1974, pp. 37-42.

4) The members of the SLAC-LBL Collaboration are: A. M. Boyarski, M. Breidenbach, F. Bulos, G. J. Feldman, G. E. Fischer, D. Fryberger, G. Hanson, D. L. Hartill, B. Jean-Marie, R. R. Larsen, D. Lüke, V. Lüth, H. L. Lynch, D. Lyon, C. C. Morehouse, J. M. Paterson, M. L. Perl, T. P. Pun, P. Rapidis, B. Richter, R. F. Schwitters, W. Tanenbaum, and F. Vannucci (Stanford Linear Accelerator Center); G. S. Abrams, D. D. Briggs, W. Chinowsky, C. E. Friedberg, G. Goldhaber, J. A. Kadyk, A. M. Litke, B. A. Lulu, F. M. Pierre, B. Sadoulet, G. H. Trilling, J. S. Whitaker, F. Winkelmann, and J. E. Wiss (Lawrence Berkeley Laboratory).

5) A. M. Boyarski et al., Phys. Rev. Letters 34, 1357 (1975).

6) It is possible that the muon sample has a small hadronic contamination. Muon identification relies on shower counter information, aided by penetration of the muon through 20 cm of iron. Collinear pion pairs could be mistaken for muon pairs.

7) A. M. Boyarski et al., to be published.

8) These values for the QED cross sections reflect the cut $|\cos \theta| < 0.6$.

9) D. R. Yennie, Phys. Rev. Letters 34, 239 (1975).

10) There is no strong evidence for the decay mode $\psi \to \pi^+ \pi^- \pi^+ \pi^- \gamma$. Aside from the missing mass peak being located slightly above a mass of zero, these events show a strong $\omega \pi \pi$ and $\rho \pi \pi$ clustering.

11) F. J. Gilman, private communication.

12) G. S. Abrams et al., Phys. Rev. Letters 34, 1177 (1975).

13) There are small phase space corrections to the ratios predicted from the isospins.

14) C. Callen et al., Phys. Rev. Letters 34, 52 (1975).
    T. Appelquist et al., Phys. Rev. Letters 34, 365 (1975).
    E. Eichten et al., Phys. Rev. Letters 34, 369 (1975).

15) J.-E. Augustin et al., Phys. Rev. Letters 34, 764 (1975).

16) A. M. Boyarski et al., Phys. Rev. Letters 34, 762 (1975).

17) J. T. Dakin et al., Phys. Letters 56B, 405 (1975).
U. Camerini et al., SLAC-PUB-1591 (1975), submitted to Phys. Rev. Letters.

18) B. Knapp et al., Phys. Rev. Letters 34, 1040 (1975); ibid., p. 1044.

19) For a review of these theories, see H. Harari, SLAC-PUB-1514, unpublished; F. J. Gilman, SLAC-PUB-1537 (1975), to be published in Proc. Orbis Scientiae II, Coral Gables, Florida, 1975; O. W. Greenberg, University of Maryland Technical Report No. 75-064, to be published in Proc. Orbis Scientiae II, Coral Gables, Fla., 1975.

20) A. M. Boyarski et al., SLAC-PUB-1583 (1975), submitted to Phys. Rev. Letters.

APPENDIX I

The storage ring SPEAR circulates one bunch of electrons and one bunch of positrons in a single magnetic guide field. The bunches collide alternately in two interaction regions. The beam energies may now be varied between about 1.3 GeV and 4 GeV. The energy distribution of electrons within a beam bunch is approximately Gaussian with a width that increases approximately quadratically with energy, and has a σ of about 1 MeV at a total energy ($E_{c.m.} = 2E_{beam}$) of $E_{c.m.} = 3$ GeV. The absolute energy calibration is based on measurements of the particle orbits and the magnetic guide fields and is known to about 0.1%. The bunch shapes are Gaussian with σ's in the transverse plane of approximately 0.1 cm and longitudinally a few cm. The luminosity is about $3 \times 10^{29} \text{cm}^{-2} \text{sec}^{-1}$ at $E_{c.m.} = 3$ GeV.

The magnetic detector is shown schematically in Fig. 1. The magnetic field of 4 kilogauss is axial and within a volume about 3 meters in diameter by 3 meters long. The interaction region is surrounded by a stainless steel vacuum pipe 0.15 mm thick. Coaxial with the pipe are a pair of cylindrical plastic scintillation counters that form one element of the trigger system. Continuing radially outward are four sets of multiwire spark chambers. Each set consists of four layers of wires at $\pm 2°$ and $\pm 4°$ with respect to the beam axis. Thus, each set of chambers provides redundant azimuthal (resolution ≈ 0.5 mm) and longitudinal (resolution ≈ 1.2 cm) position information for each charged particle. Following the spark chambers are a set of 48 plastic scintillator trigger counters. These counters are used in the trigger system and in a time-of-flight particle identification system with a resolution (σ) of about 0.5 nsec, allowing π/K separation up to about 0.65 GeV/c. Next comes the aluminum coil of the solenoid with a thickness of about 1 radiation length, followed by a layer of 24 lead-scintillator sandwich electron shower counters used to identify electrons. The next element is the iron return yoke of the magnet, which also serves as a hadron filter for the final set of spark chambers which aid in muon identification.

The accuracy of single particle identification in the detector is momentum dependent. Above about 600 MeV/c the probability of mistaking an electron for a hadron is about 6%, an electron for a muon about 1%, and a muon for a hadron about 8%. In the momentum range 600 MeV/c to 900 MeV/c, the probabilities of hadrons mimicking electrons or muons are 13% and 16%, respectively.

The trigger requirement is two or more charged particles with transverse momenta greater than about 200 MeV/c. The complete detector system covers a solid angle of $0.65 \times 4\pi$. A hadronic event is defined to be one with 3 charged particles or two charged particles acollinear by $20°$ or more. The detection efficiency for hadronic events varies smoothly from 40% at $E_{c.m.}$ = 2.5 GeV to 65% at 7.0 GeV. Backgrounds have been studied using separated beams and longitudinal (z) distributions of events. The background contribution to the resonances is very small, of order 0.1%, and is roughly 5% in the nonresonant region. Normally, cross sections are normalized by measuring Bhabha scattering in the magnetic detector. However, in the vicinity of the ψ, the e^+e^- pair production cross section is strongly enhanced by "2nd order electromagnetic" decays of the resonance. Hence, the luminosity is integrated by a set of small counters monitoring Bhabha scattering at small angles (where the scattering is dominantly caused by space-like photons). The luminosity monitor is calibrated with the magnetic detector at a beam energy far from the resonances.

DISCUSSION

CHAIRMAN: M. Breidenbach

Scientific Secretaries: R. Rückl and J. Strauss

DISCUSSION

- DRELL:

Your G-parity assignment to ψ is based on the analysis of the exclusive channels

$$\psi \rightarrow 2\pi^+ 2\pi^- \pi^0$$

and

$$\psi \rightarrow 3\pi^+ 3\pi^- \pi^0 \ .$$

In these final states, you have a $\leq 20\%$ probability that the π^0 is in fact a γ. You do not have this uncertainty in 4- and 6-charged pion states. The fact that the corresponding ratios α = Ron/Roff   1 already gives you a strong hint that the G-parity of ψ is negative.

- CRONIN:

I am asking about the beautiful azimuthal distribution for hadrons with x > 0.4 in the case of polarized beams. This is presumably due to a finite value for α via

$$N(\phi, \theta) \sim (1 + \alpha \cos \theta^2) + |P^+P^-|\alpha \sin \theta^2 \cos \phi^2 \ .$$

The modulation height was ~ 0.5, so I would think $\alpha \sim 0.6$. Why do you not see the effect of this α in the unpolarized angular distribution?

- BREIDENBACH:

First the unpolarized angular distributions are at different energies. Secondly, our data are limited to $|\cos \theta| \leq 0.6$. There is not a big effect in the polar angle distribution, so the errors on α are large.

- POLITZER:

You showed data on the one pion inclusive cross-section that must be about one year old. Is there anything more recent? It would be interesting to know whether it scales below 3 GeV where R scales, or above 5 GeV where R is again flat-ish. There is a simple, naive expectation based on the idea of a threshold for a new quark species, i.e. first there is an increase at low x because the extra cross-section is predominantly two meson production; at higher energy the curve should fill out and have the shape that it had below 3 GeV, only scaled up proportionally for all x.

— *BREIDENBACH:*

The analysis is underway right now, and it should be ready very soon.

— *POLITZER:*

I would like to make a second comment. The intermediate states d between ψ and ψ', though very welcome, pose the following problem: the branching ratio of $\psi' \to \psi\gamma\gamma$ (quoted to be about 4%) is the product of two branching ratios. But, $\psi' \to d\gamma$ is itself supposed to be less than 6%. This suggests that the new states decay predominantly (like 70%) electromagnetically. But if they are related to the ψ's, they are hadrons, and this seems absurd.

— *BREIDENBACH:*

The limits on $\psi' \to \gamma + X$ are limits on a single line. If there are two (or three) monochromatic lines, and if they are separated by more than 20 MeV, then the limit applies to each one separately.

— *DRELL:*

The statement by Mr. Politzer is not true. How else but electromagnetically could the d decay? The Zweig rule suppression would imply that they decay to the ψ: and of the competing pion decays to the ψ, one π is impossible, two π's have the wrong C, and three π's have almost no phase space.

— *WILSON:*

What improvements for your detector are you planning for the future?

— *BREIDENBACH:*

The main improvements we plan for our second generation detector are:

i) a larger solid angle, say about 95% of 4π, by putting in drift chambers extending into the forward regions;

ii) a higher detection efficiency for neutrals by additional liquid argon counters;

iii) a better trigger.

EVIDENCE FOR A NEW RESONANCE P_c AND OTHER RECENT RESULTS OBTAINED AT DORIS
USING DASP

B.H.Wiik
Deutsches Elektronen-Synchrotron DESY
Hamburg, Germany

The experimental program utilizing the e^+e^- colliding ring DORIS got underway in late 1974. Since then data have been taken using three large, rather complementary detectors. In this talk the present status of the accelerator will be described and the data obtained with the double arm spectrometer DASP discussed. Most of the data have been collected for energies at or adjacent to the masses of the newly found resonances[1] at 3.09 GeV and 3.68 GeV. Here preliminary results on the following topics will be presented:

1) Inclusive distributions, particle ratios, and limits on heavy charged mesons from the decays of the J/ψ and ψ' resonances.

2) Hadronic two body decays of J/ψ and ψ'.

3) $\gamma\gamma$ final states and the observation of the $\eta\gamma$ decay mode.

4) The cascade decay: $\psi' \to (J/\psi) + X$

5) Evidence for narrow states P_c with even charge conjugation and masses between 3.7 GeV and 3.1 GeV.

The results on the decay of the resonances into lepton pairs[2] agree within the errors with the data[3] reported by the SPEAR solenoid group and will not be discussed here.

A substantial amount of data has also been collected using the two other detectors. These groups are now analyzing their data and the first results will be presented at the Stanford Conference later this year.

DORIS

The layout of the DESY accelerator complex is shown in Fig. 1. Electrons and positrons from a 400 MeV linear accelerator are injected into the 7.5 GeV synchrotron to be accelerated to the proper energy for transfer to DORIS. DORIS is in the shape of an race track with two nearly independent rings stacked one above the other. The beams cross in the vertical plane in the middle of the two long straight sections.

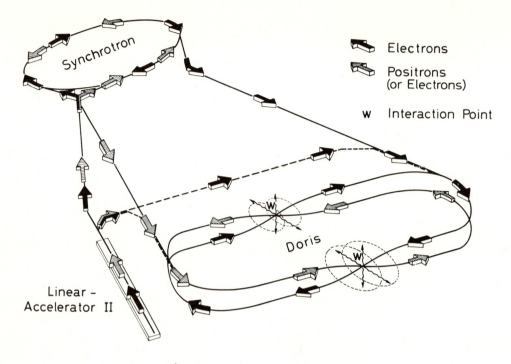

Fig. 1
Layout of the DESY accelerators

Although in principle CMS energies as high as 7 GeV can be reached with the present accelerator, all the data have been collected for energies between 3 GeV and 4.2 GeV. Upgrading the present magnet power supplies and installing all the available r.f. power in one ring will make it possible to reach a CMS energy of 10 GeV.

With the present optics the measured specific luminosity around 4 GeV (CMS) can be written as $10^{31} I^2$ (A^{-2} cm^{-2} sec^{-1}) with the circulating current I measured in Amperes. This is in good agreement with the value of $1.2 \, I^2 \, 10^{31}$ ($A^{-2} sec^{-1} cm^{-2}$) predicted from the present optics. So far luminosities in excess of 10^{30} $cm^{-2} sec^{-1}$ have been reached with stored currents of 0.3 - 0.4 Ampere in each beam. However, at these high currents, the energy spread in the beam increases due to coherent synchrotron oscillations. These oscillations can be damped for currents less than 0.2 - 0.25 Ampere by a second r.f. transmitter tuned slightly off the resonance frequency. Hence during the data taking at the 3.1 GeV and 3.7 GeV resonances, where a good mass resolution is imperative, the currents were limited to less than 0.25 Ampere. Another important parameter is the size of the interaction volume. Due to the finite crossing angle and the high r.f. frequency used, the interaction "point" is around 2-3 cm long. The transverse dimensions are less than 0.6 mm.

The main parameters of DORIS are summarized in Table I.

Table I:

Max.Energy:	7 GeV (9 GeV in the fall of 1975)
Average Circulating current:	0.200 - 0.400 Ampere / Ring
Number of Bunches:	1 - 480
Average Luminosity:	$\sim 3 \times 10^{29}$ cm^{2}sec^{-1}
Beam lifetime:	> 6 hrs.
Gas pressure:	$1 - 5 \times 10^{-9}$ torr (depending on the current)
Interaction volume:	
along the beam	2 - 3 cm
transverse to the beam	< 0.6 mm.

DASP
A schematic drawing of DASP[4] viewed along the beam direction is shown in Fig. 2. Two large H-magnets are positioned symmetric with respect to the interaction

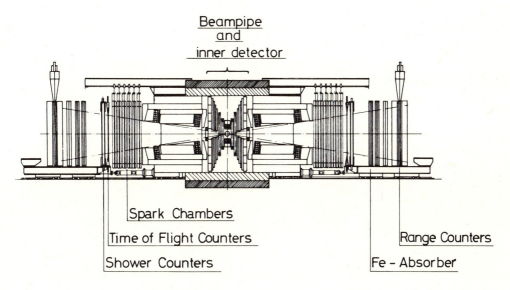

Fig. 2
DASP viewed along the beam direction

point and spaced 2.1 m apart. The geometric acceptance of each magnet is from 48° to 132° in production angle and ± 9° in azimuth resulting in a solid angle of 2 x 0.45 sterad for both magnets. The acceptance for a charged particle is smaller than this and depends on the momentum, the field strength and last detector plane required. The maximum field strength is 1.1 Tesla, the integrated field length 1.8 Tm.

A charged particle emerging from the interaction point traverses the following detectors before reaching the magnet gap: a scintillation counter adjacent to the beam pipe, a second scintillation counter which starts the time of flight measurement, two proportional chambers (3 planes in each chamber, 2 mm wire spacing) a third scintillation counter used for triggering and a wire spark chamber with magnetostrictive readout (2 planes, 1 mm wire spacing).

The momentum of a charged particle is determined from the measurement of one space point on the trajectory in front of the magnet and the knowledge of the trajectory of the particle behind the magnet. The trajectory behind the magnet is measured by 6 wire spark chambers (each has 2 signal planes, 1 mm wire spacing, 5.6 by 1.7 m^2 sensitive area). At the present a resolution of ± 0.8% is reached for a particle with 1.0 GeV/c momentum.

The particles are identified using time of flight, shower and range information:

The time of flight counters are mounted behind the spark chambers at an average distance of 4.7 m from the interaction point. The measured time of flight resolution is 0.5 nsec FWHM averaged over the 31 elements in one arm. This makes it feasible to separate pions and kaons for momenta less than 1.8 GeV/c, and kaons and protons for momenta less than 3 GeV/c, by time of flight alone. Hadrons and muons are separated from electrons using pulse height in the shower counters. These counters are made of alternating sheets of lead and scintillator, a total of 6.2 radiation lengths thick. For an incident particle with a momentum of 1.5 GeV/c the cut in the shower pulse height can be made such that 80% of the pions but only 10^{-3} of the electrons have pulse heights below the cut, that is, an electron rejection in the pion signal of 10^{-3} in one arm. To further improve the electron-pion identification a fraction of the data was collected with a spark chamber positioned just after the shower counter.

The muons are positively identified by their range. The absorbers in the range telescope are made of iron a total of 90 cm thick, subdivided into plates of different thickness in order to allow for an optimal pion/muon separation at a chosen momentum. After each plate sufficient space for either a scintillation counter hodoscope or a spark chamber is provided. The main portion of the data were taken with one wall of scintillators at depth of 70 cm in the iron. This was later reduced to 60 cm and a spark chamber was installed at a depth of 40 cm in the iron.

The inner detector shown in Fig. 3 is located in the free space between the magnets. It is made out of proportional chambers, scintillation counters, proportional tube chambers, and shower counters. This part of the detector covers more than 70% of 4π and is well suited for a measurement of the direction and energy of photons and also the direction and, in some cases, the energy of charged particles.

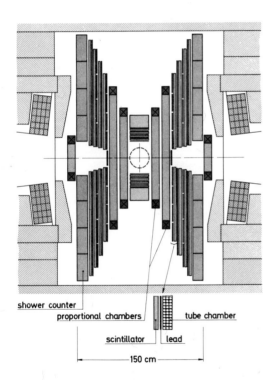

DASP — Inner Detector

Fig. 3

The inner detector of DASP viewed along the beam direction

The basic unit of the inner detector is a scintillation counter hodoscope, a sheet of lead 5 mm thick and a proportional tube chamber. The tube chambers consist of three (two) signal planes and measure the position to ± 5 mm (± 7.5 mm). A particle emitted in the direction of the inner detector first traverses one of the 22 scintillation counters surrounding the beam pipe, the proportional chambers (in a part of the azimuthal acceptance), then four of the units just described, and finally a lead-scintillator shower counter eight radiation lengths thick. The direction of a shower is determined to about $\pm 2°$, the energy resolution is around 30% for an incident photon of 1 GeV. The detector has a 50% efficiency for detecting a 50 MeV photon.

RESULTS

Inclusive spectra and particle ratios

For a measurement of the inclusive spectra the spectrometer was triggered on a single charged particle traversing one of the arms of the spectrometer. The trigger efficiency is thus independent of the final state and does not introduce any systematic uncertainties between pions, kaons and protons. Hadrons are separated from electrons and muons by the information from the shower counters and the range counters as discussed above. Pions, kaons and protons are identified using the time of flight information shown in Fig. 4. Here the events are displayed in a scatter plot with $(1/\beta^2 - 1)$ as measured by the time of flight system versus $1/p^2$. Since $(1/\beta^2 - 1) = M^2/p^2$ particles of mass M will cluster along a straight line in this plot. The particles are seen to be well separated at all momenta up to the kinematical boundary.

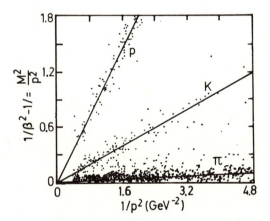

Fig. 4 Time of flight spectrum measured at $2E_{CM}$ = 3.1 GeV

EVIDENCE FOR A NEW RESONANCE Pc

In Fig. 5 the invariant cross section for the decay $J/\psi \to h^{\pm}X$ is plotted in arbitrary units versus the particle energy for pions, kaons and antiprotons (x2). The invariant cross section for antiprotons seems to be larger than the cross section for pions and kaons. Note, however, that the cross sections are all within a factor of 3 and tend to approach each other with increasing particle energy. To a good approximation, the cross sections are decreasing exponentially, for the pions the energy dependence is approximately e^{-6E}.

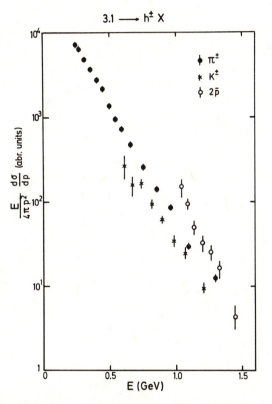

Fig. 5
The invariant cross section versus the particle energy for the decay
$$J/\psi \to h^{\pm}X$$

In Fig. 6 the invariant cross section for the decay $\psi' \to h^{\pm}X$ is plotted in arbitrary units versus the particle energy for pions, kaons and antiprotons (x2). The cross section has the same general behaviour as the cross sections measured in the decay of the J/ψ resonance. This similarity might trivially have been anticipated, since the decay $\psi' \to J/\psi + X$ accounts for roughly 60% of all the ψ' decays. The decay mode $\psi' \to \psi + \pi^{+}\pi^{-}$ is also seen directly as a step in the inclusive pion cross section at low pion energies.

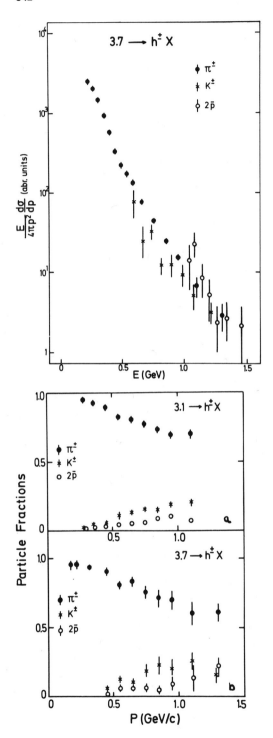

Fig. 6
The invariant cross section versus the particle energy for the decay $\psi' \to h^{\pm} X$

The fraction of pions, kaons and antiprotons observed in the decay of the J/ψ and ψ' resonances are plotted in Fig. 7a,b as a function of momentum. Similar relative particle fractions are found at both resonances; practically all the particles at low momenta are pions, but the relative yield of kaons and protons increases with momentum. At a momentum of 1.0 GeV/c the relative fractions are 0.2 for kaons and 0.1 for antiprotons (x2).

Fig. 7
Fractions of pions, kaons and antiprotons (x2) observed (a) at the J/ψ resonance and (b) at the ψ' resonance

EVIDENCE FOR A NEW RESONANCE Pc

It has often been conjectured[5] that the relative yield of kaons might increase above 4 GeV due to the production and decay of charmed particles. No such increase was observed at a CMS energy of 4.15 GeV. This is seen in Table II where the relative particle fractions observed for momenta between 0.45 GeV/c and 1.3 GeV/c are listed and compared to the results obtained at the resonances.

Table II:

$2E_{CM}$	$\pi^{\pm}$/all	$K^{\pm}$/all	$2\bar{p}$/all
3.1	0.78 ± 0.008	0.14 ± 0.008	0.062 ± 0.006
3.7	0.82 ± 0.02	0.114 ± 0.013	0.065 ± 0.009
4.15	0.84 ± 0.09	0.10 ± 0.05	—

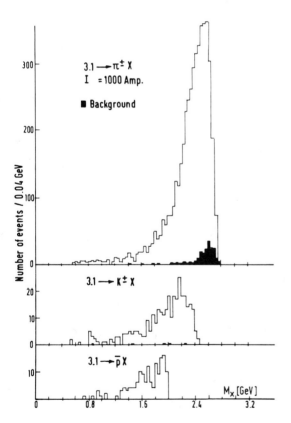

Fig. 8
Missing mass spectrum measured in the decays $J/\psi \to h^{\pm} X$

Any quasi-two body decay of the type $J/\psi(\psi') \to h + R$ will show up as a peak in the missing mass spectrum of states recoiling against the particle h. The missing mass spectra for the decay channels

$$J/\psi \to \begin{matrix} \pi^{\pm} X \\ K^{\pm} X \\ \bar{p} X \end{matrix}$$

are plotted in Fig. 8. Except for the existence of a possible $KK^*(890)$, no pronounced quasi-two body decays are observed. The background, shown as the black histogram in Fig. 8, was measured simultaneously with the data by selecting events originating outside of the interaction volume. The estimated resolution in missing mass varied between 50 MeV and 10 MeV for missing masses between the ρ mass and 2 GeV.

The missing mass spectra for the decay $\psi' \to h^{\pm}X^{\mp}$ are plotted in Fig. 9. No prominent structure with a confidence level of more than 4σ is observed.

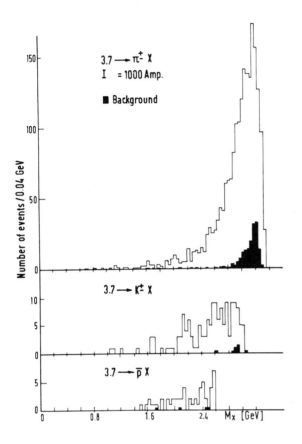

Fig. 9
Missing mass spectra measured in the decays $\psi' \to h^{\pm}X^{\mp}$

According to the color[6] interpretation of the new resonances, the ψ' is expected to decay with the emission of a single pion into a colored meson ρ_c with a mass around 3.1 GeV. Therefore, if the ρ_c exists, it will manifest itself as a peak in the missing mass spectrum observed in the decay $\psi' \to \pi^{\pm} X^{\mp}$. To search for this state a fraction of the data was collected at a reduced magnet current to ensure a uniform mass acceptance around 3.1 GeV. There is no evidence for such a peak in the measured spectrum plotted in Fig. 10. An upper limit on the decay $\psi' \to \pi^{\pm} \rho_c^{\mp}$ is derived from these data assuming the ρ_c to be less than 50 MeV wide. We find with 90% confidence that

$$\frac{\Gamma_{\psi' \to \pi^{\pm}\rho_c^{\mp}}}{\Gamma_{\psi' \to all}} < 5 \times 10^{-2} \quad .$$

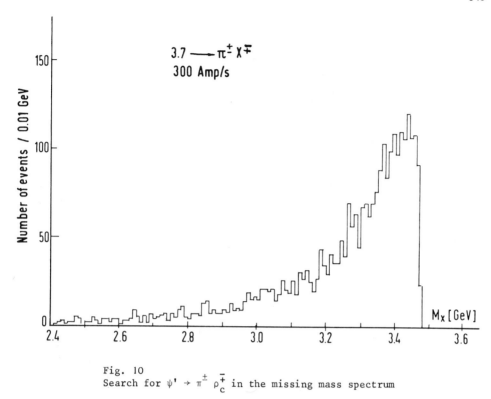

Fig. 10
Search for $\psi' \to \pi^{\pm} \rho_c^{\mp}$ in the missing mass spectrum

HADRONIC TWO BODY DECAYS

The good particle identification properties of DASP make it possible to measure the cross section for pion and kaon pair production[7]. To select these events and to separate them from the much more abundant muon and electron pairs the following criteria were used:

1) The geometrical reconstruction required tracks of opposite charges in the two spectrometer arms, collinear within 0.15 rad.

2) The production vertex was within ± 5 cm of the nominal interaction point. These criteria selected pair events. To identify the pairs as pions or kaons the criteria 3-5 were used.

3) Momentum of the track should be within ± 50 MeV of the nominal momentum.

4) Pulse height in the shower counter should be less than 4-times the most probable value for a minimum ionizing particle.

5) No range counter should fire.

At the J/ψ resonance no events, at the ψ' resonance one event satisfying the criteria above were found. The corresponding 90% upper confidence limits are listed in Table III. These limits were derived using the leptonic widths measured[3] by the SPEAR solenoid group.

Table III:

Mode	J/ψ	ψ'
$\Gamma_{\pi^+\pi^-}/\Gamma$	$< 3 \times 10^{-4}$	$< 9 \times 10^{-4}$
$\Gamma_{K^+K^-}/\Gamma$	$< 6 \times 10^{-4}$	$< 1.6 \times 10^{-3}$

The absence of $\pi^+\pi^-$ events support the isospin and G-parity assignment $I^G = 0^-$. A small K^+K^- rate is expected if the new resonances are SU_3 singlet states.

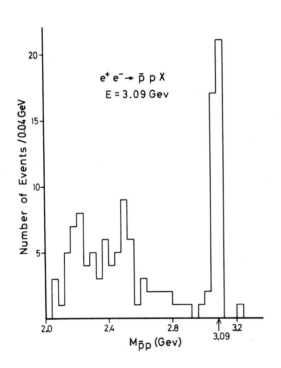

Fig. 11
Effective mass spectrum of $p\bar{p}$ pairs from the decay of the J/ψ resonance

The upper limits given on $\Gamma_{\pi^+\pi^-}/\Gamma$ and $\Gamma_{K^+K^-}/\Gamma$ for the J/ψ resonance are near the level where decay into $\pi^+\pi^-$ or K^+K^- pairs should occur via the second order radiative diagram. Extrapolating either the available data[8] or using the rho pole formula leads to $\Gamma_{\pi^+\pi^-}/\Gamma \sim 5 \cdot 10^{-4} - 5 \cdot 10^{-5}$ and a similar result for the K^+K^- decay.

The $p\bar{p}$ pairs were uniquely identified by time of flight. The effective mass spectrum of the $p\bar{p}$ pairs from the decay of the J/ψ resonance is plotted in Fig. 11. The elastic peak J/ψ → $p\bar{p}$ is clearly seen centered at a mass of 3.09 GeV. To evaluate the branching ratio the angular distribution must be known.

Assuming the angular distribution to be given by $1 + \cos^2\theta$ we find:

$$\Gamma_{p\bar{p}}/\Gamma = (0.0023 \pm 0.0006)$$

This result is based on a fraction of the events shown in Fig. 11. No candidates for the decay $\psi' \to p\bar{p}$ were found and the upper limit is:

$$\Gamma_{p\bar{p}}/\Gamma < 0.001 \ .$$

γγ - FINAL STATES

The cross section for two collinear or nearly collinear photons has been measured[9] by demanding a coincidence between the sectors of the inner detector mounted above and below the beam pipe. Bhabha scatters and cosmic ray events were rejected by requiring that no more than two of the two upper and two lower layers of scintillators between the interaction point and the first lead converters be fired. This allows for the possibility that one of the two photons converts in the beam pipe. The remaining background from beam-gas interactions and cosmic ray events was removed by cuts in the energy deposited, number of proportional tubes fired and the time of flight.

Collinear photon pairs are produced in the two photon annihilation $e^+e^- \to \gamma\gamma$. Superimposed on this smoothly decreasing cross section there might be a peak at the mass of the resonances due to any of the following processes:

1) Direct decay of the resonances into a pair of collinear photons. This decay is strictly forbidden for a spin 1 state. It is allowed for all other values of the spin provided C is even or not conserved.

2) $e^+e^- \to \psi/J \ (\psi') \to \pi^0 + \gamma$. In this case there will always be two photons collinear to within an angle of m_π/E_π. If the second photon from the π^0 decay is not detected the event will masquerade as a γγ event.

3) $e^+e^- \to \psi/J \ (\psi') \to (X \to \gamma\gamma) + \gamma$

   The resonance decays with the emission of a single photon into a heavy particle X which in turn decays into two photons. If the mass of X is close to 3.1 GeV or 3.7 GeV, X will be produced nearly at rest and the two photons from its decay will be nearly collinear.

The measured energy dependence of the γγ cross section is plotted in Fig. 12 and Fig. 13 for energies in the vicinity of 3.09 GeV and 3.68 GeV. No resonance effect is observed and the absolute value of the cross section is in good agreement with the predicted QED cross section shown as the dotted line.

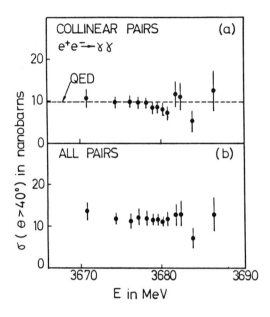

Fig. 12
Cross section for $e^+e^- \to \gamma\gamma$ as a function of CMS energy in the vicinity of 3.09 GeV. The dotted line is the rate predicted from QED

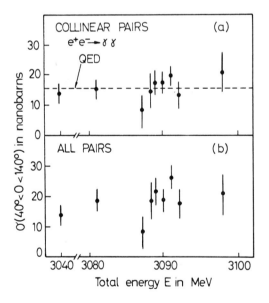

Fig. 13
Cross section for $e^+e^- \to \gamma\gamma$ as a function of CMS energy in the vicinity of 3.68 GeV. The dotted line is the rate predicted from QED

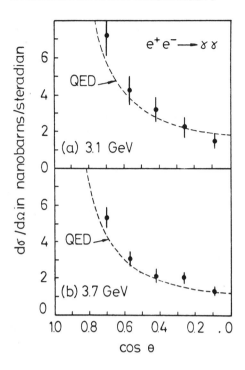

The angular distributions, plotted in Fig. 14a, b are also consistent with QED predictions, plotted as the solid line. Upper limits on the various resonance contributions were derived from the observed energy dependence of the cross sections by fitting the data to a non resonant background plus a Gaussian peak with the mass and the width observed in the total hadron cross section using the same detector. The 90% confidence upper limits are listed in Table IV:

Fig. 14
Angular distributions of $e^+e^- \to \gamma\gamma$
a) for CMS energies near 3.09 GeV
b) for CMS energies near 3.68 GeV
The dashed lines are the angular distributions predicted from QED

Table IV:

Mode	J/ψ	ψ'
$\Gamma_{\gamma\gamma}/\Gamma$	0.003	0.008
$\Gamma_{\pi\gamma}/\Gamma$	0.01	0.01
$(\Gamma_{X\gamma}/\Gamma)(\Gamma_{X\to\gamma\gamma}/\Gamma_{X\to all})$	0.003 (2.99GeV<M_x<3.09GeV)	0.008 (3.58GeV<M_x<3.68GeV)

THE ηγ DECAY MODE

Using the inner detector the decay of the resonances into η + γ has been investigated identifying the η by its 2γ decay mode. Demanding a final state with just three photons, coplanar to within 5°, eliminated the much more abundant beam-gas, cosmic ray and multihadron events. Note that the event is completely determined by a measurement of the direction of the three photons. In particular, possible π^0 and η' events are excluded using the opening angle and invariant mass of the secondary photons.

The results are:

$J/\psi \rightarrow \eta + \gamma$ $0.1 \text{ keV} < \Gamma_{\eta\gamma} < 2 \text{ keV}$

$\psi' \rightarrow \eta + \gamma$ $\Gamma_{\eta\gamma} < 0.8 \text{ keV}$

THE CASCADE DECAY: ($\psi' \rightarrow J/\psi + X$)

The yield of muon pairs produced at the ψ'-resonance was measured by demanding one or both of two oppositely charged particles accepted by the magnets to penetrate 70 cm of iron. These criteria provide a clean sample of muon pairs; the contamination due to hadron pairs can be completely neglected. The effective mass spectrum plotted in Fig. 15 shows two peaks. The first peak is centered at the mass of the ψ' and is the sum of the decay of the ψ' into a pair of muons and the muons directly produced by QED. The second peak at a mass of 3.09 GeV is from the decay of the J/ψ resonance into a pair of muons and is direct evidence[10] for the decay $\psi' \rightarrow J/\psi + X$. The ψ' decays roughly 60% of the time via this channel.

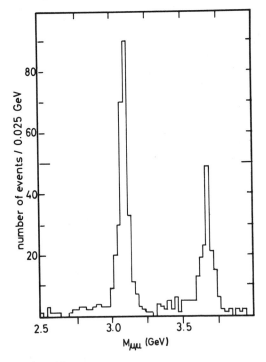

Fig. 15
The yield of muon pairs produced at the ψ' resonance a function of the effective pair mass

EVIDENCE FOR A NEW RESONANCE P_c

The following decay modes X are allowed by energy, charge and C-parity conservation

$$\psi' \to (J/\psi) + \pi^0$$
$$\pi^+\pi^- \text{ or } \pi^0\pi^0$$
$$\pi^+\pi^-\pi^0 \text{ or } \pi^0\pi^0\pi^0$$
$$\eta$$
$$2\pi^+\pi^- \text{ or } 2\pi^0\pi^0$$
$$\gamma\gamma$$

The $\gamma\gamma$ mode can arise from two successive transitions through a new intermediate state P_c. Evidence for such states has now been obtained and will be presented below.

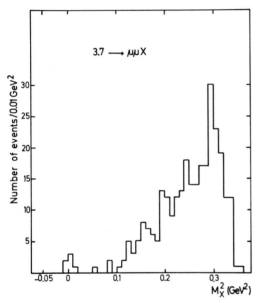

Fig. 16
The spectrum of missing masses in the reaction $e^+e^- \to \mu^+\mu^- + $ missing mass produced at the ψ' resonance. The $\mu^+\mu^-$ effective mass is required to be between 2.9 GeV and 3.2 GeV

The cascade decays were investigated in more detail by selecting events with a muon pair mass between 2.9 GeV and 3.2 GeV. For these events, the missing mass recoiling against the $\mu^+\mu^-$ pair was computed from the known muon momenta assuming the pairs result from the decay of the J/ψ resonance. The spectrum, plotted in Fig. 16 as a function of missing mass squared, peaks strongly at large values of missing mass with only a few events below the two pion threshold. In particular there is no evidence for decays with the emission of a single π^0.

The decay $\psi' \to J/\psi + \eta$ is observed in the missing mass spectrum as a narrow peak centered around 0.3 GeV^2. The observed width is consistent with the value expected from the experimental resolution alone. The existence of this decay mode proves that the ψ' and the J/ψ have the same isospin and G-parity.

The measured branching ratio

$$\frac{\psi' \to J/\psi + \eta}{\psi' \to \text{all}} = (0.037 \pm 0.015)$$

is surprisingly large considering the small phase space available.

The other decay modes listed above were disentangled using additional information from the inner detector. The results are:

1) No $2\pi^+\pi^-$ or $2\pi^0\pi^0$ events were found. The phase space available for these decay modes is exceedingly small.

2) Events with mixed charged and neutral tracks were found. These events are all consistent with either η decay or $\pi^0\pi^0$ decays with photons converting in the beam pipe. We have therefore no evidence for direct decays with the emission of $\pi^+\pi^-\pi^0$.

3) $\psi' \to J/\psi + \pi^+\pi^-$ is the dominant decay mode[10]. The measured branching ratio is

$$\frac{\psi' \to J/\psi + \pi^+\pi^-}{\psi' \to \text{all}} = 0.36 \pm 0.06$$

4) The decay $\psi' \to J/\psi + \pi^0\pi^0$ is directly observed with a branching ratio

$$\frac{\psi' \to J/\psi + \pi^0\pi^0}{\psi' \to \text{all}} = 0.18 \pm 0.06$$

The measured ratio of $\pi^0\pi^0/\pi^+\pi^-$ is consistent with 0.5 as expected for a $\Delta I = 0$ transition. Pure $\Delta I = 1$ or 2 transitions would lead to 0 and 2 for this ratio and seem to be excluded by the data.

EVIDENCE FOR NEW NARROW RESONANCES

The decay mode $\psi' \to J/\psi + \gamma\gamma$ was found[11] in two nearly independent experiments. In the first experiment the $\gamma\gamma$ cascade was identified by measuring the two electrons from the decay of the J/ψ resonance in coincidence with the photons. All four particles were observed in the inner detector only. In a second experiment, done concurrently with the first, the reaction was identified by observing the decay of the J/ψ into a pair of muons with the magnetic spectrometer and selecting events with just two photons in the inner detector. The two experiments will be discussed separately.

EVIDENCE FOR A NEW RESONANCE Pc

The decay $\psi' \to (J/\psi \to e^+e^-) + \gamma\gamma$ was measured by first selecting events with two electrons and two photons in the inner detector. For this purpose electrons and photons were defined as follows:

Electron: A charged particle, originating within the interaction volume, and producing a electromagnetic shower with an energy greater than 700 MeV. Electrons were accepted in the top and bottom part of the inner detector or in the horizontal part convered by the proportional chambers P1 and P2. The angles were determined to $\pm 1°$.

Photon: The appropriate beam counters and front counters in the inner detector should not have fired. The photon should convert "in" the detector as evidenced by either a pulse in the shower counter (the energy loss must be greater than 0.3 of that for a minimum ionizing particle) or at least one proportional tube and one scintillation counter fired. Photon showers which were within $15°$ of the axis of the electron shower were not accepted. The angular resolution for photons is $\pm 2°$ if they convert in front of a proportional tube chamber or $\pm 8°$ in ϕ and ± 20 in θ if they are detected by the shower counters only.

Restricting the noncollinearity angle of the electrons to less than $45°$, 71 such events were found. From the measured directions of the four particles and the known initial energy a OC calculation was made yielding all the momenta. From the momenta so determined the effective mass distribution of the electron pairs was computed and is plotted in Fig. 17a. The plot shows a clear peak less than 200 MeV wide centered at 3.1 GeV. This is in good agreement with the results of the Monte Carlo computation for the process $e^+e^- \to \psi' \to (\psi \to ee) + \gamma\gamma$, which yield a peak at 3.1 GeV, 130 MeV (FWHM) wide.

The most serious background is caused by the decay $\psi' \to (J/\psi \to e^+e^-) + \pi^0\pi^0$ with only two of the four photons detected. The contribution from this decay way evaluated by a Monte Carlo computation, using as input:

1) (The $\pi^0\pi^0$ mass spectrum) = 1/2 (the $\pi^+\pi^-$ mass spectrum observed in $\psi' \to J/\psi + \pi^+\pi^-$). This relationship is found experimentally and is predicted for a $\Delta I = 0$ transition.

2) The acceptance of the detector computed from the known geometric acceptance and the measured detection efficiency for a photon.

3) The measurements errors as listed above.

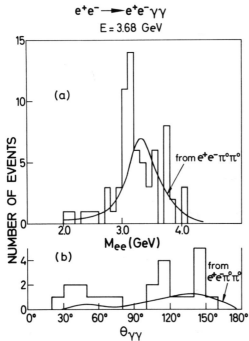

Fig. 17
a: Effective mass distribution of the e^+e^- pairs observed in the final state $e^+e^- \gamma \gamma$ at the mass of the ψ' resonance. The background predicted from $\psi' \to (J/\psi \to e^+e^-) + \pi^o\pi^o$ is shown as the solid line

b: The distribution of $e^+e^- \gamma \gamma$ events with 3.2 GeV > m_{ee} > 3.0 GeV as a function of the opening angle

The resulting mass distribution of the e^+e^- pairs is plotted as the solid line in Fig. 17a. The distribution peaks at an effective mass of 3.3 GeV and is 600 MeV wide (FWHM) in disagreement with the peak observed at 3.1 GeV.

The decay $\psi' \to (J/\psi \to e^+e^-) + (\eta \to \gamma\gamma, 3\pi^o)$ with two photons detected will also contribute to the background. The calculation of the $3\pi^o$ decay mode leads to a very broad mass distribution of the corresponding e^+e^- pairs and since the absolute magnitude is also small it is neglected. The contribution from the $\eta \to \gamma\gamma$ decay mode corresponds to electron pairs with a well defined mass centered at 3.1 GeV. However, since the η is produced nearly at rest, the angle $\theta_{\gamma\gamma}$ between the two photons from its decay will be larger than 140°. The $\theta_{\gamma\gamma}$ distribution for events in the peak between 3.0 and 3.2 GeV is shown in Fig. 17b together with the Monte Carlo prediction for

EVIDENCE FOR A NEW RESONANCE Pc

the background from $e^+e^- \pi^0\pi^0$. It follows that only 4 events in the narrow e^+e^- peak are due to the $\eta \to \gamma\gamma$. We therefore conclude that we have observed the cascade decay $\psi' \to J/\psi + \gamma\gamma$.

The decay[12] $\psi' \to (J/\psi \to \mu^+\mu^-) + \gamma\gamma$ was determined by detecting two muons as discussed above in coincidence with two photons in the inner detector. 33 such events were found.

There is a background resulting from:

$$\psi' \to (J/\psi \to \mu^+\mu^-) + \eta$$
$$\to (J/\psi \to \mu^+\mu^-) + \pi^0\pi^0$$

The η background was removed by restricting the missing mass recoiling against the $\mu^+\mu^-$ pair to values less than 520 MeV. To remove the $\pi^0\pi^0$ contribution, a 3C fit to the remaining sample of 29 events was made assuming these events to be genuine $\gamma\gamma$ events. The measured momenta and angles of the muons and the direction of the photons were used as an input to the computation together with their errors. The events were then plotted as a function of the χ^2 of the fit and candidates for genuine $\gamma\gamma$ events were required to have a χ^2 of less than 12. Altogether 9 events satisfied these criteria. The $\pi^0\pi^0$ background remaining in this sample was estimated, using the Monte Carlo computation described above, to contribute less than 0.8 events.

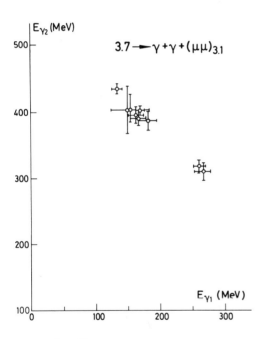

Fig. 18
Scatter plot of the two photon energies for candidates for the decay $\psi' \to (J/\psi \to \mu^+\mu^-) + \gamma\gamma$

The events are plotted in Fig. 18 as a function of the fitted photon energies.

The events cluster in one or possibly two groups demonstrating the existence of one or two narrow resonances P_c and P'_c with even charge conjugation.

The first group is centered at
$$E_1 = (169 \pm 7) \text{ MeV} \quad \text{and} \quad E_2 = (398 \pm 7) \text{ MeV}.$$
At present the order of the photon emission cannot be determined, so two values of the mass are possible:
$$M_{P_c} = (3.507 \pm 0.007) \text{ GeV} \quad \text{or}$$
$$M_{P_c} = (3.258 \pm 0.007) \text{ GeV}.$$
The total branching ratio:
$$\frac{\psi' \to P_c + \gamma_1}{\psi \to \text{all}} \cdot \frac{P_c \to J/\psi + \gamma_2}{P_c \to \text{all}} = (0.04 \pm 0.02)$$

A second group, consisting of two events, is centered at
$$E_{\gamma 1} = (263 \pm 8) \text{ MeV} \quad \text{and} \quad E_{\gamma 2} = (315 \pm 8) \text{ MeV}$$
This implies a mass
$$M_{P'_c} = (3.407 \pm 0.008) \text{ GeV} \quad \text{or}$$
$$M_{P'_c} = (3.351 \pm 0.008) \text{ GeV}.$$
The two events correspond to a branching ratio of 1.5×10^{-2}.

States with even charge conjugation were predicted[13] by the charm model in the mass range between 3.7 GeV and 3.1 GeV. This model further predicted these states to decay with a large branching ratio into the J/ψ resonance plus a photon. From a comparison of the branching ratio measured for the $\gamma\gamma$ cascade in this experiment and the limits obtained[14] on a single mono-energetic photon line, one concludes that indeed $P_c \to J/\psi + \gamma$ must be a major decay mode of this new resonance.

It is therefore tempting to associate the new P_c-resonances with the states predicted in the charmonium model. However, one should bear in mind that since these states might also be accomodated in a color model[6], their mere existence does neither prove nor disprove charm or color.

REFERENCES

1) J.J.Aubert et al., Phys.Rev.Lett. 33, 1404 (1974)

   J.E.Augustin et al., Phys.Rev.Lett. 33, 1406 (1974)

   G.S.Abrams et al., Phys.Rev.Lett. 33, 1453 (1974)

2) DASP Collaboration, Physics Lett. 53B, 393 (1974)

   DASP Collaboration, Physics Lett. 56B, 491 (1974)

   G.Wolf, Rapporteur talk at the EPS Int. Conference on High Energy Physics, Palermo, Italy, June 1975

3) A.M.Boyarski et al., Phys.Rev.Lett. 34, 1357 (1975)

   V.Lüth et al., SLAC-PUB 1617, LBL-4211 - August 1975

4) The following Institutions and physicists are involved in the DASP-collaboration:
   W.Braunschweig, U.Martyn, H.G.Sander, D.Schmitz, W.Sturm, and W.Wallraff, I.Physikalisches Institut der RWTH Aachen,
   K.Berkelman, D.Cords, R.Felst, E.Gaderman, G.Grindhammer, H.Hultschig, P.Joos, W.Koch, U.Kötz, H.Krehbiel, D.Kreinick, J.Ludwig, K.-H.Mess, K.C.Moffeit, A.Petersen, G.Poelz, J.Ringel, K.Sauerberg, P.Schmüser, G.Vogel, B.H.Wiik, and G.Wolf - Deutsches Elektronen-Synchrotron DESY and II. Institut für Experimentalphysik der Universität Hamburg, Hamburg,
   G.Buschhorn, R.Kotthaus, U.E.Kruse, H.Lierl, H.Oberlack, K.Pretzl, and M.Schliwa - Max-Planck-Institut für Physik und Astrophysik, München,
   S.Orito, T.Suda, Y.Totsuka and S.Yamada, University of Tokyo, Tokyo.

5) M.K.Gaillard, B.W.Lee and J.L.Rosner, Rev.Mod.Phys. 47, 277 (1975)

6) For a recent review see:
   O.W.Greeberg, University of Maryland - Technical Report No. 75-064 (1975) and references given in this report

7) The results on the decays $J/\psi \to \pi^+\pi^-$, K^+K^-, $p\bar{p}$ have been published

   DASP Collaboration, Physics Lett. 57B, 297 (1975)

8) M.Bernardini et al., Phys.Lett. 46B, 261 (1973)

9) The results obtained from a fraction of the data reported here have already been published

   DASP Collaboration - Phys.Lett 53B, 491 (1975)

10) G.S.Abrams et al., Phys.Rev.Lett 46B, 261 (1973)

11) The existence of a new state P_c was first announced at this meeting. The results are published in:

    DASP Collaboration - Phys.Lett. 57B, 407 (1975)

12) The discussion of this decay channel includes the analysis of additional data. The status at the time of the meeting can be found in Ref. 11.

References continued:

13) C.G.Callan et al., Phys.Rev.Lett 34, 52 (1975)
    T.Appelquist et al., Phys.Rev.Lett. 34, 365 (1975)
    E.Eichten et al., Phys.Rev.Lett. 34, 369 (1975)
    B.J.Harrington et al., Phys.Rev.Lett., 34, 706 (1975)

14) J.W.Simpson, Phys.Rev.Lett. 35, 699 (1975)

DISCUSSION

CHAIRMAN: B. Wiik

Scientific Secretary: R.K. Ellis

DISCUSSION

- *BREIDENBACH:*

Experiments similar to those done at DESY have been carried out at SPEAR. We are in basic agreement with the results found at DESY.

- *DYDAK:*

Your identification of the P_c depends crucially on your ability to separate the 2γ background coming from $\psi' \to \psi + \pi^0 + \pi^0$ with two undetected γ's. I would therefore like to ask what is your threshold and resolution of low-energy photons?

- *WIIK:*

The response of our detector as a function of angle and energy has been measured using a tagged photon beam. At 50 MeV there is a 50% probability of detecting a photon; at 150 MeV, the photon detection efficiency is 100%. The energy resolution is bad, but is nowhere used in the reconstruction of the event.

- *DYDAK:*

Do you have the coplanarity-energy plot for the twenty electron events?

- *WIIK:*

If you do a zero-C fit they are forced to be coplanar.

- *FERBEL:*

I would like to ask a series of questions about the plot of the acoplanarity versus the energy imbalance. Firstly, how do you interpret the events which do not lie near the origin?

- *WIIK:*

These are $3.7 \to 3.1(\pi^0\pi^0)$ events, where two of the four γ's were detected. In making the plot, these events were interpreted at $3.7 \to 3.1(\gamma\gamma)$, and as a consequence, there is an apparent violation of energy conservation and coplanarity.

- *FERBEL:*

Did you estimate the number of background events in the central re-
by extrapolating from the outside?

- *WIIK:*

No. The $\pi^0\pi^0$ distribution was assumed to be the same as the measured
$\pi^+\pi^-$ distribution and half as large. Using a Monte Carlo calculation and
the measured response of the apparatus, we estimated a background of less
than 0.2 events. The extrapolation from the non-central region was only
an extra check to convince ourselves that we had made no gross errors.

- *FERBEL:*

The scale which you must use to determine whether there is clustering
around the origin or not is your resolution. What is your resolution?

- *WIIK:*

About 100 MeV in energy imbalance, and 10° in acoplanarity.

- *MORPURGO:*

Are you sure that with so few events you can distinguish a transition
via an intermediate state from the direct electromagnetic process?

- *WIIK:*

For the process you mention there is no reason to expect the events
to be clustered in 20 MeV. The spectrum should be continuous.

- *MORPURGO:*

If the results which have been reported by Wiik are confirmed by
larger statistics, they would give a striking confirmation of some bound
state model. In the charmonium model there are three P states which are
predicted by Eichten et al. to lie about 200 MeV below $\psi(3.7)$. However,
the experimental partial width of 8 keV based on the branching ratio

$$\Gamma(\psi' \not\rightarrow P_c \not\rightarrow \psi)/\Gamma(\psi' \rightarrow \text{all}) = 4\%$$

would appear to be much smaller than the calculated widths $\sim$ 100 keV
(cf. the table in my lectures).

- *WIIK:*

I should like to make some speculative remarks about this discovery.
Firstly, I think that this result should not be interpreted as the discovery of charm. To discover charm, one has to observe states with non-
zero charm and watch them decay.

Secondly, tentatively we find:

DISCUSSION

$$\Gamma(3.7 \xrightarrow{\gamma} P_c \xrightarrow{\gamma} 3.1)/\Gamma(3.7 \to \text{all}) \sim 4\% \ .$$

Combined with Hofstadter's result that there is no single γ transition of magnitude greater than about 5% of the width of the 3.7, the second transition $P_c \downarrow 3.1$ must proceed principally via one photon.

- *CRONIN:*

Is that not inconsistent with the number of 4- and 6-pion decays observed at SLAC?

- *WIIK:*

No, not necessarily.

- *MORPURGO:*

I agree entirely with you on the question of charm; as you may have noticed, I used the expression "bound state".

- *DRELL:*

I should like to ask Mr. Politzer to comment on the discrepancy between the measured widths and the partial γ widths predicted by the charmonium model.

- *POLITZER:*

They appear to be about a factor of 10 wrong. In our paper, the widths should be interpreted as upper bounds. In the Cornell paper, they took models for the wave functions. One is forced to conclude that they got the wave functions wrong.

- *DRELL:*

The calculations referred to are non-relativistic. In non-relativistic calculations, one often over-estimates transition matrix elements. I am convinced that a factor of 3 in transition matrix element is a question of detail and not of principle.

- *MORPURGO:*

The structure of a two-body bound state is very simple and the principal uncertainty comes from lack of knowledge of the radius. For an electric dipole transition, the matrix element depends only on the charge and the radius. Whilst the transition matrix element may be reduced by changing the value of the radius, I suspect that there remains an unexplained suppression factor.

Do you have plans to study the high energy γ's connected with the process $\psi \to b' + \gamma$? As I explained in my lectures, this is an important process to examine.

- *WIIK:*

A study of the 3γ final state is in progress.

- *WILSON:*

In the work of Eichten et al. there are three ^3P states. The ψ' is predicted to decay into these states with widths in the ratios 5:3:1 corresponding to the $(2J + 1)$ fold multiplicity of the final states. With increasing resolution, all three states should be observed, and it will be interesting to check the ratios of the widths.

STUDIES OF NEUTRON DISSOCIATION AT FERMILAB ENERGIES[*]

T. Ferbel

University of Rochester[+], Rochester, New York 14627

Last winter I had the fervent hope to present at the Erice School our results on the dissociation of neutrons into charm-anticharm systems. Unfortunately, all we have been able to obtain in our Fermilab experiment have been rather small upper limits for the production of charm particles in high energy collisions.[1] Consequently, I will discuss in this lecture the latest results obtained in our continuing investigation of neutron dissociation into ($p\pi^-$) systems in neutron-nuclear collisions between 50 GeV/c and 300 GeV/c.[2] The experiment is a collaborative effort involving physicists from Fermilab, Northwestern University, University of Rochester and SLAC (see Table I).

Table I

Primary Contributors to Experiment

J. Biel [a]	L. Kenah [d]
E. Bleser [b]	J. Rosen [d]
T. Ferbel [a]	R. Ruchti [d]
D. Freytag [c]	P. Slattery [a]
B. Gobbi [d]	D. Underwood [a]

a) University of Rochester
b) Fermilab
c) SLAC
d) Northwestern University

---

[*] Lecture delivered at the Intl. School of Subnuclear Physics at Erice. I wish to thank Prof. Zichichi for providing me the opportunity to speak at the School.

[+] Research supported by the United States Energy Research and Development Administration.

INTRODUCTION

Two production processes are expected to contribute to neutron dissociation, namely, coulomb and diffractive excitation;[3] the characteristics of these mechanisms are indicated briefly in Table II. The cross section for coulomb excitation is proportional essentially to the square of the charge of the target nucleus (Z); at fixed ($p\pi^-$) mass, M, the yield grows logarithmically with the square of the energy in the center of mass ($\ln s$); the cross section as a function of the square of the four-momentum transfer, t (defined to be positive, i.e., $t=|t|$), has a zero in the forward direction (a consequence of current conservation), and it contains the characteristic $\frac{1}{t^2}$ fall off stemming from the γ propagator. The actual expression for the Primakoff cross section for any process $m + Z \to M + Z'$ in the high-energy limit is as follows:[4]

$$\frac{d\sigma}{dtdM} = \frac{2Z^2\alpha}{\pi\eta} \frac{M}{M^2-m^2} \frac{t-t_{min}}{t^2} \sigma(\gamma m \to M)|F(t)|^2 \qquad (1)$$

where $\eta = \frac{1}{2}$ when m is a photon and otherwise $\eta = 1$; t_{min} is the kinematically allowed minimum value of t required for producing the mass M in the m + Z collision; m is the mass of the projectile and $\sigma(\gamma m \to M)$ is the cross section for photoproduction of final state M in γm collisions at a center of mass energy E = M; F(t) is the electromagnetic form factor of the nucleus.

Expression (1) has to be modified, particularly at small s-values, for absorption in nuclear matter of the incident neutron and of the produced $p\pi^-$ system. The spin-parity (J^P) states which can be produced in the γ-exchange process are supposedly those which can be excited through the measureable reaction

$$\gamma n \to p\pi^- \qquad (2)$$

Table II

NEUTRON DISSOCIATION

$n + A \to (p\pi^-) + A'$

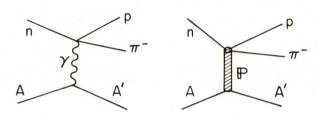

	Coulomb Excitation	Diffractive Excitation
$\sigma(s)$:	$\sim \ln s$ - constant	$\sim$ constant
$\sigma(t)$:	$\sim \dfrac{t-t_{min}}{t^2}$ (small t)	Inelastic Nuclear Form Factor
$\sigma(A)$:	$\sim Z^2$	Weak
J^P :	As in $\gamma n \to p\pi$ $(I = \tfrac{1}{2}, \tfrac{3}{2})$	Excited States* of neutron? Continuum? $(I = \tfrac{1}{2})$
$S^{1/2}$:	N(1535), Δ(1650)	N(1535) ?
$P^{1/2}$:	N(1470), Δ(1910)	N(1470) ✓
$P^{3/2}$:	N(1810), Δ(1232)	N(1810) ?
$D^{3/2}$:	N(1520), Δ(1670)	N(1520) ✓
$D^{5/2}$:	N(1670), Δ(>2000)	N(1670) ?

*The states with question marks are forbidden by the Morrison rule.

It is worthwhile to point out that although asymptotically the cross section for the coulomb process at fixed M grows logarithmically with incident momentum, typically, order of magnitude changes in production rates of low mass states occur between AGS and FNAL energies. (This can be seen by integrating expression (1) at fixed M between $t_{min} \approx (M^2-m_n^2)^2/(2p_{incident})^2$ and $t = 0.001$ GeV^2 for different values of $p_{incident} \approx s/2$.)

In principle, the coulomb-production contribution to neutron dissociation can be calculated and, consequently, can be used as a tool for the study of the more interesting process, namely the diffraction dissociation of the neutron. The diffractive excitation process is not well understood. The cross section for the process is expected to be energy independent; the t-distribution is expected to be dominated by the nuclear form factor. The dependence of the cross section on atomic number (A) is expected to be rather weak, particularly for large A values (due to the absorption of the hadron system in nuclear matter). It is not clear what sort of states can be excited in the dissociation process nor whether the helicity of the produced system can be different from that of the incident neutron. There is a famous Ansatz known as the Morrison rule[5] for diffractive production, relating the change in parity to the change in the spin of the produced object, which has not been fully tested ($\Delta P=(-1)^{\Delta J}$) and there have been suggestions[6] in the literature that, just as in ρ photoproduction, the helicity of the incident neutron is maintained by the diffractively produced object. In this lecture I will comment on what we have learned about these questions from our preliminary analysis of neutron dissociation.

The measurements I will discuss were performed in the M-3 neutral beam of the Meson Detector Building at the Fermi National Accelerator

Laboratory. A variety of nuclear targets ranging from hydrogen to uranium were employed in this study. I will first briefly discuss our nuclear coherent dissociation data; subsequently I will present new measurements of total cross secitons of neutrons on nuclei in the Fermilab momentum range: finally I will report on neutron dissociation using a hydrogen target and compare the hydrogen data with expectations from simple Deck models.

DISSOCIATION USING NUCLEAR TARGETS

The apparatus, which is sketched in Fig. 1, consists of a large aperture wire-plane spark chamber V-spectrometer. The nuclear-target box is located 400 meters downstream of the neutron-beam production target.

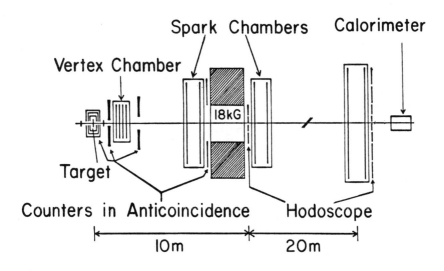

Fig. 1   Schematic of spectrometer system.

Nuclear targets, typically 0.1 radiation lengths in thickness, were used to study coherent production of $p\pi^-$ systems. The target box was surrounded with scintillator/lead sandwich veto counters to detect nuclear break-up processes and thus reduce the trigger level for the experiment. The trigger requirements were designed to suppress incoherent production as well as the production of more than two charged particles within the target box. Specifically, the logic condition $\bar{A} \cdot S \cdot H1$ was required to be satisfied for an acceptable trigger. Here, A represents the logical OR or all the veto counters, i.e. the counters surrounding the target box, the $\frac{1}{16}$ inch thick scintillation veto counter at the entry to the target box (indicating that the particle initiating the collision was charged rather than neutral), and the various baffle counters surrounding the magnet aperture (indicating presence of charged or neutral particles in addition to those entering the magnetic field volume); S represents a $\frac{1}{32}$ inch scintillation trigger counter located 2 inches downstream of the nuclear target sample. Two and only two of the six H1 hodoscope elements (immediately downstream of the magnet) were required to have signals before the acceptance requirement was fulfilled. The spark chambers were capable of being pulsed 15 times during the ~1 sec beam spill. A typical run consisted of 10,000 triggers taken over a six hour period. Quality of the performance of the apparatus was monitored using an online PDP-15 DEC computer; reconstruction of events and analysis was performed off-line. Approximately 80% of the 600,000 triggers yielded successfully reconstructed events.

As discussed in the Introduction, coherent dissociation of neutrons at high energies can proceed either through the hadronic diffractive dissociation[3] process (often referred to as Pomeranchukon exchange) or through the electromagnetic excitation of the neutron in the coulomb field of the target nucleus (Primakoff effect, or photon exchange).[4] The dissociation of neutrons has been studied at lower energies[7] where kinematic restrictions cause a severe suppression of the coherent cross section, particularly at large values of $p\pi^-$ mass (M). In this lecture I will present the general characteristics of the $p\pi^-$ mass spectra and the momentum-transfer distributions obtained in the coherent dissociation of $n \rightarrow p\pi^-$ for Pb, Cu, C and Be targets in the 120 GeV/c-300 GeV/c momentum band.

Before reporting the results of our experiment I will briefly discuss the geometrical acceptance of the apparatus. Although the magnet aperture is only 8 inches in the vertical plane and 24 inches in the horizontal bending plane, the nature of the coherent production process is such that the acceptance of the apparatus is not seriously compromised for neutron momenta in excess of 50 GeV/c. The efficiency for the data to be presented varies smoothly with variables of dynamic interest and is typically ~70%; the inefficiency is mainly a result of the geometrical losses in azimuth angle due to the limiting vertical aperture of the bending magnet. (The magnet provides approximately a 1 GeV/c transverse momentum bend.) Geometrical acceptance, as well as small losses in reconstruction, have been calculated using a Monte Carlo program which simulated the experimental arrangement; the results of the Monte Carlo have been used to weight the data and thus greatly reduce instrumental biases.

Figure 2 displays distributions of the square of the four-momentum transfer (t) between the incident neutron and the produced $p\pi^-$ systems for three regions of M: (1) The $\Delta(1236)$ mass region, defined as M<1.28 GeV; (2) The N*(1400) region, defined as 1.35<M<1.45 GeV; and (3) The N*(1688) region, defined as 1.55<M<1.80 GeV. (We use the variable $t'=t-t_{min}$.) Corrections for target-empty measurements and for small variations in the geometrical acceptance of the apparatus (calculated using the above mentioned Monte Carlo program) have been applied to the data. The steep peaks observed at small t', with values of diffractive slopes characteristic of the sizes of the nuclear targets, give us full confidence that

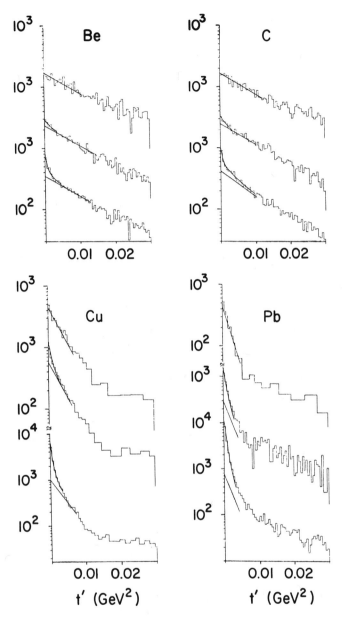

Fig. 2. Distributions in momentum transfer for neutron dissociation off Be, C, Cu and Pb nuclei. In each case we display the t spectra for $p\pi^-$ mass ranges <1.28 GeV, 1.35-1.45 GeV and 1.55-1.80 GeV (data for the lowest mass band are shown at the bottom of each set). The curves indicate the contribution of diffractive production (bottom curves) and the sum of diffractive and coulomb production (top curves).

the inelastic production process takes place coherently over the entire neucleus. (It should be recalled that we do not measure the momentum of the recoil nucleus and consequently the reaction n+A→(pπ⁻)+A has no constraints. We believe, however, that background from other coherent or incoherent channels is not important for the data which I present in this paper.[8])

The distributions in t', for each target, appear to show similar dependences on M. Namely, for the Δ(1236) region all the distributions display a sharpening of the t' spectrum at t'≲0.001 GeV^2. This excess contribution can be attributed to coulomb production of pπ⁻ systems, which because of the known large Δnγ coupling is dominated by Δ°(1236) production. Taking our experimental resolution into account, the initial fall off of the cross section in t' is consistent with the theoretically expected form for coulomb production.[9]

The smooth curves drawn on the figure are superpositions of the contributions from coulomb[9] and from diffractive coherent production. The latter is based on an optical model description for the production process provided by Kolbig and Margolis.[10] Standard Woods-Saxon parameters were used to describe the nuclear shapes.[11] The total cross section of a neutron on a nucleon was taken to be a constant 39mb and the elastic forward scattering amplitude for n-nucleon and (pπ⁻)-nucleon was taken to be imaginary. Our experimental resolution was folded into the overall prediction for the shape of the t' spectrum. The value for the (pπ⁻)-nucleon total cross section was also taken to be 39mb, a value consistent with the shape observed for the t' distribution and consistent with similar measurements made by Carithers et al[7] at incident neutron momenta of ~12 GeV/c.

Figure 3 displays the M spectra for two regions of t': (1) t' < 0.001 GeV^2, a region where coulomb production is important; and (2) 0.005 < t'

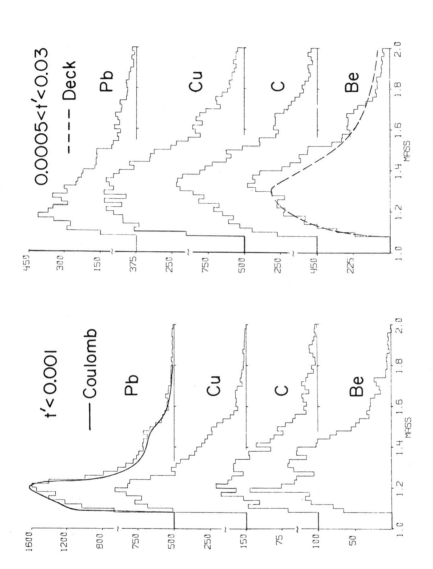

Fig. 3. Mass spectra for coherently produced $p\pi^-$ pairs for $t'<.001$ GeV^2 and $.005<t'<0.03$ GeV^2. The curve on the Pb data is the shape expected for the mass distribution assuming dominance of the Primakoff production amplitude. The curve on the Be data indicates the shape expected on the basis of a reggeized Deck model.

< 0.03 GeV^2, where diffractive production dominates. The data in Fig. 3 indicate substantial Δ(1236) production, particularly at small t'; a shoulder is evident at the N*(1520) and a small enhancement is observed at the N*(1688), both mainly at the larger t' values. The difference in the various M-spectra can be attributed largely to the known dependence of the coulomb production process on t' and Z (the nuclear charge). The cross section for Δ°(1236) production, in particular, is approximately proportional to Z^2 for t' < 0.001 GeV^2. The curve superimposed on the Pb data is the predicted shape of the mass spectrum expected on the basis of coulomb production. (The absolute normalization is also consistent with the data.) The curve displayed on the Be data is based on a calculation of the reggeized Deck effect.[12] The shapes of the predicted mass distributions are in reasonable agreement with the data.

We have used the dominance of the coulomb cross section for production on Pb to extract the momentum spectrum of the incident neutron beam. The observed pπ⁻ momentum spectrum in Pb for M < 1.28 and t' < 0.001 was first corrected for small diffractive-event background. The subsequent unfolding of the resolution and of the known coulomb production process[9] provided the corrected momentum spectrum (Fig. 4), which above 50 GeV/c is consistent with the directly measured neutron spectrum obtained using calorimetry.[13]

The energy dependence of the cross section for several mass intervals for Be, C, Cu and Pb targets is displayed in Fig. 5. Only Pb appears to shown an increase of the cross section with momentum. This rise is consistent with the dominance of the coulomb process in Pb (even for large M-values). Production cross sections using the other elements are constant to within 10% possible systematic uncertainties.

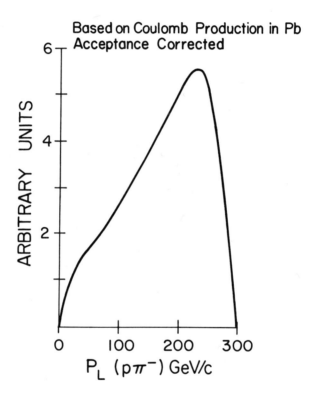

Fig. 4.   Neutron momentum spectrum in the M-3 beam at Fermilab.

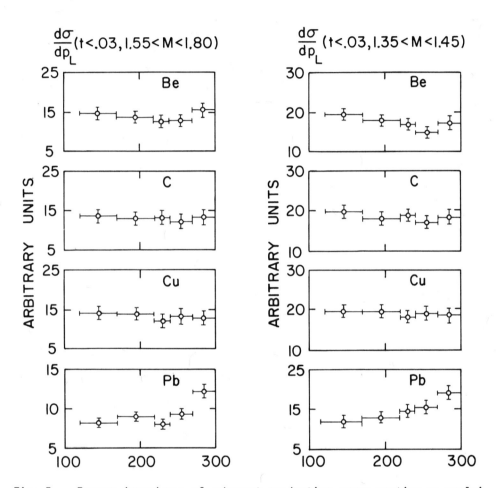

Fig. 5. Energy dependence of coherent production cross section on nuclei.

NEUTRON-NUCLEAR TOTAL CROSS SECTIONS

The cross section for neutron dissociation into $p\pi^-$ systems on Pb is ~1% of the total n-Pb cross section. Therefore, our V spectrometer can be considered as a neutron detector with excellent positional (about ±0.3mm transverse to beam axis) and energy (~±1%) resolution. We have utilized this unique aspect of the spectrometer to perform a precision measurement (limited only by statistics) of neutron-nuclear total cross sections. Using nuclear transmission targets located ~200 meters upstream of the spectrometer Pb target, we measured cross sections as a function of momentum.

The basic idea of the good geometry measurement is indicated in the sketch below. We measure the rate for $(p\pi^-)$ V production with and without a nuclear transmission target located 200 meters upstream of the spectrometer. The log of the ratio of the V counts obtained without transmission target to counts obtained with a transmission target is proportional to the total cross section for neutrons on the specific nuclear target material. (A good V count is defined to have a coherently produced V in the spectrometer with M<1.6 GeV and t<0.005 GeV^2.)

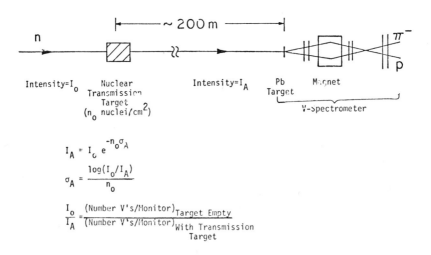

$$I_A = I_0 e^{-n_0 \sigma_A}$$

$$\sigma_A = \frac{\log(I_0/I_A)}{n_0}$$

$$\frac{I_0}{I_A} = \frac{(\text{Number V's/Monitor})_{\text{Target Empty}}}{(\text{Number V's/Monitor})_{\text{With Transmission Target}}}$$

Neutrons can scatter elastically in the transmission target and if the scattering angle is small enough the scattered neutron can produce a V in the spectrometer which can be mistaken for a V caused by a transmitted (unscattered) neutron. To illustrate the magnitude of this bias I present in Fig. 6 the distribution of the radial positions of the vertex points of the V's with and without the presence of the upstream transmission target. Neutrons that are elastically scattered in the transmission target produce a beam halo at large R detected in the downstream Pb "converter." (The unscattered neutron beam is very well collimated and the intensity falls rapidly with R.) The greatest corrections for elastic scattering occurs for the largest nuclei. However, with our additionally imposed requirement that R<0.4 cm, we see that the correction even for Pb is very small.

Transmission targets of approximately one interaction mean free path were cycled automatically, typically every ten minutes (a target empty position was included in the cycling). A check on systematic errors was provided by using two carbon targets of different length (no effect was observed). Counter telescopes were used to monitor the neutron flux throughout the data taking. The small (<0.5%) corrections for elastic scattering of neutrons in the transmission targets were extracted from the sort of distributions provided in Fig. 6 and were applied to the data. In Fig. 7 we display our measurements of the total cross sections of neutrons on C, Al, Cu and Pb targets as a function of beam momentum. These measurements connect up very smoothly with measurements at lower energies but appear to be somewhat inconsistent with other recent measurements at Fermilab.[14]

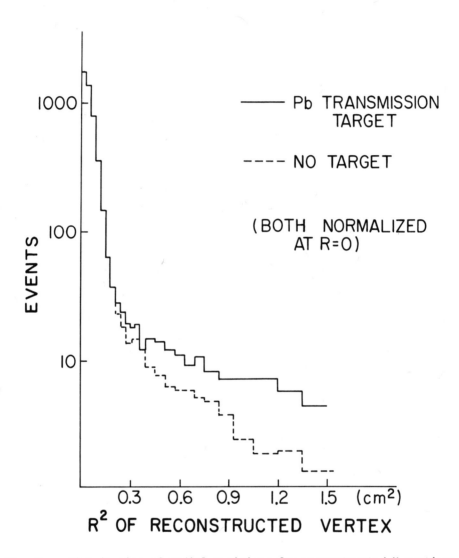

Fig. 6. Distribution of radial positions for reconstructed V vertices.

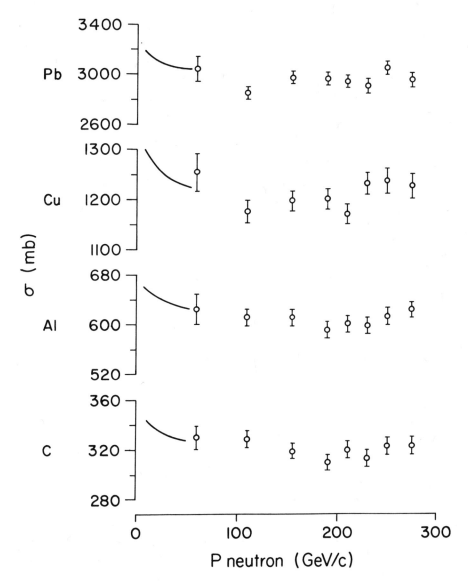

Fig. 7. Energy dependence of total neutron-nuclear cross sections. The curves indicate an averaging of measurements at lower energies.

DIFFRACTIVE PRODUCTION ON HYDROGEN

I will now discuss our results pertaining to neutron dissociation from hydrogen in the 50 GeV/c to 300 GeV/c momentum range, the specific reaction is:

$$n + p \rightarrow (p\pi^-) + p \tag{3}$$

The apparatus, except for the substitution of a hydrogen target for the nuclear-sample target box, is identical to that already described. The hydrogen target consists of a high-pressure vessel capable of withstanding gas pressures of $\lesssim 100$ atmospheres. The active target region is 20 inches in length; this region is defined by a thin veto-counter at the upstream end, and by a trigger counter at the downstream end of the high-pressure volume. A set of 16 plastic scintillator strips, each 1/4 inch thick, 11/16 inches wide and 20 inches long, surround azimuthally the active target region at a distance of $1\frac{3}{4}$ inches from the neutron beam axis (see Fig. 8). The sixteen scintillator elements are each viewed at both ends, with the aid of plastic light pipes, using photomultiplier tubes located outside of the high-pressure vessel.[15] The trigger requirement for an acceptable $np \rightarrow (p\pi^-)p$ event consists of the following: no signal in the upstream veto-counter contained within the target vessel, a signal in the downstream trigger counter, two and only two charged particles exiting the spectrometer magnet, and a signal in only one of the 16 azimuthal counters.

Following the spatial reconstruction of the forward $p\pi^-$ system (V) a check is made of the correlation between the transverse direction of the V system and the position of the activated azimuth counter for that event. The distribution in the angle (Φ) between the transverse direction of the forward $p\pi^-$ system and the transverse direction of the recoil proton, as specified by the center of the activated azimuthal counter, is peaked at

180° (see Fig. 8). This peak, which signifies the presence of our signal, has the expected width of ~22.5°; the peak to background ratio is ~5/1. In the data to be presented we have imposed a cut-off band on Φ centered at 180° (i.e., 180° ± 15°). To account for contamination of our $p p \pi^-$ final state, we have subtracted from the data in the signal band of $\Phi = 180° \pm 15°$ the data in the sum of the background bands: $\Phi = 195°$ to 210° and $\Phi = 150°$ to 165°. A small subtraction (~2%) for target-empty was also applied to the data. The present sample of reaction (3) is derived from a total of ~180,000 triggers of which ~60,000 passed all geometric and kinematic acceptance criteria. The largest background is from neutron dissociation into $(p\pi^-\pi^0)$ systems. We believe this source of background to be at a level of $\lesssim 10\%$ of the $(p\pi^-)$ signal. (Background from $(n\pi^+\pi^-)$ systems is greatly suppressed by removing events in which a neutron signal is detected in a small calorimeter downstream of the apparatus. The calorimeter subtends an angle of about ±2 mrad.)

Figure 9 displays the mass spectrum of the forward-produced $p\pi^-$ system in reaction (3) corrected for acceptance of the spectrometer. Data are shown for three t' intervals: (a) 0.02 to 0.08 GeV^2, (b) 0.08 to 0.20 GeV^2, and (c) 0.20 to 1.0 GeV^2. (The cut-off at $t' \simeq t = 0.02$ GeV^2 was chosen to assure an unbiased sample of events. The losses below $t \simeq 0.015$ GeV^2 are quite severe because, under our typical running conditions of 50 atmospheres of pressure for the hydrogen gas target, recoil protons do not have sufficient energy to reach one of the azimuthal counters to trigger the interaction.) Ignoring the contribution from coulomb production in the nuclear data, we note that the mass spectrum for comparable t', namely t'<0.08 GeV^2, is similar to those observed for nuclear targets. All mass distributions appear to display fine structure at ~1.7 GeV mass

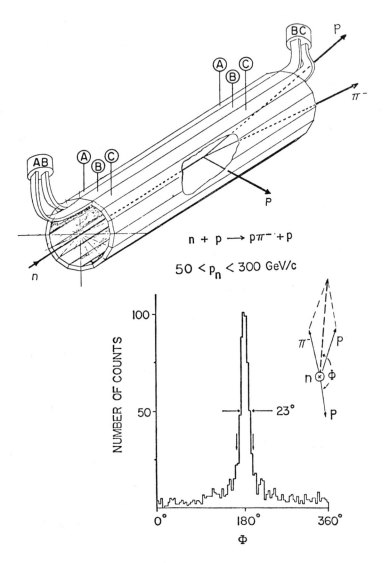

Fig. 8. Schematic of the sixteen scintillation strips placed azimuthally about the hydrogen target region. The angle Φ is between the center of an activated strip (recoil proton) and the transverse momentum of the forward produced pπ⁻ system (V).

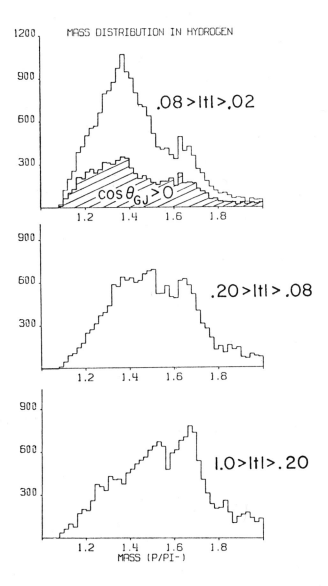

Fig. 9. Mass distribution of $p\pi^-$ systems for several regions of t. The cross hatched insert at the top corresponds to events with $\cos\theta > 0$ (Gottfried-Jackson angle of the forward proton). At the larger t values there is only a weak dependence on $\cos\theta$.

values and possible rapid changes near 1.5 GeV. Cross sections for the 1.7 GeV enhancement are more prominent at larger momentum transfers.

To investigate in more detail the correlation between t' and M for production in hydrogen we display in Fig. 10 the cross section in t' for various regions of M. A strong dependence on M is observed for the t' spectrum, particularly for M<1.4 GeV.[16] In Table III we display the results of fits of the form exp(-bt') to the data at $0.04 < t' < 0.16$ GeV^2. It is observed that the t' spectra have strong curvature (or "breaks") near $t' \simeq 0.2$ GeV^2 at small M values. For $t' > 0.4$ GeV^2 the data appear to take on a universal form [~exp(-4.5 t')], essentially independent of M. The change of slope near $t' \simeq 0.2$ GeV^2, and the disappearance of this effect at large M, has been interpreted in terms of the dominance of non-flip amplitudes for M≲1.4 GeV and spin-flip contributions for M>1.4 GeV.[17] As will be discussed in the following sections our data indicate the presence of large flip as well as non-flip helicity amplitudes (both in the t-channel as well as s-channel) for M<1.4 GeV.

To demonstrate the richness of the angular distributions we present in Fig. 11 a scatter plot of the cosθ vs φ decay angles of the proton in the Gottfried-Jackson (GJ) frame. We note a severe depletion in the region of cosθ ≈ 0.6 and φ ≈ 0°. It is interesting that a suppression of a similar nature is expected in this region of (cosθ,φ) space if production is dominated by the π-exchange Deck mechanism (see Fig. 12 for the Deck diagram in question).[9] In fact, as will be shown, the low-order decay moments in the Gottfried-Jackson frame display a mass dependence which can be reproduced surprisingly well by the Deck model. (The enhancement observed in Fig. 11 for cosθ<0 and φ ≈ 0 can also be reproduced in a qualitative way by a Deck model calculation involving proton rather than π-exchange.)

Table III

Slope Parameter Values

Mass Interval (GeV)	Slope[a] (GeV^{-2})
<1.25	19.1 ± 0.9
1.25 - 1.35	15.0 ± 0.7
1.35 - 1.45	12.8 ± 0.6
1.45 - 1.55	6.8 ± 0.7
1.55 - 1.65	4.9 ± 0.6
1.65 - 1.75	4.2 ± 0.6
1.75 - 2.00	4.2 ± 0.8

[a] Value of the parameter b from a fit of the data to the form exp(bt). The range of $|t|$ in the fit was 0.04 to 0.16 GeV^2 for the first four mass bands and .04 to 0.5 GeV^2 for the last three mass regions.

The fact that the contribution from the π-exchange Deck diagram is expected to peak near $\cos\theta \simeq +1$ has prompted a group at the ISR to examine the dependence of the diffractively produced two-body mass on $\cos\theta$.[18] As displayed previously in Fig. 9, at Fermilab energies there does not appear to be as dramatic a dependence of the pπ$^-$ mass on $\cos\theta$ as was observed at the ISR. In Fig. (13) we present t distributions for M<1.35 GeV, separately for negative and for positive values of $\cos\theta$. The distributions are observed to display a strong dependence on $\cos\theta$, qualitatively similar to that expected from a π-exchange Deck model. (Recall, however, that for $\cos\theta<0$ proton exchange contributes significantly to reaction 3.)

If the Deck mechanism is dominant in our reaction, it is expected that the azimuth ϕ_H of the proton in the helicity frame of the pπ$^-$ system will be particularly useful in distinguishing the Deck contribution from other (perhaps resonant) production channels.[19] In Fig. 14 I display the ϕ_H distribution for the data of reaction (3). A clear peak is observed near $\cos\phi_H \simeq -1$ corresponding to π-exchange, but an equally

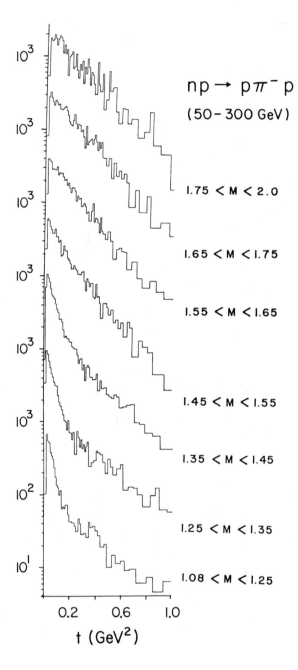

Fig. 10. Momentum transfer distributions for forward-produced $p\pi^-$ systems as a function of $p\pi^-$ mass.

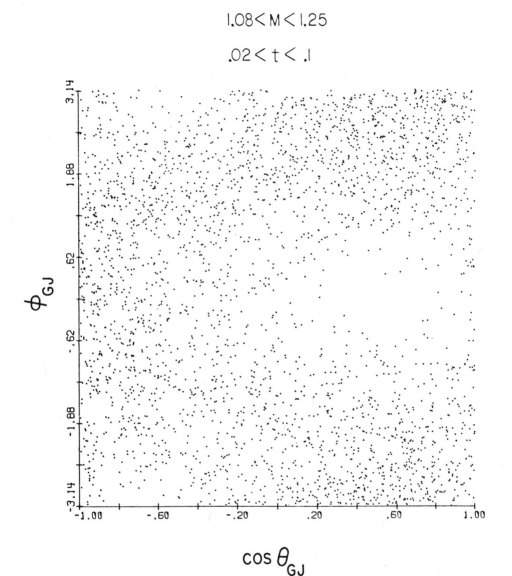

Fig. 11. Correlation between the cosθ and φ of the forward proton in the Gottfried-Jackson frame.

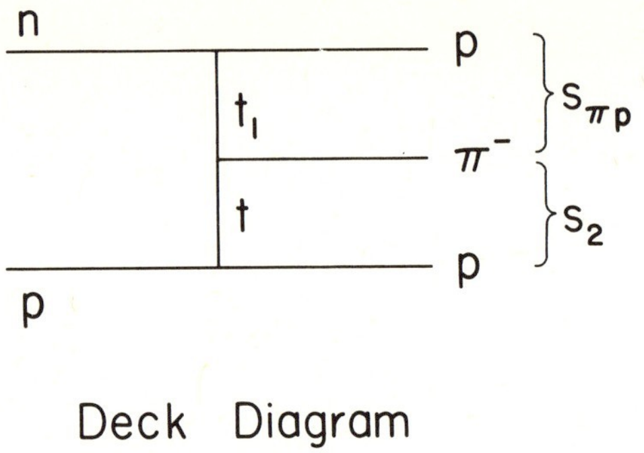

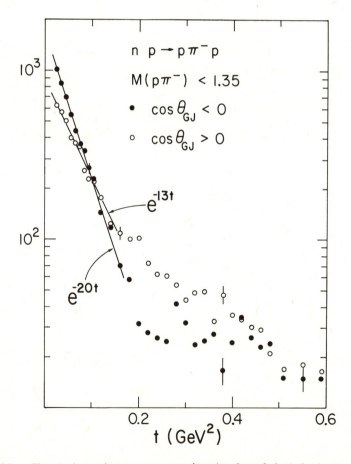

Fig. 12. The simple π-exchange Deck diagram for reaction (3).

Fig. 13. The t-dependence on cosθ in the Gottfried-Jackson frame.

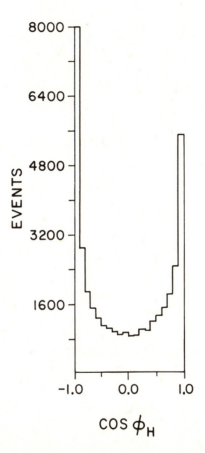

Fig. 14. Distribution in the cosine of the proton's azimuth angle calculated in the s-channel helicity frame of the $p\pi^-$ system.

clear peak is also present near $\cos\phi_H \simeq +1$ which corresponds to a Deck proton-exchange contribution to the reation. In Fig. (15) we display distributions in $p\pi^-$ mass for the proton's $\cos\phi_H > 0.9$ and $\cos\phi_H < -0.9$. A clear difference is apparent between the two mass spectra, particularly at large M. A possible conclusion one may draw from these data is that two Deck-like production processes contribute to reaction (3). It is not clear however, why the two mass spectra differ so much. (The data for $|\cos\phi_H| < 0.9$ are similar to the total ϕ_H spectrum.)

In Fig. (16) we present the energy dependence of the cross section for $p\pi^-$ production at small M (integrated over t). We observe that just as in the case of production from nuclear targets, these cross sections are also energy independent between 50 and 300 GeV/c. Consequently, it appears that the gross properties of reaction (3) for low M-values are those expected for a diffractive production process of the kind indicated in Fig. 12.

COMPARISON WITH DECK MODEL

I will now present the decay angular distributions of the $(p\pi^-)$ system in order to compare our data at small M with the Deck-production model indicated in Fig. 12. The square of a simple π-exchange Deck-type matrix element can be written as

$$|M|^2 \approx \left[-t_1 e^{2(t_1-\mu^2)}\right] \frac{[\tfrac{1}{2}(s_{\pi p}-u_1)]^{2\alpha_\pi}}{\alpha_\pi^2} (s_2-u_2)^2 e^{10t}, \qquad (4)$$

where t_1 and t are four-momentum transfers squared; $s_{\pi p}$ and s_2 are squares of the πp invariant masses as indicated in Fig. 12; u_1 and u_2 are respectively the squares of the four-momentum transferred to the pion from the incident neutron and target proton; $\alpha_\pi = 0.9(t_1-\mu^2)$ is the pion Regge

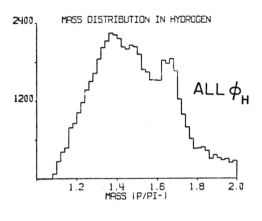
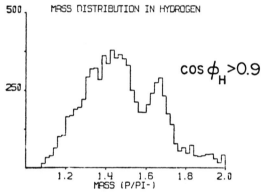
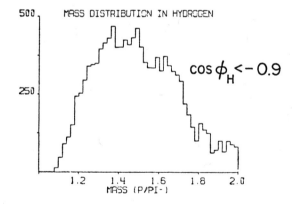

Fig. 15. Mass spectra for two regions of ϕ_H of the forward proton in the s-channel helicity frame. The Deck π-contribution is expected to dominate $\cos\phi_H < -0.9$.

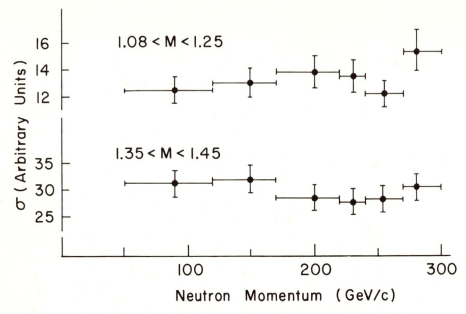

Fig. 16. Energy dependence of the cross section for neutron dissociation off a hydrogen target. The data have been integrated over t.

trajectory, and μ is the pion mass. The π^-p elastic differential cross section is taken proportional to $\exp(10t)$. The decay moments of the $(p\pi^-)$ system were extracted by numerical integration of the above expression. In addition, a variant of the Deck model was examined. This variant consisted of expression (4) but with the first bracketed omitted. This can be regarded as a Deck model with additional absorption.

Figure 17 displays the normalized low-order moments $\langle Y_{\ell m}\rangle$ versus t for fixed mass in the Gottfried-Jackson (GJ) frame and in the helicity frame. The data are in rough agreement with the trend of the π-exchange Deck calculation. A similar level of agreement is available for other $(p\pi^-)$ mass values. In Fig. 18 we display the variation of the same $\langle Y_{\ell m}\rangle$ versus mass at fixed t in the GJ frame. Again, the data only roughly follow the Deck calculations.

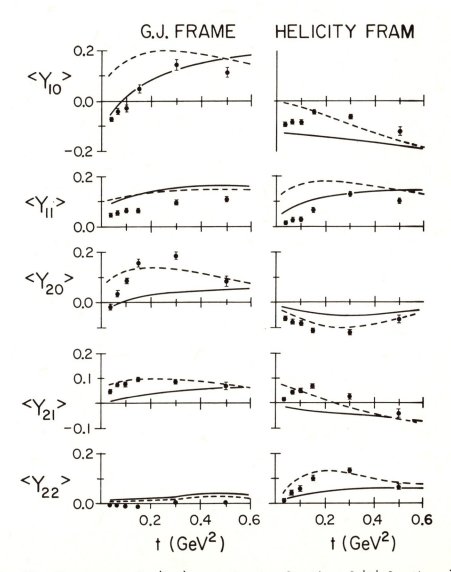

Fig. 17. The average $\langle Y_{\ell m}(\theta,\phi)\rangle$ moments as a function of $|t|$ for the $p\pi^-$ mass interval 1.300-1.375 GeV. The solid curves are predictions of equation (4). The dashed curves correspond to equation (4) but without the $(-t_1 e^{2(t_1-\mu^2)})$ factor in front of the expression. The latter can be thought of as a Deck model with additional absorption.

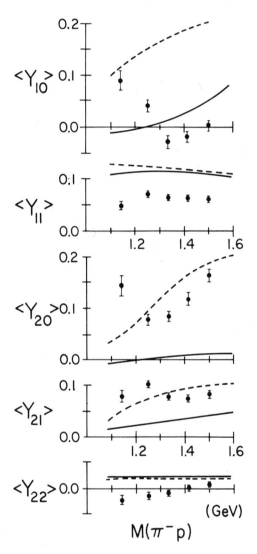

Fig. 18. The low order $\langle Y_{\ell m}(\theta,\phi)\rangle$ moments as a function of $p\pi^-$ mass for the t interval 0.08 to 0.12 GeV^2. The solid curves are predictions of equation (4). The dashed curves correspond to equation (4) but without the $(-t_1 e^{2(t_1-\mu^2)})$ factor in front of the expression. The latter can be thought of as a Deck model with additional absorption.

Although the predictions from expression (4) do not agree in detail with the results of reaction (3), this is not surprising in view of the fact that expression (4) can only be expected to describe the data in a rather restricted kinematic domain of reaction (3). The regime in which the π-exchange Deck mechanism is expected to dominate is characterized by small t_1,[12] large $\cos\theta_{GJ}$,[18] or as emphasized most recently, negative $\cos\phi_H$.[19] To examine whether the Deck diagram of Fig. 12 can account for the data in the region $\cos\phi_H < -0.9$ we have compared the predictions of expression (4) with our results applying this restriction on $\cos\phi_H$ to both the model and data.

The results of the comparison are displayed in Figs. 19 and 20. It is clear that in the region of expected applicability the spin structure of the Deck amplitude is in very close agreement with the data. (A similar comparison of results for $\cos\phi_H > 0.9$ with a proton-exchange Deck calculation indicates a comparable level of agreement between model and data.[20]) This agreement is particularly remarkable in view of the uncertainty in the origin of the M-t correlation in the data.[21] I wish to caution, however, that the severe cut we have imposed on $\cos\phi_H$ has a substantial distorting effect on the $<Y_{\ell m}>$ moments and consequently it is not likely that the improved agreement between the model and data can be attributed entirely to the success of the Deck model. A more detailed quantitative comparison is now in progress to evaluate the extent of the improved agreement resulting from $\cos\phi_H$ cuts. Although it is uncertain whether the simple Deck model can account in detail for the properties of reaction (3), it is clear nevertheless that the qualitative features of the data can be understood on the basis of a superposition of π and proton-exchange Deck contributions to neutron dissociation.

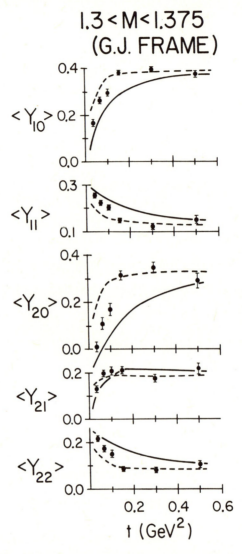

Fig. 19. The average $<Y_{\ell m}(\theta,\phi)>$ moments as a function of $|t|$ for the $p\pi^-$ mass interval 1.300-1.375 GeV for $\cos\phi_H < -0.9$. The solid curves are predictions of equation (4). The dashed curves correspond to equation (4) but without the $(-t_1 e^{2(t_1-\mu^2)})$ factor in front of the expression. The latter can be thought of as a Deck model with additional absorption.

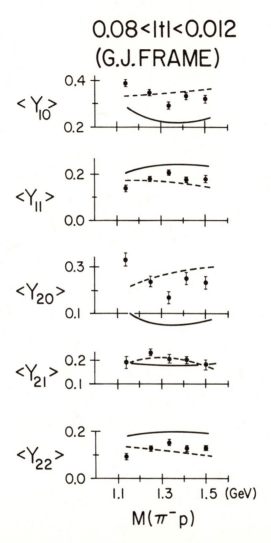

Fig. 20. The low order $<Y_{\ell m}(\theta,\phi)>$ moments as a function of $p\pi^-$ mass for the $|t|$ interval 0.08 to 0.12 GeV^2 for $\cos\phi_H < -0.9$. The solid curves are predictions of equation (4). The dashed curves correspond to equation (4) but without the $(-t_1 e^{2(t_1-\mu^2)})$ factor in front of the expression. The latter can be thought of as a Deck model with additional absorption.

HELICITY AMPLITUDES AT LOW MASS

The origin of the observed correlation between the mass M of an inelastic system produced in diffraction dissociation and the square of the four-momentum transferred to that system (t) has been the object of extensive investigation.[16] One attractive model for understanding the t-M interdependence in these highly peripheral reactions is based on the assumption that s-channel helicity amplitudes for small masses (M≲1.3 GeV) are dominantly helicity non-flip.[17] Consequently one expects a steep differential cross section for small t and a dip or sharp break near t~0.2-0.3 if the helicity non-flip system is produced peripherally (i.e., near an impact parameter b~1 Fermi). The contributions from the helicity flip amplitudes are hypothesized to become more important as the mass and spin of the diffractively produced system increases, thus leading to a substantial broadening of the t-distributions with increasing M values.[17] The $<Y_{11}>$ moment in the helicity-frame consists of interference terms proportional to an helicity non-flip amplitude and a unit helicity-flip amplitude. In terms of the s-channel peripheral model discussed above,[17] one therefore expects $<Y_{11}>$ in the helicity frame to pass through zero at the t value where a dip appears in the differential cross section. Similarly, one also expects a zero in $<Y_{11}>$ near t~0.6 where the single helicity-flip amplitudes are predicted to have a zero. The absence of these predicted zeroes in $<Y_{11}>$ of Fig. 17 implies that the s-channel peripheral model cannot be the dominant production process.

It is expected that the spin structure for $p\pi^-$ masses below ~1.5 GeV is sufficiently simple to allow an extraction of the production amplitudes. In Fig. 21 we display the preliminary results of an amplitude analysis of the data. $<Y_{\ell m}>$ are negligible for $\ell > 4$ and (in the GJ frame) for

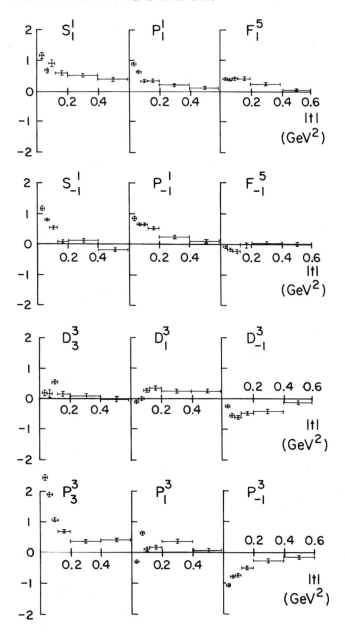

Fig. 21. Preliminary t-channel helicity amplitudes for the $p\pi^-$ mass range 1.3-1.375 GeV. The amplitudes are specified through the orbital wave (L), the total spin (J) and the helicity flip ($\Delta\lambda$) as $L^{2J}_{2\Delta\lambda+1}$. There is an ambiguity in the overall sign of the amplitudes. There is also an overall ambiguity in the relative parity of all even and odd states, e.g., the amplitudes labeled S^1, D^3, F^5 can be relabeled respectively as P^1, P^3, D^5 and vice versa.

m > 2. Therefore spins up to J=5/2 and helicity-flips ±1 saturate the moments. (The m ≤ 2 cutoff in $<Y_{\ell m}>$ is a consequence of Pomeranchukon exchange in the s_2 subsystem of Fig. 12.) It should be recalled that there is an ambiguity in the overall sign of the amplitudes. There is also an overall ambiguity in the relative parities of all even and odd states: the amplitudes labeled S^1, D^3, F^5 can be relabeled respectively as P^1, P^3, D^5 and vice versa.

The t-channel helicity amplitudes shown in Fig. 21 indicate that both helicity flip and non-flip terms are of comparable size and of comparable degree of peripherality. In addition, the presence of large S^1 and P^3 states (or alternately P^1 and D^3 states if all parities are reversed) speaks against the validity of the Morrison selection rule[5] in diffraction production.

SUMMARY

In summary, we have examined the characteristics of diffraction dissociation of neutrons into $p\pi^-$ systems at high energies. A substantial correlation is observed between the mass and the t of the produced system. The spin structure of the $p\pi^-$ amplitudes at low mass is very complex, but is described surprisingly well by the simple Deck mechanism. Both π-exchange and proton-exchange contributions are evident in diffractive production. The t-channel and s-channel helicity amplitudes contain comparable contributions from flip and non-flip terms and the produced states are not restricted to those expected on the basis of the Morrison rule.

I thank J. Biel, L. Kenah and D. Underwood for providing the latest neutron dissociation data, I also thank D. Duke and P. Slattery for extensive discussions pertaining to the analysis of the results presented in this report. Finally, I wish to acknowledge helpful conversations with E. Berger and G. Fox.

REFERENCES

1. E. Bleser et al, Phys. Rev. Letters 35, 76 (1975).

2. My lecture is based on a summary of our data presented at the Palermo Conference (to be published).

3. For the original ideas on this subject see, for example, E. L. Feinberg and I. Ia. Pomeranchuk, Suppl. Nuovo Cimento 3, 652 (1956); M. L. Good and W. D. Walker, Phys. Rev. 120, 1855 (1960); ibid, Phys. Rev. 120, 1857 (1960).

4. A. Halprin, C. M. Anderson and H. Primakoff, Phys. Rev. 152, 1295 (1966); M. Gourdin, Nucl. Phys. B 32, 415 (1971); L. Stodolsky, Phys. Rev. Letts. 26, 404 (1971).

5. D. R. O. Morrison, Phys. Rev. 165, 1699 (1968).

6. F. Gilman et al, Phys. Letts. 31B, 387 (1970).

7. D. O'Brien et al, Nucl. Phys. B77, 1 (1974); W. Carithers et al, University of Rochester Report submitted to London Conference (1974).

8. The contribution from $n \to p\pi^-\pi^0$ dissociations, for example, which we consider to be the largest source of background, is estimated to be <15% of the $p\pi^-$ dissociations for $t' < .03$ GeV. Compared to the $p\pi^-$ signal events the t-dependence of this background is rather flat.

9. We have used the formulations containing absorption corrections provided by G. Fäldt, Nucl. Phys. B43, 591 (1972); C. Bemporad, et al, Nucl. Phys. B51, 1 (1973). The $\gamma n \to p\pi^-$ cross sections were obtained from the compilation of H. Genzel and W. Pfeil, Bonn Univ. PI B1-168 (1972).

10. K. Kolbig and B. Margolis, Nucl. Phys. B6, 85 (1968).

11. B. Hahn et al, Phys. Rev. 101, 1131 (1956); H. R. Collard et al, Landolt-Börnstein new series I/2 nuclear radii.

12. See, for example, E. L. Berger, Phys. Rev. Letters 21, 701 (1968).

13. M. Longo, et al, University of Michigan Report UM-HE-74-18 (1974).

14. L. Jones et al, Phys. Rev. Letters 33, 1440 (1974) and P. Murthy et al, University of Michigan Report UM HE-75-2 (1975).

15. A report describing the properties of the high-pressure gas target is in preparation.

16. See, for example, the following: J. Bartsch et al, Phys. Letters 27B, 336 (1968); B. Y. Oh and W. D. Walker, Phys. Letters 28B, 564 (1969); M. S. Farber et al, Phys. Rev. Letters 22, 1394 (1969); H. I. Miettinen and P. Pirila, Phys. Letters 40B, 127 (1972); the review of J. Rushbrooke, Proc. III Intl. Colloquium on Multiparticle Dynamics, Zakopane (1972), A. Bialas et al, eds.; H. Lubatti and K. Moriyasu, Lett. Nuovo Cimento 12, 97 (1975).

17. See, for example, S. Humble, Nucl. Phys. B76, 137 (1974). For an earlier discussion of helicity flip contributions in Pomeranchukon processes see G. Kane, Acta Phys. Pol. B3, 845 (1972).

18. E. Nagy et al, CERN preprint 74-1321 (June 1974). This group investigated the dissociation of the proton rather than neutron; in particular, they examined the reaction $pp \to pn\pi^+$.

19. For a recent summary of the phenomenology see the excellent review of E. Berger, ANL Report ANL-HEP-PR-75-06 (1975).

20. Private communication from D. Duke.

21. For a critique of the M-t interdependence available in the Deck model see H. I. Miettinen, Proc. of the Zakopane Colloquium (1972). A. Bialas et al, eds. For a re-examination of this question see V. A. Tsarev, Phys. Rev. D11, 1864 (1975).

ACCELERATION OF HEAVY IONS TO RELATIVISTIC ENERGIES
AND THEIR USE IN PHYSICS AND BIOMEDICINE

Milton G. White
Department of Physics: Joseph Henry Laboratories
Princeton University, Princeton, N. J. 08540

1. Introduction

A. The spectacular progress in our understanding of nuclei and elementary particles was made possible by the invention and development of cyclotrons, Van de Graafs, synchrotrons, electron linear accelerators and colliding beam machines which, in 45 years, have increased the Center-of-Mass energies from 100 keV, or so, to 50 GeV. Each major energy increase has revealed facets of nature totally outside the prevailing theoretical framework; though, once discovered, theorists have been very agile in adjusting theory to fit the new facts.

B. This pattern -- advance in energy, new major discoveries, new theories -- has become so traditional that it is almost impossible nowadays to build a major new accelerator unless it promises to enter a new energy regime and guarantees major discoveries. For some time the protagonists of relativistic heavy ion accelerators encountered apathy, amounting to outright opposition from some quarters, on the ground that the collision of a 100 GeV nucleus, consisting, say, of 200 nucleons, would be no more exciting than a two nucleon, 500 MeV collision, and a lot messier. Feynman is quoted as saying that if he wanted to study nucleon interactions with nuclei he wouldn't throw a Swiss watch at another Swiss watch. He would rather scatter nucleons from a single Swiss watch. However, in so doing he would never find out that the Swiss watch contained a mainspring, a balance wheel, and various gears. In fact, he wouldn't even know that the target was a Swiss watch, let alone what time of day it read. The collision of two Swiss watches will, at least, throw out a lot of interesting time pieces and maybe one can even tell the time of day!

C. When the Omnitron was proposed[1] in Berkeley in 1966 it was designed to produce heavy ions with a maximum of 500 MeV/amu. The scientific motivation was largely nuclear physics, nuclear chemistry and radiation therapy for cancer. Unfortunately funding for the facility, to cost $28 million, was not forthcoming.

D. When it was proposed[2] to the AEC in 1968 that the Princeton-Pennsylvania Accelerator, a 3 Bev fast cycling proton synchrotron, be converted to heavy ions there was little scientific support, and that chiefly from the cosmic ray scientists. Unsuccessful attempts were made to secure scientific support from nuclear physicists, elementary particle physicists, solid state and atomic physicists. A major effort was mounted to interest cancer radiation therapists in using relativistic heavy ions, e.g. neon, based on earlier work of Tobias, Todd[3] and others who had shown that the high linear energy transfer rate (LET) of heavy ions held promise of overcoming the so called Oxygen Enhancement Ratio (OER) which makes anoxic cancer cells harder to kill by γ-rays than normally oxygenated healthy cells. Finally, through the courage and generosity of the Fannie E. Rippel Foundation, a private foundation, we were able to convert the PPA to heavy ions and to perform experiments for nine months before our money finally ran out.

E. The PPA obtained[4] its first beam of 3.9 Bev nitrogen ions on July 15, 1971 and the Bevatron[5] also succeeded a few weeks later. Now the PPA is in moth balls, intact but gathering dust. Luckily for science the Berkeley Bevalac continues to live, though on a very tight budget. Before describing the technical problems which face a relativistic heavy ion accelerator, and affect the chances of going to very high energies and very heavy ions, I would like to sketch, very superficially, a few of the many fields of science in which progress may lie in the domain of relativistic heavy ions.

2. Special Properties of High Energy Heavy Ions (HZE's) and their Fields of Applications

High energy heavy ions are characterized by not only the large total energy E, but also the large number of electrically charged protons Z and nucleons A in the nucleus. The following presentation discusses the special properties of HZE's which depend on Z, A, and E.

A. High Nuclear Electric Charge Z:

i) Loss of energy of a charged particle by ionization in passing through matter is proportional to $Z^2\beta^{-2}$ where $\beta = v/c$. For example, a 300 MeV/amu neon ion, with 10 cm residual range in water (or tissue), transfers energy at the initial rate (LET) of 400 MeV/g cm^{-2} or 40 keV/μm, rising to 1000 keV/μm at the Bragg peak, while a proton with the same residual range loses energy initially at 0.7 keV/μm rising to 90 keV/μm at the peak. Ions of higher Z would, of course, ionize at a higher rate proportional to Z^2. It is this marked difference in LET between electrons, or protons, and heavy ions which is the major reason for expecting heavy ions to be important in solid state physics and biomedical applications.

ii) There are a number of very interesting and possibly fundamental questions concerning the intense electric fields surrounding a fully ionized high Z nucleus. For example a study of the Lamb shift for hydrogen-like, high Z nuclei may reveal departures from QED at very high electric fields. The best way to fully ionize a nucleus, such as uranium, is to shoot it with very high energy, 100 MeV/amu, through a thin foil. Theorists[6] have also conjectured that there may be a limiting electric field characterized by spontaneous emission of electron-positron pairs for nuclei with Z > 170. To produce nuclear matter with Z > 92 one may hope either to produce stable, heavy nuclei lying in the region of the predicted islands of stability at Z = 112 or higher or, more speculatively, to produce Z = 184 nuclei by means of the abnormal states mechanism proposed by Lee and Wick. Regardless of how highly speculative these various theories may be it is surely in the historical spirit of physics to push ahead with experiments involving higher and higher nuclear electric charge.

B. Multi-nucleon Character of an HZE:

    i) The sharing of the large energy E among A nucleons may result in qualitatively new phenomena. When two, relativistic multi-nucleon particles collide one may see collective, or coherent interactions between nucleons. Dar[7] has pointed out that when a highly relativistic, and therefore Lorentz contracted, nucleus collides with a proton at rest at impact parameter b, the proton actually collides simultaneously with all the nucleons at impact parameter b. When b = o the proton collides simultaneously with $A^{1/3}$ nucleons. The "collision energy" is therefore $A^{1/3}$ times the energy per nucleon in the incident nucleus. Heavy ion colliding beams with an energy per nucleon equal to a proton colliding beam machine would produce "energy densities" $A^{1/3}$ times larger and perhaps reach CM energies sufficient to produce new, more massive particles. In addition, the simultaneous participation of several nucleons may introduce new and interesting features.

    ii) Greiner[8] and others, using a hydrodynamical model, predict that when the velocity of the incoming nucleons of a heavy nucleus, making a head-on collision with a stationary heavy nucleus, exceeds the velocity of sound in the nucleus, (about 1/5 c) there may be formed a shock-front which propagates through both nuclei. In the shock front matter and energy density should far exceed anything now known with consequences which may be extremely exciting. Greiner, and also Teller[9] expect, or speculate, that the vacuum state degenerates into an infinite number of $\pi^{\pm}$ pairs and π^{o}'s created without the expenditure of energy. In Greiner's picture these pions are all in bound states with a double layer formed of π^{-} inside, π^{+} outside and π^{o}'s throughout the nuclear volume. This "pionization" of matter, if it is possible, would be very exciting to study.

    iii) T. D. Lee and G. C. Wick[10], using a field theoretic approach which assumes only that the field consists of 0^{+} strongly interacting particles, find that at some critical density (about twice normal) nuclear collapse can occur leading to the formation of stable nuclei in abnormal states characterized by the nucleons having lost nearly all their rest mass. These nuclei could absorb thermal neutrons and grow to large size, even to $A \sim 10^4$. They speculate that a head-on collision of two uranium nuclei with 200-500 MeV/amu could produce a doubling of the density as the nuclei interpenetrate each other with the subsequent formation of abnormal nuclear matter. The vacuum state in their model would have extraordinary properties and might show profound departures from present understanding.

iv) Even if none of these speculations come to reality it is clear that the multi-nucleon character of the collision between relativistic heavy ions brings a new dimension into the study of nuclear matter and therefore merits serious study. Perhaps there are real surprises in store if one succeeds in putting a lot of energy and mass into a <u>large</u> volume of space i.e. a large nucleus, in contrast with the present thrust of elementary particle physics where one strives to put ever more energy into a smaller and smaller region of space. The "cooking" time of a relativistic heavy ion collision is $\sim 2 R/c \sim 10^{-22}$ secs compared with 10^{-24} secs for a proton-proton collision thus simulating, to some degree, conditions[11] prevailing in the Primeval Fireball a few milliseconds after T = 0.

v) Several authors[12] have considered the question of the optimum energy for the production of shock waves and abnormal states of nuclear matter. Because the mean free path of a single nucleon traversing nuclear matter increases with energy, rising from 1.5×10^{-13} cm at 200 MeV/amu to 3.0×10^{-13} at 1000 MeV/amu it appears that above about 2 GeV/amu the nucleus may be too transparent to develop shock waves. Possibly the optimum energy is 200-500 MeV/amu for this type of study. However, our knowledge of heavy nuclei interacting at high energy is so scanty that no one should be deterred from building accelerators capable of

10-15 GeV/amu.

3. Applications of Relativistic Heavy Ions

   i) Heavy Ions in Cancer Therapy

   I have already alluded to the possible value of fast heavy ions in cancer therapy and I will now present a few more details even though this subject is not relevant to sub-nuclear physics. Cancer is relevant to all of us individually since one fourth of this audience will probably die of cancer, very likely prematurely, unless some real advances are made in early detection and treatment. At the present time the annual death rate in the USA from cancer treated with X-rays is 175,000, of which 58,000 die because of failure to achieve local control of the tumor. It is this 58,000 deaths per year which might be reduced by more effective radiation therapy. Most of these 58,000 patients have cancer of an unoperable form, and while one day there may be developed satisfactory drug treatment methods for all of the several hundred known types of cancer there is no guarantee that this will soon come to pass. Meanwhile, radiation therapy promises a worthwhile measure of control. The attractive features of high energy heavy ions compared with X-rays are: 1) Well defined particle range and region of irradiation; 2) Very large linear energy deposition rate (LET) along the track and especially at the Bragg peak; 3) Improved ratio of dose at tumor depth to skin dose; 4) Relatively low exit dose. The large LET of heavy ions has been shown by Todd,[13] Tobias and others to be much more effective, per unit of energy transfer, than low LET X-rays in stopping the reproduction of cells. This improvement relative to X-rays is significantly greater for cells low in oxygen, a situation believed to exist for cancerous cells in a tumor mass. It has been shown that anoxic cells require three times more X-ray dose than normally oxygenated cells, while heavy ions, pi-minus stars and to some extent the heavy ion recoils from neutron bombardment all show substantially less of this so-called Oxygen

Enhancement Ratio (OER). At the present time there is an active clinical program to evaluate the merits of neutron therapy at a number of cyclotrons, pi-minus therapy at LAMPF and, eventually, heavy ions at the Bevalac.

ii) In order to achieve a worthwhile improvement in the OER it is necessary to use ions with a Z = 10 (neon) or, possibly argon (Z = 20). Since the desired range in tissue is 10-15 cm an energy for neon of 300-400 MeV/amu is required, while for argon one needs 400-500 MeV/amu. Todd predicts that 2.5×10^8 neon ions stopping within a cubic centimeter of tumor, at 10 cm depth, would reduce the cells to the 10^{-10} level of survival while delivering only 90 rem to the skin. By comparison the same type of analysis shows that X-rays (1.3 MeV), neutrons (25 MeV) and π^- mesons (45 ± 5 MeV) would deliver to the skin 3,800, 1,200, and 105 rem respectively for the same reduction to the 10^{-10} cell survival level. The calculation assumes anoxic, human kidney T-1 cells. Research at the PPA on Chinese hamster cells and other biological material using nitrogen ions of 280 MeV/amu confirmed[14] Todd's predictions as far as we were able to go in the short time available for research. In view of the large amount of money currently being spent in the USA for clinical trials of marginally superior neutron therapy it would make scientific and economic sense to increase substantially the heavy ion therapy program which promises to be much more effective. However, before human trials can begin a substantial program of basic research must be completed.

iii) <u>Nucleon Structure, Elementary Particles</u>

- Coherent production of mesons.
- Production of excited states of nucleus through multiple internal collisions.
- Phase separation of matter, anti-matter in a micro Primeval Fireball.
- Possible production of new particles by the high CM energy.
- Test of Feynmans' limiting fragmentation picture.

- Test of elementary particle ideas at large hadron number.

iv) Nuclear Structure, Abnormal States, Shock Waves

- Very high angular momentum states of nuclei and nuclear molecules.
- Highly excited collective modes of nuclei.
- Nuclear fragmentation.
- Nuclear shock waves.
- Pionization of nuclear matter.
- Production of abnormal states of nuclear matter.
- Production of new shapes of nuclear matter.
- Superheavy nuclei.
- Production of nuclear matter far from the valley of stability.
- Hypernuclei containing several Λ particles i.e. Superstrange nuclei.
- Thermodynamics of hot, nuclear matter.

v) Cosmic Rays

- Determination of HZE fragmentation cross-sections and interpretation of origin of cosmic rays.
- Electronic charge exchange cross sections of relevance to the origin of K- capturing nuclei e.g. ^{7}Be-.
- Calibration of fossil cosmic ray tracks in meteorites and lunar material.
- Search for superheavy and abnormal state nuclei in cosmic rays.

vi) Solid State Physics

- Radiation damage studies, color centers, internal damping.
- Formation of voids in solids.
- Track formation in solids.
- Macromolecules, chain breaks.
- Ion implantation.
- Channeling and blocking in crystals.

vii) General Physics

- Production of fully ionized, large Z nuclei for study of QED at very high electric fields.

- Production of abnormally large Z through collisions of high Z nuclei.

vii) Biological Studies and Biomedical Applications

- Use in cancer therapy.
- Study of the dynamics of cell division.
- Study of brain organization and brain damage.
- Radiation effects of high LET heavy ions on living tissue.
- Macromolecules and DNA-chain breaks.
- Effects on nerve tissue.
- Use as a micro surgical tool.

4. Desirable Ions and Energies for Physics and Biomedical Application

i) In a field so new and fertile as relativistic heavy ions almost any ion at almost any energy is worth studying. However the accelerator designer needs some guidance, or at least a set of reasonable parameters for which to strive. As a practical matter the state of present day ion sources and the existence of accelerators capable of conversion to heavy ion acceleration will have a dominant effect on what will actually be done in the next few years.

ii) From the view point of the most exciting basic physics so far suggested i.e. shock waves, pionization of matter, abnormal states of nuclei it seems clear that one ultimately requires uranium ions with 100-1000 MeV/amu and at currents as high as technically feasible, with 10^9 particles per second being desirable. However, even much smaller currents can certainly be usefully employed. While uranium is the ultimate goal, much lighter ions, even down to argon or neon should certainly be employed in searches for unusual behavior.

iii) The biomedical applications, especially cancer therapy, are best satisby beams of neon, or possibly argon, with sufficient energy to penetrate 20 cm of tissue which requires, for neon 500 MeV/amu and for argon 650 MeV/amu. Since 2.5×10^8 neon ions/cm^3 of tumor will reduce cell survival to the 10^{-10} level it is clear that quite small beam currents will be adequate. For example, a 1000 cm^3 tumor, treated in

10 fractionations, requires only 2.5×10^{10} neon ions per fraction. If given over a period of 100 seconds the required beam current is 2.5×10^8 particles per second. A fast cycling synchrotron, such as the PPA, or the CEA converted to heavy ions need deliver only 10^7 ions per pulse.

5. Requirements for a Relativistic Heavy Ion Accelerator

   i) It is obvious that, today, only a synchrotron can produce the required energies at a reasonable cost. Linacs are limited to 10-20 MeV/amu for an accelerator 70 meters in length, and electron ring accelerators have yet to fulfill their early promise. These devices may prove to be good injectors of partially ionized, very heavy ions in to a large synchrotron, but if the goal is to inject fully ionized uranium into an existing large synchrotron then, as I will show, a rapid cycling intermediate energy booster synchrotron will be necessary.

   ii) Figure 1, taken from a paper by Prelec and van Steenbergen[15] depicts the energy capabilities of several accelerators for ions with charge to mass ratios expressed by the dimensionless parameter $\varepsilon \equiv (q/q_p)/(m/m_p)$. Notice that the Bevatron, accelerating fully stripped uranium, can reach 2 GeV/amu while the AGS could go to 10 GeV/amu; the PPA, 850 MeV/amu. The "new" CEA, proposed as a booster for the AGS, could produce uranium with 500 MeV/amu if provided with U^{+60} and of course higher for fully ionized uranium.

   iii) The major problems encountered in a heavy ion synchrotron are: achieving a very high charge state and avoiding charge changing collisions with the background gas in the acceleration chamber. Both problems are connected with capture and loss of electrons by the ions as they pass through stripping foils or the dilute gas of the vacuum chamber. Unfortunately there are little experimental data on capture and loss of electrons by high velocity heavy ions; so theoretical extrapolation from low velocity is necessary. Calculations by Schmelzer[16], Betz[17] and Dmitriev and Nikolaev[18] are conveniently summarized in the papers by Prelec and van Steenbergen. I will give only the background

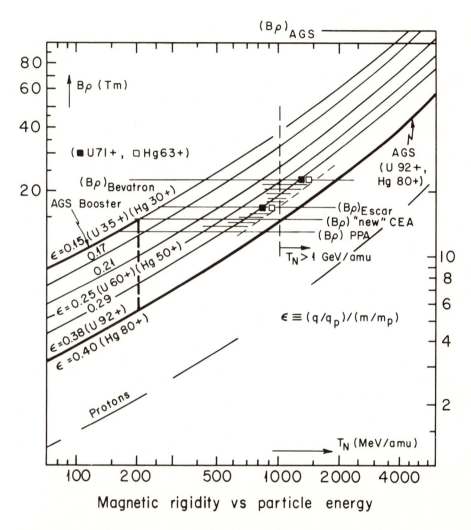

Figure 1 Particle energy versus Bρ for ions in various charge states.

for these calculations and their conclusions.

iv) The charge exchange collision rate may be expressed as $dN_{ch.ex.} \propto$

$N \cdot \sigma(\beta) \cdot \beta c dt \cdot Ng$ where dN is the number of ions suffering a change of ± 1, in time dt; N the total number of ions traveling with velocity $\beta \cdot c$; $\sigma(\beta)$ is the total charge changing cross section $\sigma_{tot} = \sigma_{cap} + \sigma_{loss}$ and Ng the number of background gas atoms per unit volume. Integration leads to

(1) $\quad N/N_o = \exp\left[-10^{27} P_{vac} \int_o^T \sigma(\beta) \cdot \beta \, dt\right] \quad P_{vac}$ in Torr

v) The method employed to evaluate $\sigma(\beta)$ is as follows. In passage through a dilute stripper, in the equilibrium state, it holds that

(2) $\quad N(q + 1 \to q)_{capt} = N(q \to q + 1)_{loss}$

Ignoring multiple electron exchange processes, it follows that

(3) $\quad F(q + 1) \cdot \sigma_{capt}(q+1) = F_q \sigma_{loss}(q)$;

where F is the fraction of the beam having charge states q and $q + 1$. This quantity may be approximated by the assumption of a Gaussian charge state distribution around the average charge $\bar{q}$.

(4) $\quad F(q) \approx d_q (2\pi)^{-1} \exp\left[-(q - \bar{q})^2 / 2 \, d_q^2\right]$

The quantity $d_q = 0.32 \, Z^{0.45}$.

Consequently

(5) $\quad \sigma_{loss}(q) = \sigma_{capt}(q + 1) \exp\left[-2(q - \bar{q}) + 1\right] / 2 \, d_q^2$

To evaluate $\sigma_{loss}(q)$ requires knowing $\sigma_{capt}(q + 1)$ and $\bar{q}$ as a function of T_n, the energy per nucleon. For σ_{capt} several authors have proposed a relationship which scales from experimentally determined proton capture at the same velocity v and in the same medium.

(6) $\sigma_c \simeq q^2 (v/2v_0)^3 \cdot \sigma_c^{proton}$; v_0 = electron velocity in lowest state of hydrogen

(6') $\sigma_c \simeq q^2 (\beta/2\alpha)^3 \sigma_c^{proton}$; $\alpha = \frac{e^2}{hc} = 1/137$

$\beta = v/c$

Using the experimental relationship for σ_{capt}^{proton} in N_2.

(7) $\sigma_c^{proton} = 1.4 \times 10^{-15} (v/2v_0)^{-8.6}$

one finally obtains for an ion of charge q and energy Tn:

(8) $\sigma_c(q) \simeq 2.2 \times 10^{-18} q^2 Tn^{-2.8}$ Tn = MeV/amu

Note how rapidly σ_c drops with Tn, therefore emphasizing the importance of strong acceleration early in the acceleration cycle. Before one can calculate σ_{loss} from Eqt's. 5, 8 it is necessary to find a relationship between $\bar{q}$, the equilibrium charge, and Tn, Z_{ion} and $Z_{stripper}$. It is here that one encounters the greatest uncertainty in predicting the behavior of a heavy ion synchrotron.

vi) Two semi-empirical relationships have been proposed by Betz and Schmelzer[16,17] and by Dmitriev and Nikolaev[18]. These relationships are displayed in Fig. 2&3 for both gaseous and solid strippers. The physical arguments behind these semi-empirical relationships have a long history dating back to early calculations of Bohr, Bethe, et.al and will not be reviewed here. There is a real need for experimental data since the various theoretical estimates vary by at least 10-20%.

vii) Since the relationship of Dmitriev and Nikolaev appears to fit the experimental data the best, and is the more conservative, we will use it in the following discussions. From Eqts. 5, 8, and the equation of Dmitriev and Nikolaev we arrive at the total charge changing cross-sections $\sigma_{tot} = \sigma_{capt} + \sigma_{loss}$ as a function of Tn, Z_{ion}, Z_{gas}. Some confidence in this approach may be gained from experimental work at the PPA[4] which, for nitrogen ions, roughly confirmed the predictions up to

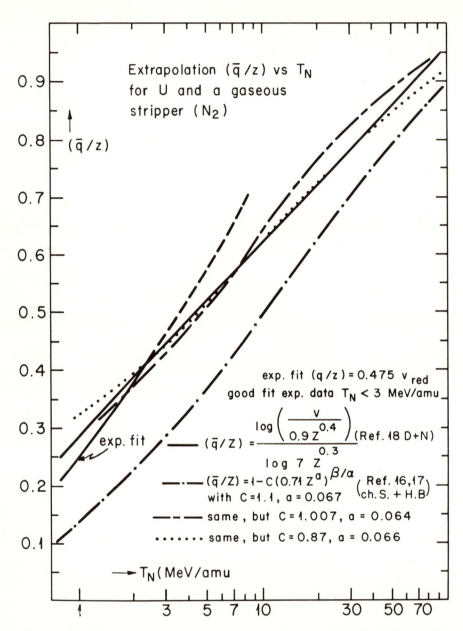

Figure 2   Semi-empirical relationship for the ratio of most probable charge to nuclear charge, q/Z, as a function of energy per nucleon T_N.

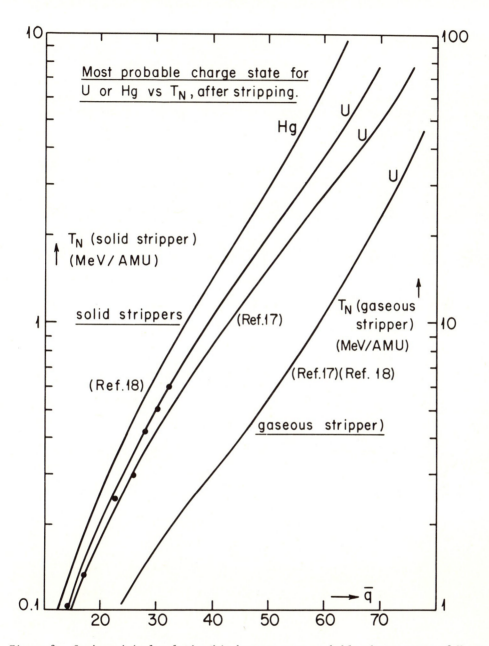

Figure 3 Semi-empirical relationship between most probable charge state of U or Hg versus T_N after stripping in a foil or in N_2.

Tn = 278 MeV/amu and for argon up to 400 Me /amu.

viii) Figure 3 gives the average charge state $\bar{q}$ for U and Hg as a function of the energy per nucleon Tn in solid and gaseous strippers. It is clear from these semi-empirical extrapolations that to produce fully ionized uranium by foil stripping an energy in excess of 100 MeV/amu will be required.

6. AGS Parameters

i) Figure 4 shows the minimum energy at which one could inject uranium with charge q into the AGS. The minimum here refers to the lowest magnetic field of suitable quality for stable orbits. $T_N(max)$ refers to the top energy to which the AGS magnet could carry uranium in charge state q.

ii) However, because of the capture and loss of electrons which we have been describing; for U^{+40} and Tn > 0.1 GeV/amu, we find $\sigma_{tot} \cdot \beta \approx 10^{-17}$ cm^2 leading to N/No $\simeq \exp\left[-10^{10} t_r P_{vac}\right]$ where t_r is the magnet rise time. For better than 1/e transmission the vacuum must be better than $P_{vac} < (10^{-10}/t_r)$ mm of Hg (Torr). For the AGS $t_r \simeq 0.5$ sec thus requiring a vacuum better than 2×10^{-10} Torr. This is a very severe requirement; therefore Prelec and van Steenbergen propose injecting only fully ionized U^{+92} which they would inject at Tn = 100 MeV/amu. Their proposal is to use the CEA fast cycling synchrotron magnet in front of the AGS to boost the energy sufficiently for efficient stripping of uranium to U^{+92}.

iii) Figure 5 shows how the energy per nucleon varies with time for the "old" and "new" CEA where "new" differs from "old" by reduction of the radius from 36 to 20 meters; the change being dictated by the desirability of using PPA-like radiofrequency cavities. Figure 5 also shows how the particle loss factor $\sigma \cdot \beta$ (cm^2) varies along the acceleration cycle for uranium in charge state +30. While the rate of ion loss is greatest early in the acceleration cycle the time spent there is the least. It turns out that for a 1/e transmittal a vacuum of about 10^{-9}

ACCELERATION OF HEAVY IONS TO RELATIVISTIC ENERGIES

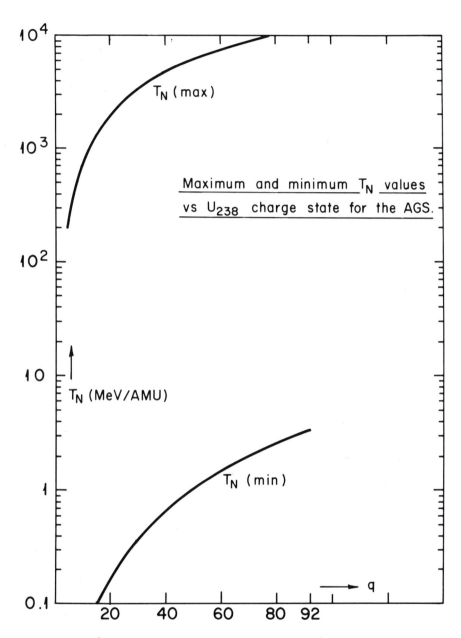

Figure 4 AGS injection energy as function of the charge state.

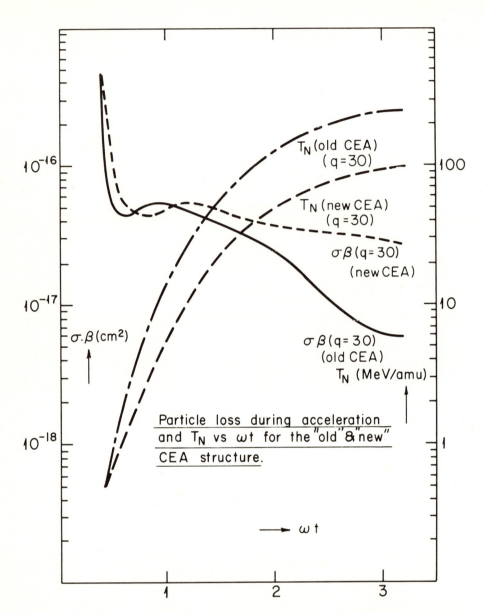

Figure 5 The CEA as a heavy ion booster accelerator. Shown is the energy per nucleon for U^{+30} as function of ωt where $\omega = 2\pi \cdot f$ (f = cycling frequency).

mm of Hg is required, a not difficult requirement.

iv) After acceleration of U^{+30} to 100–200 MeV/amu it is extracted and sent through a foil which strips to U^{+92} for insertion into the AGS. Table I summarizes the various parameters. Note that under the assumptions stated, i.e. the preacceleration of U^{+11} by 10.8 MV and stripping to U^{+30} before insertion into the booster, the CEA space charge limit is about 5×10^8 particles per pulse (at 30Hz the average current is thus 1.5×10^{10} p/sec) leading to an AGS intensity of about 10^9 U^{+92} per cycle with a maximum energy of 12 GeV/amu. Several booster cycles of injection are employed to reach this current. There is essentially no particle loss in the AGS due to charge changing collisions at a vacuum of around 5×10^{-8} Torr, a readily obtained vacuum.

v) The maximum number of heavy ions per pulse is limited by well known incoherent transverse space charge instabilities; which, for the AGS, is 10^{13} protons/pulse at 200 MeV injection energy according to the Laslett formula. The space charge limit for any other particle with atomic weight A, charge state ε and corresponding $\beta\gamma$ is:

$$N_{sp\ ch\ ion} = \left[(\beta^2 \gamma^3)_{inj\ ion} \bigg/ (\beta^2 \gamma^3)_{200\ proton} \right] (1/A\ \varepsilon^2)_{ion} \cdot 10^{13} ppp$$

The most noteworthy feature in this relationship is the $1/\varepsilon^2$ dependence of the space charge limit. The ability to accelerate without charge exchange loss in a low charge state will lead to higher ion currents. Even with U^{+92} the space charge limit in the AGS main ring, with Tn(min) = 200 MeV/amu, is 2.8×10^{11} ions/pulse. The real bottleneck is the CEA booster for which the above consideration leads to a space charge limit of 1.8×10^9 ions/pulse for U^{+38} injected at 0.925 MeV/amu. The only way to raise this limit is to inject into the CEA at a much higher energy. Use of a lower charge state would raise the current, but would decrease the final energy below that required for efficient production of U^{+92} by foil stripping. The Unilac and Superhilac are clearly outstanding as preaccelerators into a booster. As Table I shows, the CEA booster particle output per pulse, taking into account charge changing collisions, is 8×10^8 (column 3), assuming a vacuum of 1.0×10^{-9} Torr. However, in view of the high CEA pulse rate of 30 cycles per second it now becomes feasible to inject several CEA cycles into the AGS. Putting all these considerations together one finally predicts an AGS current of 2.1×10^9 U^{+92} ions per pulse at 2 GeV/amu. Though the AGS magnet could go to 12 GeV/amu the AGS cycling rate would have to be reduced to keep the rf accelerating voltage within the range of the present proton cavities; also harmonic jumping would have to be employed. The design of Prelec and van Steenbergen starts with the assumption that the addition of heavy ion capabilities must in no way compromise the present AGS proton operation. For this reason they have had to make some adjustments to the various parameters which would not be necessary for a fully dedicated heavy ion machine.

7. The Bevalac Parameters

i) As a practical matter the only relativistic heavy ion accelerator now operating, or likely to operate in the forseeable future, is the Bevalac at the Lawrence Berkeley Laboratory. Since its present performance has been rather extensively reported in the literature I will limit my remarks to a recent proposal to increase its capabilities to permit acceleration of U^{+72} to 1.1 GeV/amu at an intensity of 10^8 ions/cycle.

ii) The approach taken at Berkeley is somewhat different from that outlined above and is made possible by the existence of the Superhilac as an injector. However, even the Superhilac can not reach energies for uranium above 8.5 MeV/amu, therefore, according to Figure 2, it will be difficult to obtain a charge state much above U^{+72} by stripping. Because of the long acceleration time in the Bevatron is therefore essential to achieve a vacuum of at least 10^{-9} - 10^{-10} Torr to avoid excessive beam loss. An ingenious, bold proposal[20] has been made to add a vacuum liner to the existing chamber, which currently achieves only

TABLE I

Preaccelerator - Booster - AGS System

							Speculative source
SOURCE							
equiv. volt. MV	10.8	10.8	10.8	20	4	5.8
source ion	U238	U238	U238	U238	U238	U238
q	11+	11+	11+	11+	11+	38+
emA	1	1	1	1	1	-
post strip q	30+	30+	30+	38+	20+	no stripper
intensity eA	$0.4 \cdot 10^{-3}$	$0.4 \cdot 10^{-3}$	$0.4 \cdot 10^{-3}$	$0.52 \cdot 10^{-3}$	$0.27 \cdot 10^{-3}$	no attenua.
emit. μrad-m	15π	15π	15π	11.2π	25π	no dilution
BOOSTER						
structure	"old CEA"	"old CEA"	"new CEA"	"new CEA"	"old CEA"	"new CEA"
T inj MeV/AMU	0.50	0.50	0.50	0.925	0.184	0.924
β inj	0.0325	0.0325	0.0325	0.044	0.020	0.044
B inj Gauss	306	306	459	494	279	494
n inj turns	3	3	3	3	2	-
N capt. p/p	$(4.6 \cdot 10^9)$	$(4.6 \cdot 10^9)$	$(3.4 \cdot 10^9)$	$(2.5 \cdot 10^9)$	$(2.5 \cdot 10^9)$	(-)
N sp.ch. p/p	$2.1 \cdot 10^9$	$2.1 \cdot 10^9$	$2.1 \cdot 10^9$	$1.8 \cdot 10^9$	$1.5 \cdot 10^9$	$1.8 \cdot 10^9$
rep.rate Hz	15	15	30	30	15	30
V max rf kV	236.4	236.4	205.0	235.8	236.4	235.8
rf freq. MHz	1.03 - 13.56	1.03 - 19.71	1.41 - 18.57	1.92 - 24.62	0.63 - 13.56	1.9 - 24.6
harmonic h	24	24	24	24	24	24
accel.range ωt	π/8-1.75	π/8-π	π/8-π	π/8-π	π/8-2.5	π/8-π
P vac* ≤ Torr	$1 \cdot 10^{-9}$	$8 \cdot 10^{-10}$	$1 \cdot 10^{-9}$	$3.8 \cdot 10^{-9}$	$5 \cdot 10^{-10}$	$3.8 \cdot 10^{-9}$
T eject. MeV/AMU	100	260	100	200	100	200
B max Tesla	(0.759)	0.759	0.669	0.765	(0.759)	0.765
N out* p/cycle	$8 \cdot 10^8$	$8 \cdot 10^8$	$8 \cdot 10^8$	$7 \cdot 10^8$	$5 \cdot 10^8$	-
post strip. q	92+	92+	92+	92+	92+	92+
inten. p/cycle	$4.8 \cdot 10^8$	$6.4 \cdot 10^8$	$4.8 \cdot 10^8$	$5.6 \cdot 10^8$	$3 \cdot 10^8$	-
emit. μrad-m	1.5π×9π	0.9π×5.5π	1.5π×9π	1.1π×6.5π	3.4π×9π	-
AGS						
β inj	0.428	0.622	0.428	0.566	0.429	0.566
B inj Gauss	449	752	449	650	449	650
n inj turns cycles	1 4	1 4	1 6	1 6	1 4	-
N capt. p/p	$1.5 \cdot 10^9$	$4.1 \cdot 10^9$	$2.3 \cdot 10^9$	$2.7 \cdot 10^9$	$0.9 \cdot 10^9$	-
N sp.ch. p/p	$1.2 \cdot 10^{11}$	$3.9 \cdot 10^{11}$	$1.2 \cdot 10^{11}$	$2.8 \cdot 10^{11}$	$1.2 \cdot 10^{11}$	$\sim 3 \cdot 10^{11}$
rep.rate c/s	1	1.25	1	1.25	1	-
rf freq. MHz	1.9-2.5	-	1.9-2.5	-	1.9-2.5	-
V max rf kV	~ 75		~ 75		~ 75	
rf freq. MHz	2.5-4.2	2.8-4.2	2.5-4.2	2.5-4.2	2.5-4.2	2.5-4.2
V max rf kV	avail.	avail.	avail.	avail.	avail.	avail.
harmonic h	12	12	12	12	12	12
P vac ≤ Torr	$1.3 \cdot 10^{-8}$	$1 \cdot 10^{-7}$	$1.2 \cdot 10^{-8}$	$0.7 \cdot 10^{-7}$	$1.3 \cdot 10^{-8}$	$0.7 \cdot 10^{-7}$
N out p/p	$1.3 \cdot 10^9$	$3.7 \cdot 10^9$	$2.1 \cdot 10^9$	$2.4 \cdot 10^9$	$8 \cdot 10^8$	-
p/sec	$1.3 \cdot 10^9$	$4.6 \cdot 10^9$	$2.1 \cdot 10^9$	$3 \cdot 10^9$	$8 \cdot 10^8$	-
T out** GeV/AMU	2	2	2	2	2	2

* in Booster $(N/N_0) = 1/e$, in AGS $(N/N_0) = 0.9$; see text.

** Maximum value in AGS $\sim$ 12 GeV/AMU, at lower cycling rate.

10^{-7} Torr, and also cryopanels cooled to 20° K. Also titanium pumps will be liberally distributed around the ring. A vacuum in the desired range is predicted, though thermal outgassing of the panels prior to cool down may be required to achieve 10^{-10} Torr. Figure 6 is a cross-section of the magnet and vacuum chamber. Figure 7 is a schematic representation of the present and proposed capabilities of the Bevalac. Figure 8 puts the Bevalac into perspective with other existing heavy ion accelerators.

iii) On the same representation I have indicated the PPA's ultimate limits if provided with a 10^{-9} Torr vacuum chamber, a suitable injector, and a storage ring to permit the attainment of fully ionized uranium. Figure 9 indicates how a very modest storage ring could be added to permit cycling ions twice through the PPA. The idea is to build a combination booster-synchrotron/storage ring (PPB) which would operate as follows: A U^{+11} beam is injected into the PPB, which accelerates by 10-20 MV equivalent. The ejected beam is then stripped to U^{+30}, inserted into the PPA and accelerated to 150 MeV/amu. After ejection and stripping to U^{+92} the beam is reinserted into the PPB for storage until the PPA magnet has been cycled back to low magnetic field whereupon the beam is ejected from the PPB back into the PPA and accelerated to full energy of 850 MeV/amu. The predicted time average currents are in the 10^9 pps range.

Conclusion:

Now that accelerator builders have opened up for research a new area of the physical world we may anticipate many exciting discoveries. Probably the most important ones will bear little resemblance to anything I have sketched above, but at least they will have served the purpose of getting the research underway.

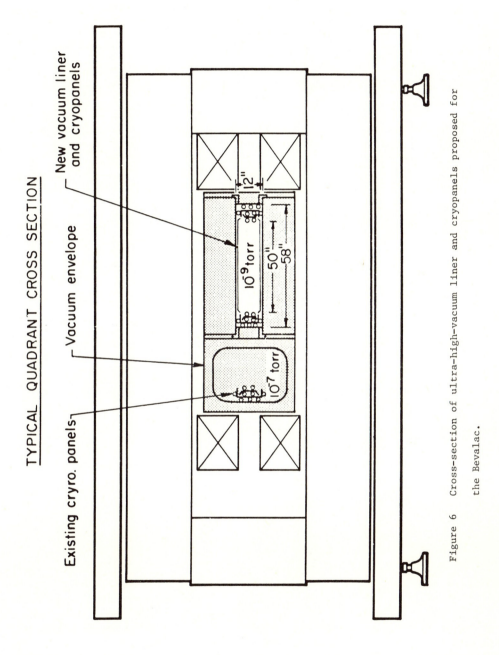

Figure 6 Cross-section of ultra-high-vacuum liner and cryopanels proposed for the Bevalac.

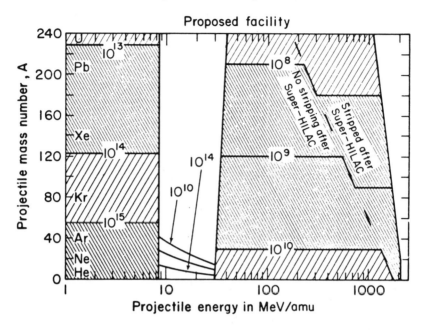
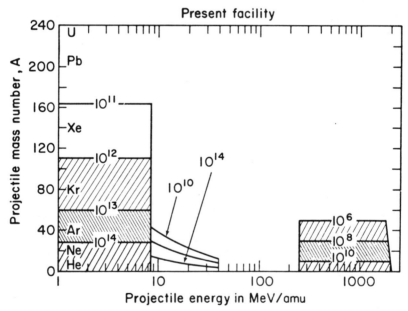

Figure 7 Schematic representation of the ions and energies obtainable with the Bevalac.

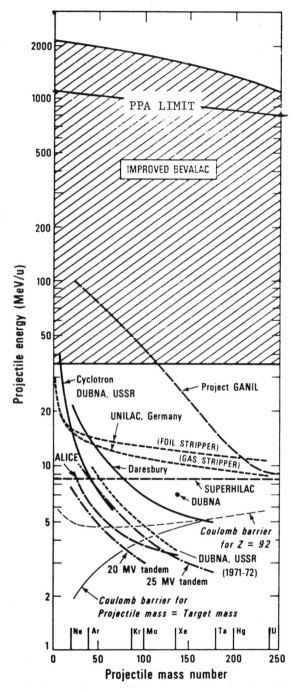

Figure 8 Comparison of the heavy ion capabilities of several existing or proposed heavy ion accelerators.

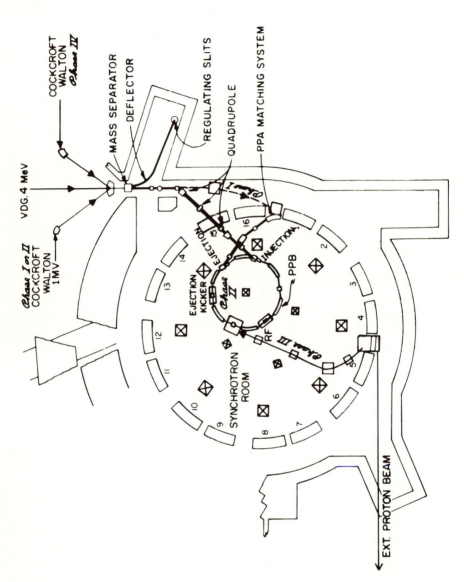

Figure 9 Proposed heavy ion booster/storage ring for the PPA.

References:

1. Omnitron Proposal UCRL-16828 (1966).

2. Heavy Ions and Higher Proton Currents Proposed for the Princeton-Pennsylvania Accelerator (M. Isaila, J. Kirchgessner, K. Prelec, F. C. Shoemaker, M. G. White) Particle Accelerators $\underline{1}$, 79 (1968); also PPAD 646 D.

3. The Prognostic Value of Cellular Studies in Evaluating New Accelerator Beams for Therapy (Paul Todd) LA-4653-MS, TID-4500; Physical and Biological Aspects of High LET Radiations with Reference to Radiotherapy (M.R. Raju) LA-5041-MS (1972).

4. Heavy Ion Acceleration (M. G. White, M. Isaila, K. Prelec, H. Allen) Science $\underline{174}$, 1121 (1971).

5. H. A. Grunder, W. D. Hartsough and E. J. Lofgren, Science $\underline{174}$, 1128 (1971).

6. Greiner et.al. Lett Nuovo Cimento $\underline{8}$, 37 (1973); Phys. Rev. Lett. $\underline{32}$, 554 (1974).

7. High Energy Collisions Between Complex Nuclei - Why? (A. Dar) Invited Talk Presented at the Topical Meeting on High Energy Collisions Trieste Sept. 9-13 (1974).

8. W. Greiner and W. Scheid, J. de Physique $\underline{32}$, C6-91 (1974).

9. G. F. Chapline, M. H. Johnson, E. Teller and M. S. Weiss, Phys. Rev. $\underline{D8}$, 4302 (1973).

10. T. D. Lee and G. C. Wick, Phys. Rev. $\underline{D9}$, 2291 (1974).

11. E. R. Harrison, Physics Today $\underline{21}$, 21 (1968).

12. Bear Mountain Workshop on Abnormal States of Nuclear Matter (1974)(to be published.)

12A. High Energy Heavy Ions: A New Area for Physics Research (Harry Heckman) 5th International Conference High-Energy Physics of Nuclear Structure: 1st and 2nd High Energy Heavy Ion Summer Study LBL (1973 and 1974); also specific papers from LBL on High Energy Heavy Ion Collisions.

13. P. Todd (loc. cit).

14. Cellular Effects of Heavy Charged Particles (P. Tod, C. B. Schroy, W. Schimmerling, and K. G. Vosburgh) Life Sciences and Space Research XI Akademie-Verlag,

Berlin p. 261-269 (1973); Radiobiology of Heavy Particle Radiation Therapy; Cellular Studies (E. J. Hall) Radiology Vol. $\underline{108}$, 119 (1973); Inactivation of Cultured Human Cells and Control of C3H Mouse Mammary Tumors with Accelerated Nitrogen Ions (D. Q. Brown, H. G. Seydel and P. Todd) Cancer Vol. $\underline{32}$, 541 (1973).

15. Search for "Abnormal Nuclear States" Heavy, A > 200 Ion Acceleration in the AGS., (K. Prelec and A. van Steenbergen) I.E.E.E. Particle Accelerator Conference (1975); also BNL Report AGS Div. 74-1 (1974).

16. Special Problems in Heavy Ion Accelerators (Ch. Schmelzer) in "Linear Accelerators" Lapostelle and Septier Edo (1970).

17. H. D. Betz et.al. Phys. Lett. $\underline{22}$, 643 (1966); Rev. of Mod. Phys. $\underline{44}$, 465 (1972).

18. Dmitriev and Nikolaev, Sov. Phys. JETP $\underline{20}$, 409 (1965); Phys. Lett. $\underline{28A}$, 277 (1968).

19. K. Prelec and A. van Steenbergen (loc. cit.).

20. High Intensity Uranium Beams from the Superhilac and Bevalac - Proposal No. 32 Lawrence Berkeley Laboratory (May 1975).

MINI BLACK HOLES

Steven C. Frautschi

California Institute of Technology
Pasadena, California 91125

The condition for a black hole is that the potential energy of a particle at its surface be of the same order of magnitude as the particle's rest energy mc^2:

$$V \simeq -\frac{G\,Mm}{R} \simeq -mc^2 . \tag{1}$$

Thus the radius and mass of a black hole are linearly related:

$$R \simeq GM/c^2 . \tag{2}$$

The possible spectrum of black holes covers an enormous range (Table 1).

Table 1

Spectrum of possible black holes

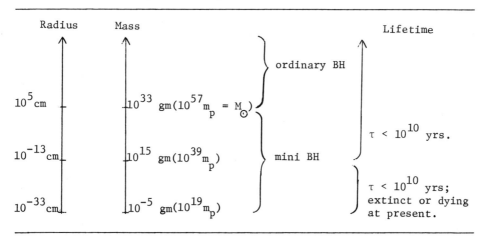

The spectrum divides naturally into several parts. First, there is a lower limit of 10^{-5} gm set by quantum mechanics. One cannot compress a wave packet to less than the Compton wavelength

$$\lambda = \hbar/Mc . \tag{3}$$

Thus the size of the black hole must exceed the Compton wavelength:

$$R > \lambda, \tag{4}$$

which implies $(G M/c^2) > (\hbar/Mc)$, i.e.,

$$M > \left(\frac{\hbar c}{G}\right)^{1/2} \simeq 10^{-5} \text{gm}. \tag{5}$$

As a result, the danger of electrons, protons, etc., collapsing spontaneously into black holes is negligible.

The spectrum of <u>ordinary black holes</u> begins at approximately the solar mass ($M_\odot \simeq 10^{57} m_p \simeq 10^{33}$ gm, corresponding to radii of a few kilometers) and runs up through possible large black holes at the cores of galaxies, etc. Black holes in this range may be forming all the time by collapse of stars and larger objects.

The spectrum below 10^{33} gm we shall call <u>mini black holes</u>. This part of the spectrum comprises an enormous range of masses (10^{33} gm to 10^{-5} gm) and radii (a few km. to 10^{-33} cm). At the present time, any mass in this range is stable against collapse by self-gravitation into a mini black hole. The stability is ensured by degeneracy pressure (of neutrons for neutron stars, electrons for the sun, earth, and smaller bodies) even when the matter is cold.

Thus any mini black holes that exist now are primordial. In the first 10^{-4} sec of the big bang, matter was compressed to a much higher density than in the sun or even in neutron stars, making gravitational collapse into mini black holes possible. For example, at $t = 10^{-23}$ sec in the conventional big bang the density was

$$\rho \simeq \frac{10^{15} \text{gm}}{(10^{-13} \text{cm})^3}. \tag{6}$$

If at this time there were large density fluctuations of order 10^{-13} cm across, the denser regions could have collapsed into mini black holes.

Whether such a dense region actually collapsed into a black hole depends on the competition between its gravitational self-attraction and the outward pressure. The pressure depends on the equation of state. The densities involved are far beyond any we can study in the lab, so a reliable equation of state is out of the question, but we can illustrate the possibilities by comparing two contrasting models of hadrons as worked out by Carr[1] and Chapline[2].

In the <u>elementary particle model</u>, matter consists of a finite number of varieties of free quarks, gluons, and leptons. The temperature of the early big bang then has the time-dependence

$$T \simeq m_\pi \left(\frac{10^{-4} \text{ sec}}{t}\right)^{1/2} . \qquad (7)$$

At such high temperatures the number density of particles is high ($n \sim T^3$). Spontaneous fluctuations in density are therefore small ($\delta\rho/\rho \sim n^{-1/2}$). Furthermore, at such high temperatures the density of kinetic energy greatly exceeds the density of rest energy, so the pressure is high. Even if a high-density fluctuation does form, the high pressure tends to block its further growth. Thus few mini black holes form in the elementary particle model <u>unless</u> large fluctuations are present in the initial boundary conditions, in which case copious formation is possible.

In the <u>statistical bootstrap model</u>, the number of varieties of hadron rises exponentially with mass. The temperature does not rise above $T_o \simeq m_\pi$. At this relatively low temperature, the number density of particles is lower than in the elementary particle model, so spontaneous fluctuations in density are larger[3,4]. Furthermore the low kinetic energy of order T_o per particle implies that there is little pressure to resist gravitational collapse once a high-density region has formed. Thus many mini black holes form in the statistical bootstrap model (in fact some cosmologies based on the model can be ruled out[1,2] on the grounds that they would lead to too many black holes, in conflict with present observational limits).

Depending on the initial conditions and the model of hadrons, then, primordial mini black holes may be extremely rare or very common. Conversely, if mini black holes were discovered they would provide important information on conditions during the early big bang - especially during the first second, concerning which we have no data at present.

We now come to the important question of <u>black hole decay</u>. The classical result was that a black hole is stable; its mass either remains constant or grows as more matter falls in. The new quantum mechanical result of Hawking[5,6] is that pairs form in the gravitational field surrounding the black hole, and the outward-moving member of the pair may escape, thus reducing the mass of the black hole. Note that this spontaneous emission does not contradict the usual statement that no particle can escape from inside the Schwarzschild radius: in fact the pair creation takes place outside the Schwarzschild radius.

To understand Hawking's result, let us recall the famous <u>Klein paradox</u> of relativistic quantum mechanics. Consider a potential V(x) which changes by ΔV in an interval Δx. If the change is sufficiently large and abrupt, pairs are produced. The requirements are

$$\Delta V \gtrsim mc^2, \qquad (8)$$

$$\Delta x \lesssim \hbar/mc^2 \qquad (9)$$

for pairs of mass m (more generally $\Delta V \gtrsim E$, $\Delta x \lesssim \lambda$ for particles of energy E, wavelength λ). In our case we apply these conditions to the gravitational potential near a black hole of mass M and radius R. The gravitational potential, of course, couples to photons and neutrinos as well as electrons, mesons, and all other forms of matter. The potential strength acting on a body of mass m at r > R is

$$V(r) \simeq -\frac{G\,Mm}{r} \simeq -mc^2\,\frac{R}{r}, \qquad (10)$$

where eq.(2) has been used. Thus if we take an interval $R \leq r \leq 2R$, i.e., $\Delta r = R$, we get $\Delta V \simeq mc^2$, so the first Klein condition eq.(8) is satisfied for any mass (more generally, for any energy). The second Klein condition, eq.(9), is more restrictive and only allows emission of particles with wavelength

$$\lambda \gtrsim R. \qquad (11)$$

Eq.(11) has the remarkable implication that smaller black holes emit shorter wavelengths, i.e., higher energy particles! Thus they lose mass at a higher rate. The more a black hole emits, the smaller it gets; the smaller it gets, the faster it emits! Eventually an explosion occurs and the black hole disappears.

To get a feeling for the numbers involved, consider first an ordinary black hole of mass several times 10^{33} gm formed by collapse of a star. Its radius of 10 km only permits radiation of wavelengths $\lambda \gtrsim 10$ km; thus only photons, neutrinos, and gravitons are emitted. On the other hand a mini black hole of radius 10^{-11} cm emits photons, neutrinos, and gravitons of wavelengths down to 10^{-11} cm, as well as low energy electrons and positrons. A mini black hole of radius 10^{-13} cm (M $\simeq 10^{15}$ gm) emits some muons and pions as well. For radii much smaller than this, the list of radiated particles starts to depend on one's model of hadrons.

The quantitative emission rate has been worked out by Hawking. We shall just quote the results here; for the derivation and for the beautiful connection to entropy change we refer you to his original papers[5,6,7]. Hawking finds that the emission is characterized by a temperature

$$T \simeq \frac{10^{26} \,°K}{(M/gm)} \ . \tag{12}$$

It comes as a surprise, and a stimulus for further research[8], that the emission is thermal. Given this fact, however, the numerical value of the temperature in eq.(12) is just what one expects from the cutoff $\lambda \gtrsim R$ on the wavelength of emitted particles. Note that for an ordinary black hole with mass several times 10^{33} gm, and $R \simeq 10$ km, the radiated wavelengths $\lambda \gtrsim 10$ km correspond to a temperature much less than $3°K$; thus ordinary black holes presently absorb more energy from the $3°$ blackbody radiation than they emit. Eventually, continued expansion of the universe should cool the blackbody radiation to the point where ordinary black holes become net radiators. But at present only mini black holes with radii of about 10^{-2} cm or less are net emitters.

For black holes with initial $M > 10^{15}$ gm Hawking finds the evaporation time

$$\tau \simeq \left(\frac{M}{10^{15} \, gm}\right)^3 10^{18} \, sec \ ; \tag{13}$$

that is, black holes in this mass range live longer than the present age of the universe. Black holes with initial M less than about 10^{15} gm evaporate in less than the age of the universe and, since they cannot have been formed later than the early stages of the big bang, must be extinct by now. The important dividing line of 10^{15} gm corresponds to the mass of, say, a cubic kilometer of water. The radius of a black hole with this mass is 10^{-13} cm!

As the radius implies, the evaporation rate of mini black holes of less than about 10^{14} gm depends strongly on unknown details of the hadron spectrum. It is very rapid indeed in the statistical bootstrap model, leading to a violent explosion which releases 10^{35} ergs in the last 10^{-23} seconds. In the "elementary particle model," evaporation proceeds more slowly, but even here 10^{30} ergs are released in the last 0.1 second[5,6].

Can mini black holes be detected? A black hole in its final stages emits predominantly hadrons. These decay into pions, which decay into photons, electrons, and neutrinos. The most promising method of detection is to look for gamma rays, perhaps in short bursts.

The present limits have been surveyed by Carr[9] and by Page and Hawking[10]. In the first place, the number density of black holes at each mass is limited by

$$\frac{N(M)}{N_0(M)} \lesssim 1 , \qquad (14)$$

where N_0 is the number that would close the universe. For the mini black holes of mass near 10^{15} gm, which would be radiating most strongly at present, the observed average gamma-ray flux sets a limit about 10^8 times below eq.(14). A limit stronger than eq.(14) is also obtained for massive black holes[11], using the fact that when positioned between us and a star they would act as a lens for the starlight, causing us to "see double."

One very unique property predicted for black hole emissions is the *final explosion*, resulting in a short burst of gamma rays. While it is not possible to predict the number of bursts, we can easily estimate an upper limit by making the most optimistic assumption at each turn. We assume:

i) The universe is barely closed; i.e., $\rho \simeq 10^{-29}$ gm/cm^3.

ii) As much as possible of this density is in the form of mini black holes with masses distributed around 10^{15} gm. The limit on such black holes set by the average gamma ray flux is

$$\rho_{BH} = 10^{-8} \rho \simeq 10^{-37} \text{gm/cm}^3 , \qquad (15)$$

i.e.,

$$N \simeq 10^{-52}/\text{cm}^3 . \qquad (16)$$

iii) These black holes are concentrated in galaxies. Since galaxies fill about 10^{-6} of all space, we estimate for the density within our galaxy

$$N_{(\text{galactic})} \simeq 10^{-46}/\text{cm}^3 . \qquad (17)$$

The lifetime of a 10^{15} gm black hole is about 10^{10} years, so one in 10^{-10} explodes in one year's observing time. From (17) we then deduce the number of bursts within our galaxy:

$$n_{burst} \simeq \frac{10^{-56}}{cm^3 yr} \simeq \frac{10^{-3}}{(\ell t.yr.)^3 yr} \quad . \tag{18}$$

Thus the nearest burst in a typical year is 10 light years away.

Continuing our program of maximizing the effect, we work in the statistical bootstrap model where the bursts are biggest. In this model E_γ is typically several hundred MeV. Therefore explosion of a 10^{14} gm black hole gives 10^{38} γ's. The gamma ray flux at a distance R = 10 light years is $10^{38}/4\pi R^2$ or $0.1/cm^2$. These might arrive in a burst as short as 10^{-7} seconds[10].

For comparison, the observed average flux of gamma rays is about $10^{-4}/cm^2$ sec. Bursts coming from the nearest mini black holes would be resolvable[*] by detectors with an effective area of greater than 10 cm^2 and with good time resolution to distinguish bursts from gamma-ray background. Thus it should be possible to test the most extreme assumptions within the next few years. Since these probably give huge overestimates, however, detection of mini black holes by means of gamma ray bursts must be considered a very difficult enterprise.

Carr, Page, and Hawking have also considered the possibility of detecting the <u>steady emission</u> from a mini black hole which is not in the explosive phase, but happens to be nearby. In this case the unique features would be proper motion of a gamma-ray source across the sky, and the thermal spectrum at such extraordinarily high temperature. Given our maximal estimate, eq.(17), for the density of black holes within the galaxy, the nearest black hole would be 2×10^{15} cm away, several times the distance to Pluto. Its steady emission rate can be obtained by differentiating eq.(13):

$$-\frac{d(Mc^2)}{dt} \simeq 10^{17} \frac{erg}{sec} \times \left(\frac{10^{15} gm}{M}\right)^2 \quad . \tag{19}$$

For a mass of 10^{15} gm about 10% of the energy is carried off by photons[13], each having an average energy of about 10^8 eV, so some 10^{28} γ's are emitted per year yielding a flux of 1 γ/m^2 yr. from the nearest black hole. Excellent directional resolution would be essential to distinguish steady

---

[*]Some bursts have in fact been observed[12], but they are in the 100 KeV range whereas an exploding black hole would produce predominantly gamma rays of 100 MeV or more.

emission by such a weak source from background. The detector would also need to be very large - at least 10^4 m^2 in area to yield a counting rate of one photon per thousand seconds.

Collisions of black holes with the earth would provide another opportunity to detect them[1]. In this case there is no reason to restrict our attention to 10^{15} gm black holes, and in contrast to eq.(17) a galactic density of as much as $N \simeq 10^{-40}/\text{cm}^3$ (for somewhat larger black holes which radiate less) is possible. The relative velocity might be 3×10^7 cm/sec, typical of masses confined within our galaxy. The highest possible collision rate is then only

$$C = v \sigma N$$
$$\simeq 3 \times 10^7 \text{cm/sec} \times \pi R_{\text{earth}}^2 \times 10^{-40} \text{cm}^{-3}$$
$$\simeq 10^{-7}/\text{yr}. \qquad (20)$$

(Some of you may recall earlier estimates that we are being pierced by mini black holes all the time; these estimates were based on the much larger number of minimal-sized (10^{-5} gm) black holes that could be contemplated before the decay and disappearance of the low-mass end of the spectrum was discovered.)

Do mini black holes have any practical use? Suppose, in spite of the difficulties of detection, that we locate a mini black hole which has not yet exploded and is judged to have a long remaining lifetime. A group of Livermore scientists[14] has proposed that we use it to help solve our energy problems! First they would tow it (the black hole could be put into orbit around an asteroid; the asteroid could be steered by thermonuclear explosions) from where it was found to a convenient location, such as an orbit around the earth. In an orbit some thousands of kilometers above the earth, one could approach within 10 meters of the black hole before its gravitational attraction dominated the gravitational pull of the earth. The radiation from the black hole would be collected and the energy conveyed down to earth in a usable form.

The potentially available power would be large. Expressing eq.(19) in watts, we find for the spontaneous emission

$$-\frac{d(Mc^2)}{dt} \simeq 10^{10} \text{ watts} \left(\frac{10^{15} \text{gm}}{M}\right)^2, \qquad (21)$$

which (for $M = 10^{15}$ gm) amounts to about a thousandth of the present power consumption by humans. Furthermore, a sufficiently large mini black hole would continue to radiate steadily for at least as long as the sun.

The Livermore group also pointed out several other ways of getting energy from a black hole, which promise to yield even more energy than the spontaneous radiation. For example, one could take matter (radioactive wastes, for instance) up to the hole and drop it in. On its way in, the matter would develop kinetic energies of the order of its rest mass. Much of this kinetic energy would be radiated away, especially since the matter would orbit (it is impossible to hit a hole 10^{-13} cm across from a distance of 10 meters), and bump into other matter. Thus the material around the mini black hole would form a minisun, whose radiation could be collected and utilized.

Finally, what about possible <u>relations of mini black holes to particle physics</u>? It was already mentioned that both the formation and decay of mini black holes are enhanced if there are many species of hadron, as in the statistical bootstrap model.

Beyond this, there is a deep question concerning the conservation of baryon number. The formation and decay of black holes is believed to violate conservation of the baryon number, lepton number, and indeed any quantum number which does not act as source for a massless field. Only a massless field extends out to large distances, allowing measurement (by integration over a Gaussian surface far outside) of the "charge" after it has fallen through the Schwarzschild radius, and thereby allowing confirmation of charge conservation.

A particle such as the proton is stable against collapse into a black hole, but must be able to go into black holes <u>virtually</u> since black holes are another state of matter. Thus baryon number, lepton number, etc., should be violated at some level. Recently particle theorists have suggested, on completely different grounds, that leptons and quarks may belong to the same unified group, and that transitions take place among them resulting in a very slow decay of the proton. At present no one has any idea whether these two suggestions for baryon nonconservation are ultimately related.

References

1. B. J. Carr, Ap. J. 201, 1 (1975).
2. G. F. Chapline, "Hadron Physics and the Early Universe," Phys. Rev., in press (1975).
3. R. Hagedorn, Astron. & Astrophys. 5, 184 (1970).
4. R. Carlitz, S. Frautschi, and W. Nahm, Astron. & Astrophys. 26, 171 (1973).
5. S. W. Hawking, Nature 248, 30 (1974).
6. S. W. Hawking, Commun. Math. Phys., in press (1975).
7. S. W. Hawking, "Black Holes and Thermodynamics," preprint OAP-412, Caltech (1975).
8. S. W. Hawking, "Fundamental Breakdown of Physics in Gravitational Collapse," preprint OAP-420, Caltech (1975).
9. B. J. Carr, "Some Cosmological Consequences of Primordial Black Hole Evaporations." preprint OAP-415, Caltech (1975).
10. D. N. Page and S. W. Hawking, "Gamma Rays from Primordial Black Holes," Caltech preprint (1975).
11. W. H. Press and J. E. Gunn, Ap. J. 185, 397 (1973).
12. R. W. Klebesadel, I. B. Strong, and R. A. Olson, Ap. J. (Letters) 182, L85 (1973).
13. D. N. Page, "Particle Emission Rates from a Black Hole. I. Massless Particles from an Uncharged, Nonrotating Hole," preprint OAP-419, Caltech (1975).
14. L. Wood, T. Weaver, and J. Nuckolls, Annals N.Y. Acad. Sci. 251, 623 (1975).

A REVIEW OF RECENT PROGRESS AND
OLD PROBLEMS IN SUBNUCLEAR PHYSICS

A. Zichichi
CERN, Geneva, Switzerland

1. INTRODUCTION

The recent EPS Conference on Subnuclear Physics gave me the opportunity of reviewing the numerous results obtained in the field of the new particles and of the old interactions.

The purpose of this lecture is to focus your attention on the most exciting results whose detailed discussion will be developed during the various lectures of this School. This review will close with a list of outstanding problems in subnuclear physics.

2. THE NEW PARTICLES

2.1 The zero width of the J particle

Figure 1 shows the results on the J width, presented by Dr. S. Ting[1]. The experimental resolution being 5 MeV/c^2, the particle first discovered by the MIT Group is obviously consistent with zero width.

2.2 The J particle is a hadron

Immediately after the discovery of the J, some theorists thought it could have been the first of a series of intermediate weak bosons. The proof that the J particle is a hadron is obtained via the photoproduction data[1] of Fig. 2, from which it can be concluded that $\sigma_{JN} \simeq 1$ mb. Another interesting feature of the photoproduction data is the dramatic increase of the photoproduction cross-section with incident photon energy. No theoretical explanation exists for this.

2.3 $J^{PC} = 1^{--}$ for $\psi(3.1)$ and $\psi'(3.7)$

Dr. V. Lüth from SLAC[2] has reported conclusive evidence for the interference between the two Feynman diagrams:

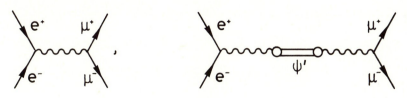

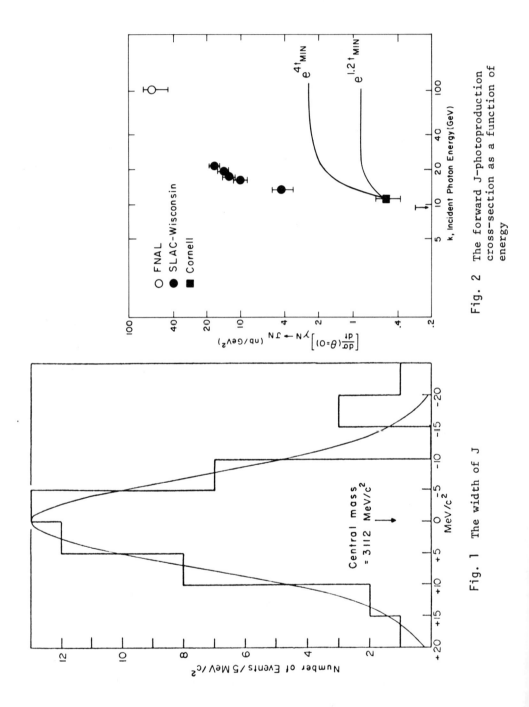

Fig. 1 The width of J

Fig. 2 The forward J-photoproduction cross-section as a function of energy

This proves (with about 5 standard deviations confidence level) that the ψ' particle has the same quantum numbers as the photon, i.e. $J^{PC} = 1^{--}$. As you know, the same authors had already observed, some time ago, the same interference pattern for the lower state ψ(3.1). The data are shown in Fig. 3.

2.4 G = -1 for ψ(3.1)

Another interesting result presented by the SLAC Group refers to the measurement of the even (4, 6) and odd (3, 5, 7) pion rates inside and outside the ψ peak. It is found that the odd pion rates are about five times larger than the even pion rates inside the ψ peak. Moreover, the even pion rates inside and outside of the ψ peak are the same. All this is shown in Fig. 4. Thus the G-parity of the ψ is obviously negative.

2.5 I = 0 for ψ(3.1)

Knowing that the G-parity and the charge conjugation of the ψ(3.1) are odd, the isotopic spin has to be even, i.e. 0 or 2. The proof of I = 0 comes from the measurement of the branching ratio,

$$\frac{\psi \to \bar{p}p}{\psi \to \text{all}} = (21 \pm 4) \times 10^{-4}$$

This value is far above the expected electromagnetic rate, which proves the existence of the direct decay mode

$$\psi \to \bar{p}p$$

without an intermediate photon. Since $(\bar{p}p)$ can only have I = 0, 1, and I = 1 is excluded because G = -1, the conclusion is that I = 0 for ψ(3.1).

2.6 ψ(3.1) is an SU(3) scalar

Since the ψ(3.1) has I = 0, the question arises whether this object is an SU(3) singlet, or the eighth member of an SU(3) octet, or a mixture. Some detailed study of the production of quasi-two-body states of strange mesons seems to support the hypothesis that ψ(3.1) is an SU(3) singlet.

2.7 Decay rates and widths of the ψ(3.1) and ψ'(3.7)

An impressive list of various decay rates of the two ψ's has been presented by the SLAC Group. These are reported in Tables 1 and 2. For the decay widths and for other details we refer the interested reader to

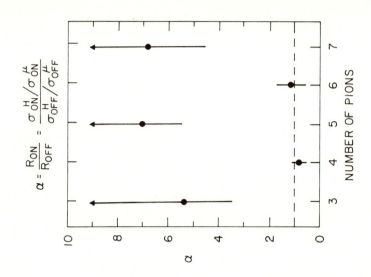

Fig. 4 Comparison of the ratio of multipion events to $\mu^+\mu^-$-pair production at the $\psi(3095)$ (ON) and at 3.0 GeV (OFF).

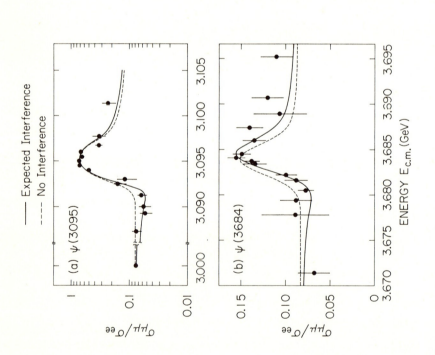

Fig. 3 The ratio of the μ-pair yield as a function of c.m. energy. The dashed line gives the expected ratio for no interference; the solid line gives the ratio for full interference.

Table 1

Decay modes of the $\psi(3095)$ (from Ref. 2)

Mode	Branching Ratio (%)	No. of Events Observed	Comments
e^+e^-	6.9 ± 0.9	ca 2000	
$\mu^+\mu^-$	6.9 ± 0.9	ca 2000	
$\rho\pi$	1.3 ± 0.3	153 ± 13	$>70\%$ of $\pi^+\pi^-\pi^0$
$2\pi^+ 2\pi^-$	0.4 ± 0.1	76 ± 9	
$2\pi^+ 2\pi^- \pi^0$	4.0 ± 1.0	675 ± 40	20% $\omega\pi^+\pi^-$ 30% $\rho\pi\pi\pi$
$3\pi^+ 3\pi^-$	0.4 ± 0.2	32 ± 7	
$3\pi^+ 3\pi^- \pi^0$	2.9 ± 0.7	181 ± 26	
$4\pi^+ 4\pi^- \pi^0$	0.9 ± 0.3	13 ± 4	
$\pi^+\pi^- K^+K^-$	0.4 ± 0.2	83 ± 18	not including $K^*(892)\ K^*(1420)$
$2\pi^+ 2\pi^- K^+K^-$	0.3 ± 0.1		
$K_S K_L$	< 0.02	≤ 1	90% C.L.
$K^0 K^{0*}(892)$	0.24 ± 0.05	57 ± 12	
$K^\pm K^{\mp*}(892)$	0.31 ± 0.07	87 ± 19	
$K^0\ K^{0*}(1420)$	< 0.19	≤ 3	90% C.L.
$K^\pm\ K^{\mp*}(1420)$	< 0.19	≤ 3	90% C.L.
$K^{*0}(892)\bar{K}^{*0}(892)$	< 0.06	≤ 3	90% C.L.
$K^{*0}(1420)\bar{K}^{*0}(1420)$	< 0.18	≤ 3	90% C.L.
$K^{*0}(892)K^{*0}(1420)$	0.37 ± 0.10	30 ± 7	
$p\bar{p}$	0.21 ± 0.04	105 ± 11	assuming $f(\theta) \sim 1+\cos^2\theta$
$\Lambda\bar{\Lambda}$	0.16 ± 0.08	19 ± 5	
$p\bar{p}\pi^0$ $\bar{n}p\pi^-$ $\bar{p}n\pi^+$	0.37 ± 0.19	87 ± 30	

Table 2

Decay modes of the $\psi(3634)$ (From Ref. 2)

Mode	Branching Ratio (%)	Comments
e^+e^-	0.97 ± 0.16	} μ-e universality assumed
$\mu^+\mu^-$	0.97 ± 0.16	
$\psi(3095)$ anything	57 ± 8	
$\psi(3095)\pi^+\pi^-$	32 ± 4	} these decays are included in the fraction for ψ + anything
$\psi(3095)\eta$	4 ± 2	
$\psi(3095)\gamma\gamma$	$<6.6*$	via an intermediate state
$\rho^0\pi^0$	$<0.1*$	
$2\pi^+2\pi^-\pi^0$	$<0.7*$	
$\bar{p}p$	$<0.03*$	

* 90% confidence limit based on a preliminary analysis

RECENT PROGRESS AND OLD PROBLEMS

Dr. Lüth's report[2]. In spite of all these results, the nature of the ψ's, in terms of colour or charm, or eventually something else, remains open.

3. NO NEW NARROW OBJECTS IN THE MASS RANGE (1.915-7.8) GeV/c² WITH AN UNEXPLORED GAP AT (2.544-2.966) GeV/c²

Dr. P. Monacelli[3] has reported a systematic search performed at Frascati, in the mass range (1.915-2.544) GeV/c², scanning with 1 MeV steps, and in the range (2.966-3.100) GeV/c², scanning with 2 MeV steps. [The machine energy resolution (FWHM) goes from 1.1 to 3.0 MeV, in the above mass range.] No peak appears, at least up to ∼ 10% of the integrated cross-section for the ψ. The fluctuations observed are in accordance with Poisson statistics.

The above search has been extended by the SLAC people[2] up to 7.68 GeV/c². Again no new narrow peaks have been observed. The data are reported in Figs. 5 and 6.

4. SEARCH FOR OTHER NEW PARTICLES (CHARM, COLOUR, HEAVY LEPTON, JETS)

4.1 No bump in the $(K^+\pi^-)$ invariant mass range (1.5-2.5) GeV

Dr. Ting[1] has reported a high-statistics search for different particle pairs produced in p-Be interactions at 30 GeV. No evidence for a bump in the invariant mass plot of $(K^+\pi^-)$ pairs has been found.

An interesting feature of this experiment is the behaviour of the cross-section for the various inclusive processes of pair production. The general features are as follows:

$$p + Be \begin{cases} \rightarrow A + \text{anything} \equiv \sigma_A^{inc} \\ \rightarrow B + \text{anything} \equiv \sigma_B^{inc} \\ \rightarrow C + \text{anything} \equiv \sigma_C^{inc} \end{cases}$$

Produced pairs	Q	B	S
$A \equiv (\pi^- p)$	0	1	0
$B \equiv (\pi^+\pi^-); (\bar{p}p); (K^+\pi^-); (K^-p)$	0	0; 1	0; ±1
$C \equiv (K^+K^-); (K^+\bar{p}); (\pi^+\bar{p}); (K^-\pi^+)$	0	0; -1	0; ±1

A, B, C, indicate the states produced, and Q, B, S, the quantum numbers of the various pairs: Q = electric charge; B = baryonic number; S = strangeness.

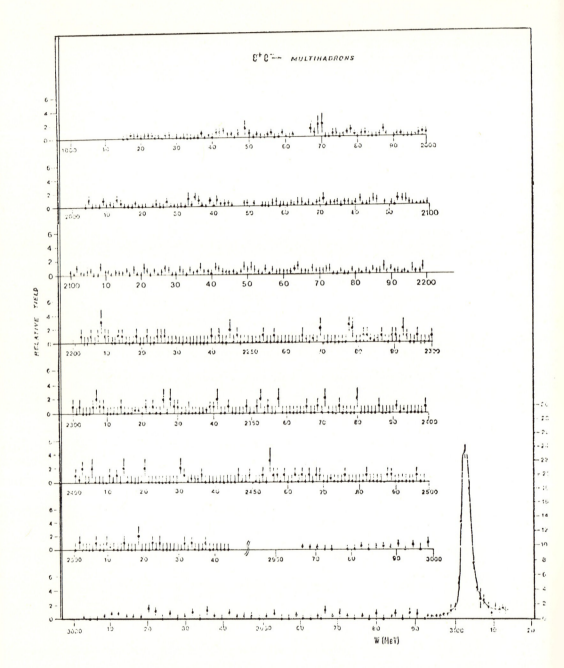

Fig. 5 Multihadron relative yield versus c.m. energy (γγ2 group). The J/ψ(3100) resonance is reported on the same scale for comparison.

RECENT PROGRESS AND OLD PROBLEMS

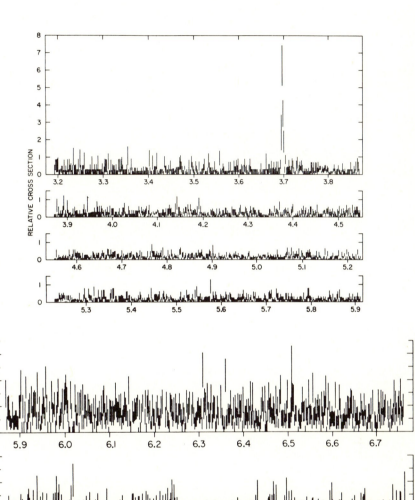

Fig. 6 Relative cross-section for $e^+e^- \to$ hadrons as obtained from a fine mesh scan. The $\psi(3684)$ stands out clearly; it was first seen in the process of this scan. The energy scale shown is known to be miscalibrated by a factor of 1.003.

The experimental results are

$$\sigma_A^{inc} : \sigma_B^{inc} : \sigma_C^{inc} \cong 10^2 : 10^1 : 10^0 .$$

These were given as preliminary but, if confirmed, they are of remarkable interest.

For example, the inclusive production of $(\bar{p}p)$ pairs is about 10 times larger than that of (K^+K^-) pairs or of pairs with and without strangeness but with negative baryonic number, such as $(K^+\bar{p})$ and $(\pi^+\bar{p})$.

4.2 Threshold effect in $R = \sigma(e^+ + e^- \to hadrons)/\sigma(e^+ + e^- \to \mu^+ + \mu^-)$

Dr. G.J. Feldman[4] has presented the SLAC data on $e^+ + e^- \to$ hadrons. More precisely, these data should be given as $e^+ + e^- \to$ tracks, because the hadronic nature of the observed multitrack events is not proved experimentally. It can only be inferred from the hadronic nature of the multitrack events, established via accurate calibration by the BCF Group[5] in their study of (e^+e^-) annihilation in the total centre-of-mass energy range (1.2-3.0) GeV. The SLAC data show a behaviour which gives two plateaux in R, one below 3.5 GeV, where $R \simeq 2.5$ and one above 4.8 GeV, where $R \simeq 5.5$, as shown in Fig. 7. These three units of R could be due to new particles being produced, as discussed below. It should be noticed that the mean charged multiplicity shows no sign of change in slope in the transition region between the two plateaux of R; it increases like ln s all over the SPEAR energy range, as can be seen from Fig. 8.

4.3 The wide bump in the (4.1-4.5) GeV range

This bump could be due to the production of pairs of charmed particles

$$e^+ + e^- \to C + \bar{C}$$

which would decay into standard hadrons; however, preliminary measurements of the (K/π) ratio show that, across this bump, this ratio remains the same.

4.4 Observation of $(e^\pm \mu^\mp)$ pairs above 4 GeV

Preliminary results of a new effect have been reported by Dr. Feldman[4]. At 4.8 GeV the SLAC Group has observed the production of electron and muon pairs, i.e. the reaction

$$e^+ + e^- \to e^\pm + \mu^\mp + \text{anything (provided it gives no detectable particles)} \quad (4.1)$$

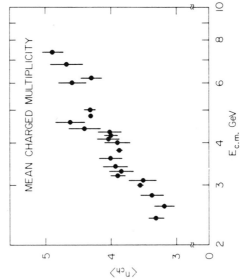

Fig. 8 The mean charged multiplicity versus centre-of-mass energy

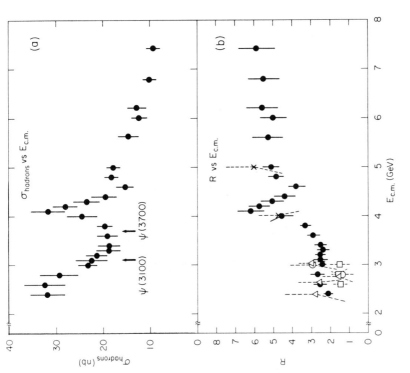

Fig. 7 a) The total hadronic cross-section versus centre-of-mass energy.
b) The ratio R of the total hadronic cross-section to the theoretical muon-pair production cross-section. The dotted points are older results from Frascati and the CEA.

This implies that the observed events (4.1) are not accompanied by other tracks. This selection is done in order to have good e and μ rejection against background from multihadron production. The interesting aspect of these preliminary results is the fact that the number of like pairs, $e^-\mu^-$, $e^+\mu^+$, is consistent with zero (background level ~ 1 event), while the unlike pairs, $e^{\mp}\mu^{\pm}$, are 21 events with 4 to 6 expected as background. The authors exclude that the effect could be accounted for by systematic effects, and the search is going on. It should be recalled that (eμ) events of like and unlike charges have been thoroughly looked for at Frascati by the BCF Group[6] and that no signal had been detected up to 3 GeV total centre-of-mass energy. The BCF search was motivated by the high-mass leptons. No charm or gauge theories were available at that time, and it is not excluded that old heavy leptons are at the origin of the (eμ) events discovered at SLAC. According to Dr. Feldman's report it seems that the threshold for this new phenomenon is at about 4 GeV, as indicated in Fig. 9.

4.5 (Spin-½)-like behaviour of the azimuthal distribution of leading hadrons

At 7.4 GeV the time needed to build up a polarization parallel and antiparallel to the magnetic guide field is ~ 14 minutes; at 3.1 GeV it is 18 hours. The time needed to build up this polarization goes like E^{-5}. Since the ring is filled every two to three hours, there is almost no polarization at the ψ's, a small amount at 4.8, but very high average polarization at 7.4. At this energy the azimuthal angle distribution of high-momentum hadrons shows a clear dependence of the type expected for μ pairs, as shown in Fig. 10.

That leading hadrons should have the same distribution as μ pairs is predicted by the (spin-½) parton model. In this picture the time-like photon produces a pair of (spin-½) partons which immediately dress up with hadrons to produce standard particles, while keeping the (spin-½) angular dependence. The dressing up should have a jet structure.

4.6 Jet-like structures in (e^+e^-) annihilation

Another interesting effect observed in the study of $e^+ + e^- \to$ tracks is the angular distribution of the produced tracks (called hadrons by the authors). On each event they find the axis which minimizes Σp_T^2, for all visible prongs. This is the apparent jet axis. Then they compute the "sphericity" for the event:

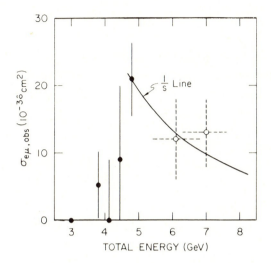

Fig. 9 The observed cross-section for events of the type e and µ, with no other particles in the SLAC-LBL magnetic detector. The two high-energy measurements (dashed lines) are preliminary. These data have not been corrected for momentum and angle cuts and for the geometry of the detector. This correction can be a factor of 2 to 10, depending on the origin of the events.

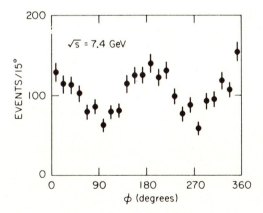

Fig. 10 Preliminary results for the azimuthal distribution of observed hadrons with $x \equiv 2p/\sqrt{s} > 0.3$ at 7.4 GeV centre-of-mass energy

$$\text{sphericity} = \frac{3\Sigma p_T^2}{2\Sigma p^2},$$

where the sum is extended to all prongs and $p^2 = p_T^2 + p_L^2$, with T and L referring to the jet axis[*]. The sphericity distributions at the various energies are compared with models, and a clear indication for jets (zero sphericity) is found at E = 7.4 GeV (Fig. 11). Note that the jet axis shows the same azimuthal dependence as for the high-energy hadrons discussed above.

4.7 Theoretical understanding of the jets

The exciting discovery of jets at SPEAR, reported by Dr. Feldman, seems to be quite welcomed by theoreticians, as remarked by Dr. G. Preparata. In e^+e^- collisions one begins to see quite distinctly the formation of very spectacular final states characterized by a jet axis distributed with the typical $(1 + \cos^2 \theta)$ law, with respect to which all final hadrons have limited ($\lesssim$ 300 MeV/c) transverse momentum. This constitutes a good indication for spin-½ hadronic constituents endowed with the type of dynamics envisaged in Preparata's massive quark model[7].

5. NEW EFFECTS IN HIGH-ENERGY NEUTRINO REACTIONS

Dr. C. Rubbia[8] and Dr. D. Cline[9] have presented a series of interesting results on high-energy neutrino reactions at Fermilab energies. They can be summarized as follows:

5.1 Threshold effect at $W_h \simeq 4$ GeV

Using neutrinos

$$\nu_\mu + N \to \mu^- + \text{hadrons}$$

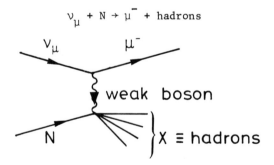

$W_h \equiv$ total hadronic energy of the system X, measured with hadron calorimeters

---

[*] i.e. to the axis for which the sphericity is minimum.

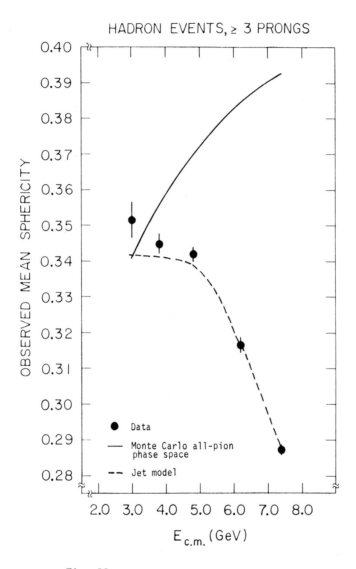

Fig. 11 Observed mean sphericity versus E_{cm}

No threshold effect in W_h is observed either for $E_\nu < 30$ GeV or for $E_\nu > 30$ GeV.

Using antineutrinos

$$\bar{\nu}_\mu + N \to \mu^+ + \text{hadrons}$$

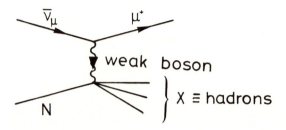

No threshold effect in W_h is observed for $E_{\bar{\nu}} < 30$ GeV. *But* a threshold effect at $W_h \simeq 4$ GeV *is observed* for $E_{\bar{\nu}} > 30$ GeV. All these data are presented in Fig. 12.

5.2 Dimuon production

A total of 61 ($\mu^+\mu^-$) events have been reported from high-energy neutrino interactions:

$$\nu_\mu + N \to \mu^- + \mu^+ + \text{hadrons}$$

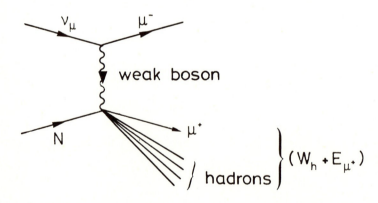

The dimuon events show the following features:

i) the dimuon mass spectrum has no resonance structure; the statistics are, however, limited;

ii) the total energy of the hadronic system W_h, plus that of the μ^+, i.e. $(W_h + E_{\mu^+})$, shows a threshold at $W_{min} \simeq 4$-5 GeV (Fig. 13).

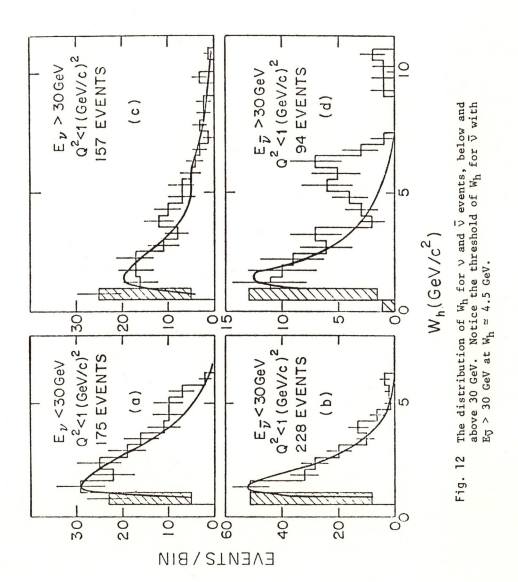

Fig. 12 The distribution of W_h for ν and $\bar{\nu}$ events, below and above 30 GeV. Notice the threshold of W_h for $\bar{\nu}$ with $E_{\bar{\nu}} > 30$ GeV at $W_h \simeq 4.5$ GeV.

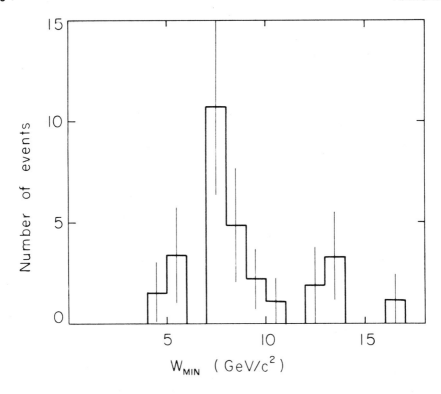

Fig. 13 Invariant mass (lower limit) of the (μ^+-hadron) system for dimuon events. Note the apparent threshold at ~ 4.5 GeV/c^2.

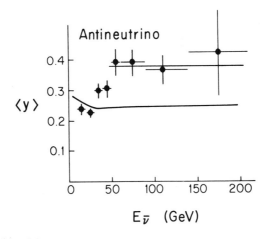

Fig. 14 Evidence for a threshold effect in $\langle y \rangle$

RECENT PROGRESS AND OLD PROBLEMS   759

The quantity W_{min} is called minimum because there must be at least
a neutrino, to conserve leptonic number, which certainly carries
extra energy.

5.3 Threshold effect in $\langle y \rangle$

When the reaction

$$\bar{\nu}_\mu + N \to \bar{\mu} + \text{hadrons}$$

is studied in terms of the $\langle y \rangle$ distribution ($y = \nu/E$; ν energy transfer,
E total energy), a discontinuity is observed in going from $E_{\bar{\nu}} \lesssim 20$ GeV
to $E_{\bar{\nu}} \gtrsim 50$ GeV, as shown in Fig. 14.

Presumably these "threshold" effects are all related.

6. NO EVENTS OF THE ($\Delta S = -\Delta Q$) TYPE FROM NEUTRINO REACTIONS

Here the problem is to see if events like the one reported some time
ago by the BNL Group, $\nu_\mu + N \to \mu^- + \Lambda + \pi^+ + \pi^+ + \pi^-$ (i.e. with $\Delta S = -\Delta Q$),
exist in nature. Dr. B.P. Roe[10] of the Michigan-Fermilab Group has
reported an extensive search for 3C candidates using the Fermilab neutrinos.

No evidence for $\Delta S = -\Delta Q$ events was found, with 50 times more neutrino
flux than in the BNL experiment; and this for $\Sigma p_X > 10$ GeV/c, where Σp_X is
the total visible momentum of an event.

The same group has reported a large rate of strange particle production:
$\sim 15\%$ of the events with $\Sigma p_X > 10$ GeV/c have a V^0-like object; therefore the
total fraction of events with strange particles must still be larger, owing
to channels (Σ^-, K^+, etc.) which produce no V^0-like events. This 15% should
be compared with the $\sim 5\%$ production in strong interactions. However, this
high rate of strange particles seems to have no anomalous origin. For
example, the number of observed Λ^0's (14 events) is in good agreement with
expectation from associated production, even if the basic number is very
low: three double ($\Lambda^0 K^0$) events. This would have not been the case if
$\Delta S = -\Delta Q$ events would have been produced at the same rate as associated
production.

7. THE THRESHOLD EFFECT IN THE INCLUSIVE e/π RATIO SERIOUSLY QUESTIONED

Dr. B.G. Pope[11] has reviewed the inclusive production of leptons,
as shown in Fig. 15.

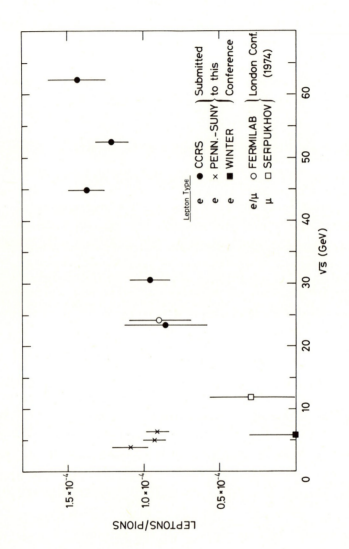

Fig. 15 The inclusive ratio lepton/pion versus total centre-of-mass energy

The new data came from Dr. M.L. Good[11a], who reported the measurement of the inclusive ratio (e^+/π^-) in hadron interactions originated by 10, 15, and 24 GeV/c protons at Brookhaven. These results, contrary to previous knowledge, show that the e/π inclusive ratio appears to remain constant and at the level of $\sim 10^{-4}$, down to the lowest energy investigated. Thus the previous belief of a threshold effect at $\sqrt{s} \simeq 25$ GeV is now very seriously questioned by these interesting results.

The origin of these leptons remains, as yet, unexplained.

8. THEORETICAL UNDERSTANDING OF THE NEW PARTICLES

Here there were two fronts: one for "charm", the other for "colour".

8.1 Cabibbo for charm

Here are Dr. Cabibbo's[12] main pro-charm points:

i) Γ_ψ and $\Gamma_{\psi'}$ are O.K. for charm; they are not too narrow.

ii) The vector mesons should be lighter than the pseudoscalar mesons, with the following masses:

$$D^* \simeq (1,900 \pm 50) \text{ MeV}/c^2 ; \quad S^* \simeq (2,000 \pm 50) \text{ MeV}/c^2 .$$

Cabibbo also thinks that the origin of the (eµ) events observed at SPEAR is

$$e^+ + e^- \to S^{*+} + S^{*-} \to \mu^\pm + e^\mp \ldots .$$

iii) The decay multiplicity for S^* and D^* is expected to be high, as is the case for (e^+e^-) and ($\bar{p}p$) annihilation at rest, where the two bodies are certainly not favoured.

iv) The leptonic branching ratios are expected to be

$$\frac{D^* \to \mu + \nu}{D^* \to \text{all}} \simeq 4\% ; \qquad \frac{S^* \to \mu + \nu}{S^* \to \text{all}} \simeq 10\% .$$

v) He also proposes the following explanation for high-p_T leptons observed in hadronic collisions:

$$p + p \to (S^*)^+ + \dots \text{ (or } D^* + \dots)$$
$$\hookrightarrow \mu^+ + \nu, \ e^+ + \nu$$

Needed is $1(S^*)^+/100 \ \pi^+$ at large p_T to explain the data.

8.2 Matthews and Yamaguchi for colour

Dr. P.T. Matthews[13] and Dr. Y. Yamaguchi[14] pleaded for the colour scheme where the colour is not an SU(3) scalar but another SU(3) symmetry group.

Dr. Yamaguchi pointed out that the radiative width of ψ is O.K.:

$$\Gamma[(3.1) \to \gamma + X] \simeq 50 \text{ keV} .$$

He remarked that a naïve quark model with charm predicts 7 MeV for $\Gamma[(3.1) \to \gamma + X]$. To reduce this width, the charm-lovers need arbitrary modifications, as do the colour-lovers.

9. DEEP INELASTIC LEPTON-HADRON SCATTERING

On this front the most exciting news came from Dr. K.W. Chen[15], who reported on the breaking of scale invariance, observed with the high-energy Fermilab muons by the Cornell-Michigan-LBL-University of California Collaboration.

9.1 Scale-breaking in νW_2

Let me briefly recall the story of the Bjorken scaling. Although the Bj limit is only achieved formally at infinite ν and Q^2, it has become commonplace to set ∞ equal to 1 or 2 (GeV)^2 in confronting the data. *The reason:* predictions, derived for infinite ν and Q^2, seemed borne out by experiment at rather small values. More precisely, it was experimentally found that, for Q^2, $\nu \gtrsim 1$ or 2 GeV^2, the structure functions W_1 and νW_2 became functions of x alone:

$$W_1(Q^2, \nu) \to F_1(x)$$
$$\nu W_2(Q^2, \nu) \to F_2(x) .$$

The experiment studied the inclusive reaction

$$\mu^+ + \text{Fe} \to \mu^+ + \text{anything}$$

at 150 and 56.3 GeV incident muon energies.

The basic numbers were: 150 GeV: Fe target, 622 g/cm^2, 1.5 × 10^9 μ^+
                          56 GeV: Fe target, 233 g/cm^2, 4 × 10^9 μ^+.

Azimuthal acceptance: full.

Polar acceptance: small angle $0.011 \leq \theta \leq 0.048$
                   large angle $0.017 \leq \theta \leq 0.065$

The spectrometer had the important feature of being scaled to preserve the acceptance and the resolution in the scaled variables x and y $\left[x = (Q^2/2\nu) = (1/\omega); \; \omega' = \omega + (M^2/Q^2); \; y = (\nu/ME); \; \nu = p \cdot q \right]$.

Two tests of scaling were performed:

i) With Monte Carlo simulation of the spectrometer acceptance and resolution, the q^2-dependence of νW_2 at fixed ω was evaluated. The Monte Carlo simulation generated events using a parametrization of $\nu W_2(\omega')$ from SLAC low-energy data, with $R = \sigma_L/\sigma_T = 0.18$. The Fe target was modelled as a collection of nucleons in Fermi motion (the results are proved to be insensitive to the parameters of the model). Radiative corrections used a peaking approximation, differing from the more exact correction by less than 3%. The correction for coherent wide-angle bremsstrahlung rose from less than 3% for $\omega < 20$ to 15% at ω near 50. The simulated muons suffered Coulomb scattering, energy loss, straggling from μ-e scattering and bremsstrahlung in the iron, and mismeasurements by the spark chambers; further analyses treat real and simulated data identically. The results are shown in Fig. 16.

The basic feature of the data is the opposite q^2-dependence at large and low values of ω's. For example, at $\langle \omega \rangle = 32.8$ the cross-section greatly increases with q^2, while at $\langle \omega \rangle = 3.5$, the cross-section drops with increasing q^2. The most synthetic way of expressing the results is via the quantity b, which is

$$b = \frac{\partial^2 \ln \nu W_2}{\partial \ln q^2 \, \partial \ln \omega'} = 0.099 \pm 0.018 \, .$$

The deviation from zero of b measures the departure of the data from Bjorken scaling. *Question:* What should be done to the data to have b = zero? *Answer:* At 150 GeV, the scattered energies should be raised by 2.6% and the cross-section simultaneously reduced by 18%. These shifts

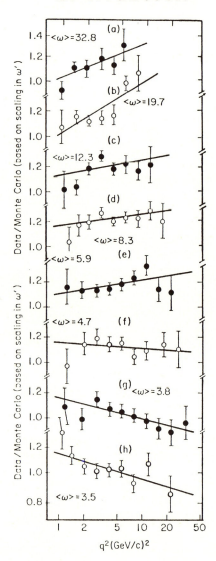

Fig. 16 Ratio of observed to Monte Carlo rate versus Q^2 for eight ranges of ω. Widths of these ranges and parameters of the straight-line fits are detailed in the Table on p. 463 of the EPS Internat. Conf. on High-Energy Physics, Palermo, 1975. Errors are statistical.

would only change the form of scale-breaking into a relative depletion of event rate at 150 GeV. Moreover, the χ^2 which measures the over-all smoothness and consistency of the data at the two energies, would rise by more than 40, compared with the value obtained avoiding the shifts.

ii) The ratio of the differential counting rates at 150 and 65 GeV is found to be neither 1 nor constant; it is a function of ω with the following value of the best fit

$$r = \left[\frac{Ed^2\sigma}{dxdy}\right]_{E=150 \text{ GeV}} \Big/ \left[\frac{Ed^2\sigma}{dxdy}\right]_{E=56 \text{ GeV}} =$$

$$= \left(\frac{\omega}{\omega_0}\right)^n = \left(\frac{\omega}{6.1^{+8.9}_{-3.6}}\right)^{0.096 \pm 0.028}$$

This shows that it is not only through a Q^2-dependent form factor that the scale-breaking mechanism can be accounted for.

Possible interpretation of the data

If expressed in terms of propagators, the observed effect corresponds to masses which are an order of magnitude greater than the proton mass -- some theorists already speak of parton form factors. Remember: partons were invented to have point-like scattering sources inside the nucleon.

Schildknecht and Steiner[16] have suggested that the rising q^2-dependence νW_2 at $\omega \geq 6$ might be attributed to the excitation of new hadronic degrees of freedom.

In the standard front, field theories with asymptotic freedom, theories with anomalous dimensions, predict breaking of scale invariance at both low and high ω, of a type similar to observation.

9.2 Scale breaking in W_1

Dr. R.E. Taylor[17] reviewed all data on deep inelastic lepton scattering and reported on electron data which show strong evidence for scale breaking, the most significant being

$$2 m W_1 \neq F(\omega').$$

As emphasized by Taylor, this effect could be cancelled by eliminating low ω, or could be blamed on non-leading terms. The results on W_1 are shown in Fig. 17.

10. ELECTROPRODUCTION AND MUPRODUCTION

10.1 Electroproduction deviates from photoproduction

As is well known, in π^0 photoproduction a dip exists in $d\sigma/dt$ at $|t| \simeq 0.4$ $(GeV/c)^2$.

This dip does not show up when virtual, instead of real photons are used to photoproduce the π^0, i.e.

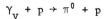
$$\gamma_v + p \to \pi^0 + p$$

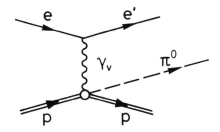

Other interesting results are as follows:

i) $d\sigma/dt$ shows an exponential decay until $|t| \simeq 0.8$ $(GeV/c)^2$, then it becomes flat.

ii) For all Q^2-values measured (i.e. $Q^2 = 0.22, 0.55, 0.85$ GeV/c^2) (Q^2 = virtual photon mass squared), $d\sigma/dt$ has nearly the same values for $|t| \gtrsim 0.8$ $(GeV/c)^2$.

iii) The values of $d\sigma/dt$ at $|t| \simeq 1$ $(GeV/c)^2$ are all about a factor of 9 below the photoproduction point, which reaches a flat maximum at $|t| \simeq 1$ $(GeV/c)^2$. The data were reported by Dr. G. Weber[18] and are shown in Fig. 18.

10.2 Muproduction deviates from photoproduction

Virtual photons can be produced either by primary electrons or by primary muons. Dr. C. Heusch[19] has reported preliminary results of his experiment on μ-production, i.e. on the study of the process

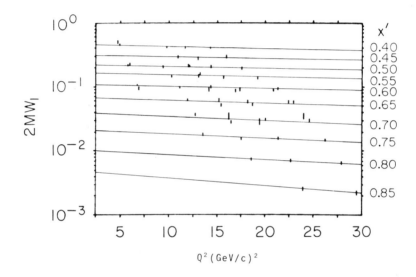

Fig. 17 $2MW_1$ versus Q^2 binned in x'. Data at the same x' are fit to a straight line in Q^2. W_1 clearly does not scale in x' since all data points at a given ω' should be constant in Q^2 and the connecting lines should be horizontal.

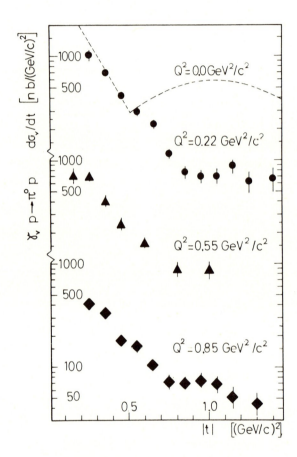

Fig. 18 The differential cross-section for the reaction $\gamma_v p \to \pi^0 p$ for different values of Q^2.

RECENT PROGRESS AND OLD PROBLEMS

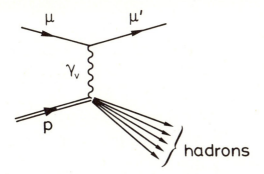

If we indicate with n (n = 1, 3, 5, 7) the multiplicity of the produced charged hadrons, with W their total energy, and with σ_n the corresponding cross-section, the main result is that all cross-sections (σ_{1-7}) are Q^2-independent, but σ_1 and σ_3 deviate and are above and below the photoproduction values, respectively, as shown in Fig. 19.

Another result worth mentioning refers to the positive-to-negative hadronic ratio $R = h^+/h^-$ which is found to increase for all x-values ($x = p_L/p_{max}$) with p_L, but becomes flatter when x > 0.3 events are selected.

10.3 Fractional charge structure of the proton constituents from $e^{\pm} + p \rightarrow e^{\pm} + \gamma +$ anything

Dr. D.O. Caldwell[20] has reported evidence for fractional charge structures of the proton, via the measurement of the ratio

$$\frac{e^+ + p \rightarrow e^+ + \gamma + \text{anything}}{e^+ + p \rightarrow e^- + \gamma + \text{anything}} = 1.31 \pm 0.10$$

which should be 1 if everything was O.K. within the proton. According to a sum rule, the difference ($e^+\gamma$) minus ($e^-\gamma$) gives the mean cubed charge of the parton. The cubic charge comes into play as it can be seen from the following diagrams

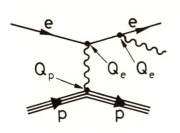 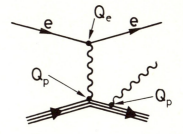

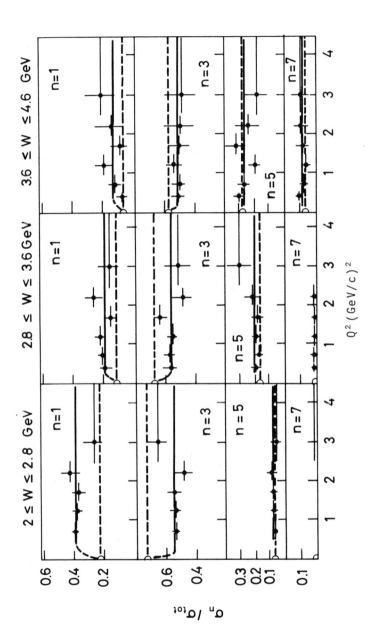

Fig. 19 Final data from the U.C. Santa Cruz/SLAC Group D Collaboration on topological cross-sections: n indicates the observed charged multiplicity of the event. The open circle followed by the dotted line shows the photoproduction point.

RECENT PROGRESS AND OLD PROBLEMS

whose interference is proportional to $(Q_e)^3 \times (Q_p)^3$, where Q_e = electron charge, Q_p = parton charge.

The same experiment, when measuring the inclusive $e^{\pm}$ scattering without γ, finds the value "1" expected on the basis of a negligible (2γ) intermediate state, i.e.

$$\frac{e^+ + p \to e^+ + \text{anything}}{e^- + p \to e^- + \text{anything}} = 1.0031 \pm 0.0024$$

for the following ranges: $(1 < q^2 < 3)$ GeV/c^2 and $(2 < \nu < 10)$ GeV.

11. HIGH-ENERGY AND HIGH-PRECISION QED CHECKS

In these exciting days of great discoveries it becomes more interesting to check if QED remains O.K. when higher values of invariant four-momentum transfers become accessible to experimental checks together with higher accuracy.

In this field there were two important contributions.

11.1 Higher precision (g-2) of the muon

After less than 15 years from the first accurate measurement of the anomalous magnetic moment of the muon at the 0.4% level, more than two orders of magnitude have been achieved in the accuracy of this important quantity in the physics of the intrinsic properties of the elementary particles. And this is more so, since the muon puzzle is there, just as it was in 1947; the muon remains the electromagnetic brother of the electron, the only difference between these two leptons being in the nature of their leptonic number, i.e. in the most unexpected corner, that of weak interactions, where nobody knows how to build up mass differences of the order of 100 MeV.

The result of the recent CERN experiment was reported by Dr. J.H. Field[21]:

$$\left(\frac{g-2}{2}\right)_\mu = [1165895 \pm 27] \times 10^{-9} .$$

This should be compared with the theoretical prediction:

$$\left(\frac{g-2}{2}\right)_\mu = \left[(1165829 \pm 2) + (73 \pm 10)\right] \times 10^{-9}.$$

↓ this includes $(\alpha/\pi)^3$, i.e. sixth-order radiative corrections.

↓ this is the contribution of the hadronic vacuum polarization calculated from $e^+ + e^- \to$ hadrons.

The present status is therefore

$$a_{exp} - a_{th} = (-7 \pm 29) \times 10^{-9},$$

where $a = (g-2)/2$. The fact that the virtual photon emitted by the muon transforms itself into hadrons and back into a photon, is established with 2.5 standard deviations safety factor. Figure 20 shows 80 (g-2) cycles. The theoretical uncertainty in the hadron vacuum polarization $(\pm 10) \times 10^{-9}$, could be improved by a better knowledge of $\sigma(e^+ + e^- \to$ hadrons) in the low-energy side of the ρ peak[22].

11.2 The energy dependence of $e^+ + e^- \to \mu^+ + \mu^-$ measured within few percent level

A precise study of the reaction

$$e^+ + e^- \to \mu^+ + \mu^- \tag{11.1}$$

in the time-like range $s = (1.44$-$9.0)$ GeV^2, where $s = (2E)^2$, with E = beam energy, has been presented by Dr. F. Palmonari[23], who has reported on a very accurate determination of the energy dependence of the time-like process (11.1) using the Frascati (e^+e^-) storage ring. These data are reported in Fig. 21. If the cross-section for the above process is parametrized,

$$\sigma(e^+ + e^- \to \mu^+ + \mu^-) = A \cdot s^{-n},$$

the experimental value of n is found to be

$$n = 1.000 \pm 0.012.$$

As is well known, the energy dependence of the purely QED process (11.1) is expected to be characterized by n = 1.000; thus, within ±1.2% the energy dependence of (11.1) follows QED predictions. The absolute value of the cross-section is within ±3.2% equal to that expected on the basis of QED.

RECENT PROGRESS AND OLD PROBLEMS

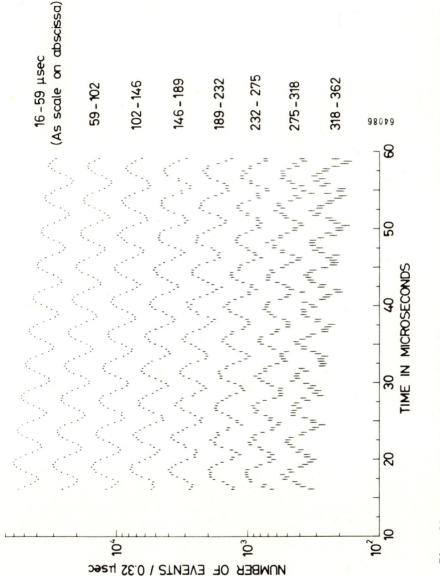

Fig. 20 Time distribution of the decay electrons from muons spiralling in the large magnet. 80 (g-2) cycles are displayed.

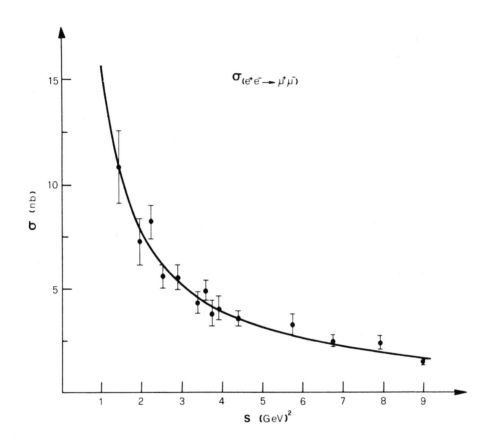

Fig. 21 The values of $\sigma(e^+e^- \to \mu^\pm\mu^\mp)$ integrated over the experimental apparatus at 14 values of the time-like four-momentum transfer s. First-order radiative corrections, using the peaking approximation have been applied to the data, after a detailed check at each energy of the acollinearity distributions (see p. 266 of the EPS Internat. Conf. on High-Energy Physics, Palermo, 1975). The solid line is the QED prediction.

12. NEUTRAL CURRENTS

12.1 Experimental results on neutral semileptonic currents

Dr. M. Rollier[24] has reviewed all experimental data on neutral semileptonic currents. By now, four independent experiments show evidence for neutral current with $\Delta S = 0$ in inclusive processes. In different experiments, some specific channels due to semileptonic neutral currents have been found. The lack of evidence for semileptonic neutral currents with $\Delta S = 1$ continues; the Columbia-CERN-NYU-BNL Collaboration has collected nine events of the type

$$K_L^0 \to \mu^+ + \mu^-,$$

which corresponds to the branching ratio value

$$\frac{K_L^0 \to \mu^+ + \mu^-}{K^+ \to \mu^+ + \nu} = 5 \times 10^{-9}$$

whose magnitude is expected from the fourth-order electromagnetic decay of K_L^0,

without any neutral current contribution.

Some indications from narrow-band neutrino experiments at Fermilab by the Caltech group show evidence for excluding S and P terms in neutral currents.

Finally, all experimental data point towards the following value of the Salam-Weinberg angle:

$$\sin^2 \theta_W = 0.39 \pm 0.05 .$$

12.2 Neutral currents - purely leptonic

Dr. H. Faissner[25] has reviewed the important area of purely leptonic neutral currents, with the following results:

i) $(\bar{\nu}_\mu e) \to (\bar{\nu}_\mu e)$ exists with 99% confidence; three *clean* events observed, with background estimated to be 0.46 ± 0.20.

ii) The data are not inconsistent with Salam-Weinberg and with $\sin^2 \theta_W < 0.5$.

12.3 Theoretical models for neutral currents

Out of the infinite number of theoretical models, Dr. M. Gourdin[26] has analysed half a dozen, with the following results:

i) the Adler-Tuan model is out by 3 standard deviations when compared with inclusive processes;

ii) the Beg-Lee model is out by 3 standard deviations in the inclusive and has difficulties with the elastic scattering;

iii) the Sakurai model is out by 2.5 standard deviations in the inclusive and has big difficulties with elastic scattering;

iv) the three Weinberg models I, II, III are all in good agreement with experimental data.

13. MESON SPECTROSCOPY

13.1 A new $J^P = 4^+$ meson in $\pi^0\pi^0$

An interesting result has been reported by Dr. W.D. Apel[27] on the studies of the reaction

$$\pi^- + p \rightarrow \pi^0 + \pi^0 + n,$$

at 40 GeV/c incoming pion momenta. Evidence for a new resonance in the $(\pi^0\pi^0)$ state is found (see Fig. 22) with the following properties:

$$\text{mass} = (2020 \pm 30) \text{ MeV}/c^2$$
$$\text{width} = (180 \pm 60) \text{ MeV}/c^2$$
$$\text{isospin} = 0 \text{ or } 2$$
$$J^P = 4^+$$
$$\text{production cross-section} \simeq 40 \text{ nb}.$$

This new meson is called h by the authors, who are the members of the Karlsruhe-Pisa-IHEP Serpukhov-Vienna-CERN Collaboration. This new meson fits nicely on the ρ, f, g, h Regge trajectory, as shown in Fig. 23.

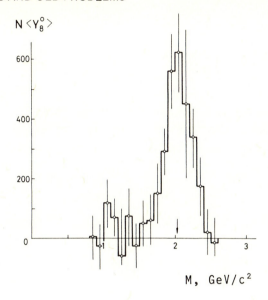

Fig. 22 Experimental evidence for the existence of a new meson decaying into two π^0's. Shown is the eighth unnormalized moment of the $\pi^0\pi^0$ system as a function of the $\pi\pi$ mass.

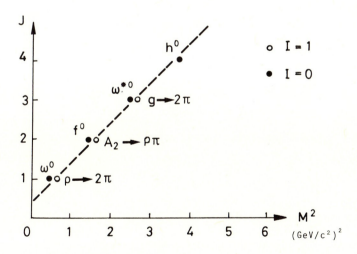

Fig. 23 Plot of spin J versus mass squared for the I = 1 and 0, 2π and 3π meson resonances

14. HADRONIC INTERACTIONS

14.1 A new way of analysing σ_{tot} of $(\pi^{\pm}p)(K^{\pm}p)(p^{\pm}p)$

Dr. H. Lipkin[28] has reported an interesting new way of studying the energy dependence of the total cross-sections for $\pi^{\pm}$, $K^{\mp}$, $p^{\pm}$, against protons. The starting point is that σ_{tot} is basically made up of two pieces, a Regge component R, which decreases with energy, and a constant or rising term f(s), i.e.

$$\sigma_{tot} = R \cdot s^{-\frac{1}{2}} + f(s) ;$$

the old way of plotting the data was by using σ_{tot} versus p_{lab}. The new way is based on the following variables:

$$\begin{cases} y = s^{\frac{1}{2}} \sigma_{tot}(s) = R + xf(x^2) \\ x = s^{\frac{1}{2}} . \end{cases}$$

Notice that R must be very large in $\sigma(\bar{p}p)$ because this cross-section decreases monotonically up to $p_{lab} \simeq 200$ GeV/c. On the other hand, R must be very small for $\sigma(K^+p)$ because this cross-section rises already at very low energy and therefore the "f" term must be dominant. The large differences which appear in the old way of presenting the data as σ_{tot} versus p_{lab} is due to the Regge term R, which is responsible for the most rapid energy variation and has different values for different processes. In the new Lipkin analysis, two processes, which differ only in their Regge contribution, will appear vertically displaced with respect to one another. The results can be seen in Fig. 24, from where it is found that all curves look rather the same in shape, and that the splitting between (K^-p) and (K^+p) is larger than $(\pi^-p)(\pi^+p)$ and smaller than $(\bar{p}p)(pp)$, i.e.

$$\updownarrow \begin{matrix} \bar{p} \\ p \end{matrix} \qquad \updownarrow \begin{matrix} K^- \\ K^+ \end{matrix} \qquad \updownarrow \begin{matrix} \pi^- \\ \pi^+ \end{matrix}$$

This behaviour of the cross-sections is expected from the Regge description with exchange degeneracy. Moreover, the fact that $(\pi^{\pm}p)$ cross-sections lie between $(K^{\pm}p)$ and $(p^{\pm}p)$ is a manifestation of the known phenomenological regularity observed in the Pomeron component of the total cross-section.

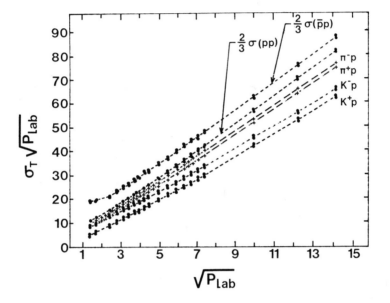

Fig. 24 The hadronic total cross-sections plotted following Lipkin's recipe. Notice the splitting, which is minimum between $(\pi^- p)-(\pi^+ p)$ and maximum between $(\bar{p}p)-(pp)$, $(K^- p)-(K^+ p)$ being in the middle.

14.2 Moving dip in dσ/dt for (pp) elastic scattering agrees with geometrical scaling predictions

Dr. K.R. Schubert[29] has reported on elastic (pp) scattering in the ISR energy range. As is well known, a dip was observed some time in dσ/dt elastic. The authors have now established that in the ISR energy range (E_{tot} = 23-62 GeV) the position of the dip goes from

$$t_{dip} = -(1.44 \pm 0.02) \, (GeV/c)^2 \text{ at } \sqrt{s} = 23 \text{ GeV}$$

to

$$t_{dip} = -(1.26 \pm 0.03) \, (GeV/c)^2 \text{ at } \sqrt{s} = 62 \text{ GeV},$$

as shown in Fig. 25.

The ratio of these quantities

$$\frac{t_{dip} (\sqrt{s} = 23 \text{ GeV})}{t_{dip} (\sqrt{s} = 62 \text{ GeV})} = 1.14$$

is in excellent agreement with the predictions of geometrical scaling. As remarked by Dr. Schubert, geometrical scaling explains the constancy of the ratio

$$\frac{\sigma(pp)_{el}}{\sigma(pp)_{tot}} \simeq 0.19$$

versus energy.

14.3 Breakdown of Feynman scaling

Evidence for breakdown of Feynman scaling in the central region at ISR energies was reported by Dr. B.G. Duff[30] from the British-Scandinavian-MIT Collaboration. This group has measured the inclusive pion cross-section

$$pp \rightarrow \pi + \text{anything}$$

at x = 0 and five ISR energies. The observed number of events increases with $\sqrt{s}$, from 23 GeV to 62 GeV, showing evidence for Feynman breaking. The remarkable feature of this breaking is that it has the same behaviour in the five bins of p_T investigated, from p_T = 50-100 MeV/c up to p_T = 300 to 375 MeV/c; this is shown in Fig. 26. The results have been presented as preliminary.

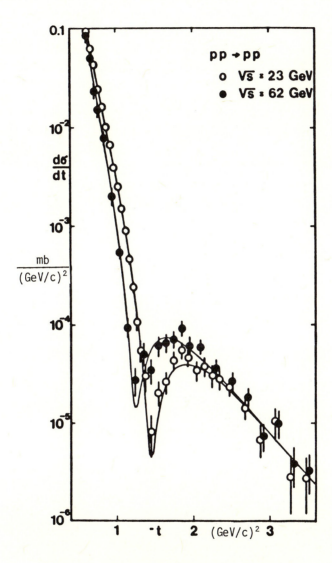

Fig. 25 The experimental proof that the position of the dip in (dσ/dt)pp is moving with energy

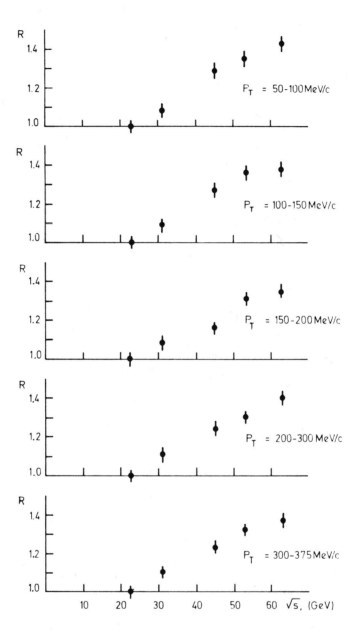

Fig. 26 Experimental results of Feynman breaking: the energy dependence of inclusive charge-averaged cross-sections for pions. Notice that the data follow the same trend for different transverse momenta. R is the ratio between the quantity (No. of events per luminosity) at $\sqrt{s}$ and the same quantity at $\sqrt{s}$ = 23 GeV; where R, by definition, is 1.

14.4 Effective slopes of $(\pi^{\pm} p)(K^{\pm} p)$ are different from $(p^{\pm} p)$

New results on the measurement of the effective slopes in the elastic scattering of hadrons in hydrogen were presented by Dr. C. De Marzo of Fermilab[31] (Batavia). The reactions investigated were

$$\left. \begin{array}{c} \pi^+ p \to \pi^+ p \\ \pi^- p \to \pi^- p \\ K^+ p \to K^+ p \\ K^- p \to K^- p \\ pp \to pp \\ \bar{p}p \to \bar{p}p \end{array} \right\} \text{from 50 GeV/c to 175 GeV/c incident particle laboratory momenta.}$$

The differential elastic cross-section $d\sigma/dt$ was fitted with the form $d\sigma/dt = A \exp(B|t| + C|t|^2)$.

The following quantity was evaluated at $t = -0.2 \ (\text{GeV/c})^2$:

$$\left[\frac{d}{dt} \ln \frac{d\sigma}{dt} \right]_{t=-0.2(\text{GeV/c})^2}.$$

The effective slopes, $b_e = B + 2C|t|$, of negative and positive pions and kaons, within errors are the same: 8 $(\text{GeV/c})^{-2}$. The remarkable fact that a t-value exists where the slopes of these cross-sections are equal, is very interesting and is expected from SU(3). The crucial point is now to see if also the absolute values coincide at $t = -0.2 \ (\text{GeV/c})^2$. The slopes of antiprotons and protons are equal, 10.5 $(\text{GeV/c})^{-2}$, but higher than the (π, K) case. The data are reported in Fig. 27.

14.5 Slope-mass correlation in double diffraction at ISR energies

An interesting effect has been reported by Dr. G.C. Mantovani[32], of the Pavia-Princeton Collaboration, for the double-diffraction process

$$p + p \to (p\pi^+\pi^-) + (p\pi^+\pi^-).$$

If we call b the slope of $d\sigma/dt$, and M the mass of the $(p\pi^+\pi^-)$ system, it is found that at $\sqrt{s} = 31$ GeV, b goes from ≥ 12 to ~ 2 $(\text{GeV/c})^{-2}$ when M goes from 1.4 to ≥ 1.8 GeV. In the same mass range, b goes from ~ 12 to ~ 7 at $\sqrt{s} = 53$ GeV, for the single-diffraction case. These slope-mass correlations are shown in Fig. 28.

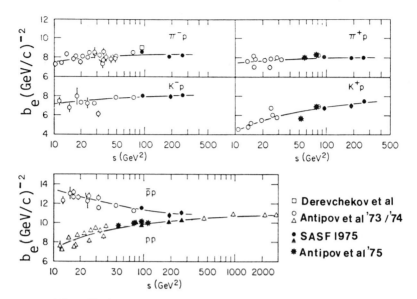

Fig. 27 The effective slope parameter b_e for the various reactions with $\pi^\pm$, $K^\pm$, $\bar{p}$, p as primary particles against protons

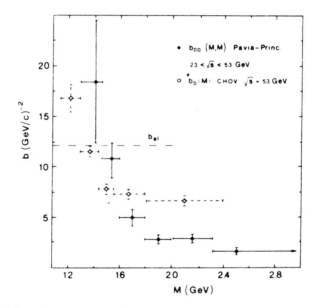

Fig. 28 The slope-mass correlation for double-diffraction processes: $pp \to (p\pi^+\pi^-)(p\pi^+\pi^-)$. The single diffraction is also shown (dashed error bars). The dashed line shows the elastic slope.

Another result worth mentioning is the value of the cross-section for the double-diffraction reaction, which is ~ 5 µb and nearly constant in the above-mentioned $\sqrt{s}$ range; of course, this constancy is essential if we believe that the observed process is entirely due to double-diffraction mechanisms. Finally, the observed mass spectra are found to be similar to the single-diffraction processes and are s-independent.

14.6 Difference in multiplicity and π inclusive spectra for 100 GeV/c antiprotons compared to protons

The first data on 100 GeV/c antiproton interactions in hydrogen were reported by Dr. W.W. Neale[33] of the Cambridge-Fermilab-MSU Group.

The multiplicity distributions are definitely higher for antiproton-induced reactions compared to protons of the same lab. momentum (100 GeV/c). For example (see Fig. 29), for charged multiplicity equal to 16, the ratio of the cross-sections is

$$\frac{\sigma(\bar{p} + p \to 16 \text{ tracks})}{\sigma(p + p \to 16 \text{ tracks})} \simeq 1.7 \ .$$

Moreover, the comparison of the two inclusive processes

$$\left.\begin{array}{l} \bar{p} + p \to \pi^+ + \text{anything} \\ p + p \to \pi^- + \text{anything} \end{array}\right\} \text{at 100 GeV/c lab. momentum}$$

in the rapidity plot shows that the excess of cross-section is concentrated in the "central region", as reported in Fig. 30.

14.7 Antibaryon annihilation seems to produce higher values of $\langle p_T \rangle$ than strangeness annihilation

The average transverse momenta of the pions in $K^- + p \to \Lambda + 3\pi$ are studied according to the diagrams

where off-shell $(\bar{p}p)$ and (K^-K^+) annihilations contribute. These data are compared with on-shell $(\bar{p}p)$ annihilation into three π's. It is found that

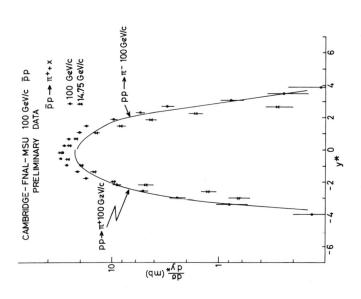

Fig. 30 Centre-of-mass rapidity distribution for π^+ inclusive production in $\bar{p}p$ collisions at 100 GeV/c and 14.75 GeV/c. The solid curves are for π^+ production in the backward and π^- in the forward hemisphere in 100 GeV/c pp collisions.

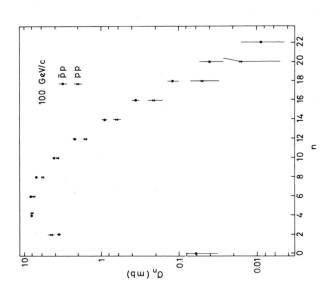

Fig. 29 Comparison of $\bar{p}p$ and pp topological cross-sections at 100 GeV/c

the average $\langle p_T \rangle$ is ~ 20% higher for the $(\bar{p}p)$ on-shell annihilation. On the other hand, "off" and "on" mass-shell $(\bar{p}p)$ agree. This would indicate a correlation between $\langle p_T \rangle$ of the produced π's and the mass of the annihilating particles, as shown in Fig. 31.

It is not easy, in fact it is impossible, to get data on this topic which are not model-dependent. However, a more extensive study of the subject is clearly of great interest. This contribution was presented by Dr. J.R. Fry[34] of Liverpool.

14.8 Increase of (K^+K^-) and $(\bar{p}p)$ yield in six-prong events from K^+p at 32 GeV/c

Dr. P.V. Chliapnikov[35] has reported on the $(K^{\pm}p)$ experiments at 32 GeV/c performed by the France-Soviet Union and CERN-Soviet Union Collaborations, in the Mirabelle bubble chamber at Serpukhov. Here is a very interesting result: the relative production of (K^+K^-) and $(\bar{p}p)$ pairs in the (K^+p) interactions is much higher in six-prong than in four-prong events. More precisely

$$\frac{\sigma(K^+p \to K^+pK^+K^-) + \sigma(K^+p \to K^+p\bar{p}p)}{\sigma(K^+p \to K^+p\pi^+\pi^-)} = 0.06 \pm 0.02 ,$$

as in low-energy data; but

$$\frac{\sigma(K^+p \to K^+pK^+K^-\pi^+\pi^-) + \sigma(K^+p \to K^+p\bar{p}p\pi^+\pi^-)}{\sigma(K^+p \to K^+p\pi^+\pi^-\pi^+\pi^-)} = 0.4 \pm 0.1 .$$

This effect could be related to the strong increase of $(K_S^0 K_S^0)$ production observed both in (K^+p) and (K^-p) interactions, when the incident K momentum goes from 16 to 32 GeV/c.

14.9 Evidence that (K^+p) diffractive dissociation is ~ 50% of the (pp)

(K^+p) diffractive dissociation is predicted by the triple Regge coupling and the Pomeron factorization to be about 50% of the (pp) case, contrary to a naive picture where the diffraction would appear as a universal feature for all particles: after all, they should exchange only a Pomeron with the target particle. Dr. Chliapnikov has presented[35] experimental evidence that the cross-section for the K^+ diffractive dissociation is

$$\sigma_D(K^+p) \approx 1 \text{ mb} ,$$

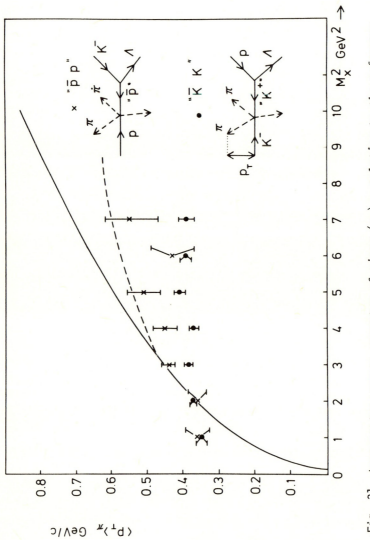

Fig. 31 Average transverse momentum of pions $\langle p_T \rangle_\pi$, relative to the reference axes shown, in the fragmentation regions of the reaction $K^- p \to \Lambda 3\pi$ at 8.25 GeV/c. The solid line is the phase-space expectation, while the dashed line shows the experimental trend from $\bar{p}p \to 3\pi$ data.

RECENT PROGRESS AND OLD PROBLEMS   789

to be compared with the proton single diffractive dissociation

$$\sigma_{SD}(pp) = (2.2 \pm 0.5) \text{ mb} .$$

14.10 Leading cluster effects in $(\pi^- p)$ at 200 GeV

Dr. V.P. Kenney[36] from Notre Dame has presented an analysis of inclusive π^+ and π^- data from $(\pi^- p)$ interactions at 200 GeV. In the entire forward c.m.s. hemisphere a large "leading cluster" effect is observed. This produces π^- inclusive cross-sections which are greater than those observed in (pp) interactions. These leading clusters are observed for all multiplicity events, i.e. from 2 to 16 charged particles.

14.11 Surprising, deep, and narrow dip in $\bar{p}p \to \bar{n}n$ at $-t \simeq m_\pi^2$

A surprising and interesting result was contributed by Bolotov and collaborators, and reported by Dr. F. Schrempp[37] in his review talk.

The differential cross-section $d\sigma/dt$, for the reaction

$$\bar{p} + p \to \bar{n} + n$$

at 40 GeV/c incident antiproton, shows a deep, narrow dip at $-t \simeq$
$\simeq 0.02 \text{ (GeV/c)}^2$, which equals m_π^2. The surprise is also due to the fact that $d\sigma/dt$ of the line-reversed reaction

$$n + p \to p + n$$

measured at 42.5 GeV/c by Babaev et al.[38] shows no sign of such a dip. The results are shown in Fig. 32.

The mechanism which comes into play is π-exchange, and Dr. Schrempp, who knew the results once he was already in Palermo, having no time to produce a theoretical model, said: '... This represents crucial information about energy and momentum dependence of the reaction mechanism of "puzzling π-exchange" ...'.

14.12 Jet-like structures observed at ISR

Dr. P. Darriulat[39] has presented evidence for an interesting effect, which could be associated with the production of hadronic jets in (pp) high-energy collisions. The reaction investigated requires a π^0 with $p_T \simeq 2.5$ GeV/c, produced at 90°, and being used as a trigger:

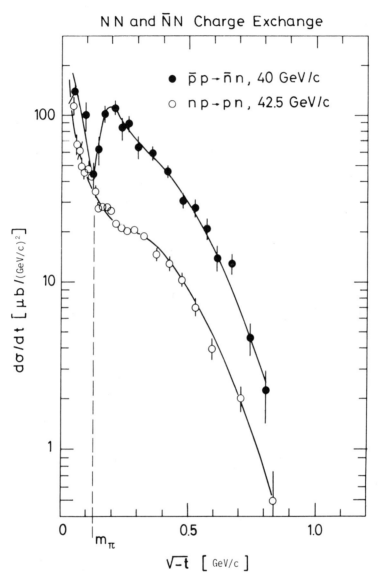

Fig. 32 $\bar{p}p \to \bar{n}n$ at 40 GeV/c, showing a narrow dip at $-t \simeq 0.02$ GeV^2. For comparison, the line-reversed reaction $np \to pn$ at 42.5 GeV/c is shown. The solid lines are to guide the eye.

RECENT PROGRESS AND OLD PROBLEMS   791

$$p + p \rightarrow \text{jet} + \pi^0 + \text{anything}$$

triggering particle ↑ (above π^0)

↳ this jet produces a particle of high momentum opposite to the π^0, and other particles with low transverse momentum

The claim is that, when the momentum of the leading particle in the jet increases, the transverse momentum remains constant, i.e.

The leading particle momenta vary in the range

$$p_a = (1.3-2.1) \text{ GeV/c}$$
$$p_b = (1.1-1.3) \text{ GeV/c}$$
$$p_c = (0.6-0.9) \text{ GeV/c}$$

and p_{out} does not follow this increase (Fig. 33). The effect needs to be carefully studied and is of great interest in establishing, in a firm and definite way, if jets are really produced at the ISR.

14.13 <u>Evidence for triple Regge failure</u>

Dr. R. Salmeron[40], on behalf of the Rutherford Lab.-Ecole Polytechnique-Saclay Collaboration, has reported an extensive analysis of (K^-p) interactions at 14.3 GeV/c. Among the large number of results, let me quote the application of a simple triple Regge model to the reactions

$$K^- + p \rightarrow K^{*-}(890) + X^+$$
$$K^- + p \rightarrow \overline{K^{*0}}(890) + X^0 \ .$$

The analysis shows that the simple triple Regge model explains the amount of *unnatural* parity exchange in the production of both $K^{*-}(890)$ and $\overline{K^{*0}}(890)$; *but* fails to explain quantitatively the ratio of amounts of natural parity exchange in both reactions. The t-interval of the data is $0.2 \leq t \leq 0.8 \text{ (GeV/c)}^2$.

14.14 <u>Observation of "relative scaling"</u>

Dr. I.A. Radkevich[41] (ITEP Moscow) has reported a study of a large number of (π^-p) and (π^-d) reactions. These reactions at high momentum transfer show a strong energy dependence. However, the ratio of different reactions, theoretically described by many different diagrams or models,

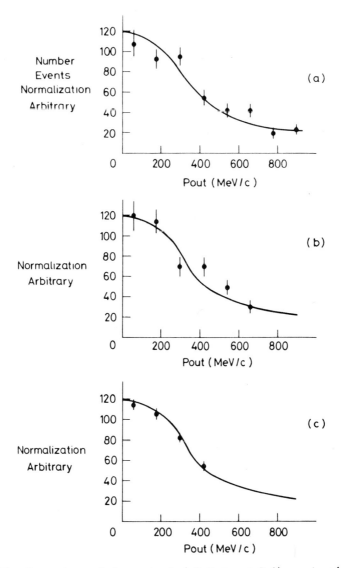

Fig. 33 Charged particle central (−2.3 < y < 2.3) production in association with a 90° π^0 with $p_T \approx 2.5$ GeV/c; at $\sqrt{s}$ = 53 GeV. Distributions of the charged particle momentum component out of the plane, p_{out}, defined by the incident protons and the π^0. Data are split according to the charged-particle momentum component along the π^0:
a) 1.3 to 2.1 GeV/c;
b) 1.1 to 1.3 GeV/c; and
c) 0.6 to 0.9 GeV/c. A same arbitrary curve is handdrawn on each figure for comparison purposes.

RECENT PROGRESS AND OLD PROBLEMS

are energy-independent. This new phenomenon is called "relative scaling", and Dr. Radkevich has presented evidence for it.

14.15 The average "spin" of diffractively produced clusters increases with mass

Dr. P. Schmid[42] has reported the results of an analysis of the reaction

$$\pi^+ + p \to \pi^+ + \pi^+ + \pi^- + p$$

at 16 GeV/c incident pion momentum. When the angular momentum of the diffractively produced $(\pi^+\pi^+\pi^-)$ system is plotted versus the value of its effective mass, $M(\pi^+\pi^+\pi^-)$, a striking correlation emerges, which is numerically given by:

$$\langle J \rangle = 0.55 + 1.1 \, Q_{eff}$$

as shown in Fig. 34; $Q_{eff} = M(3\pi) - M_{threshold}$; $M_{threshold}$ = the mass of a two-body resonance plus a pseudo-scalar meson, for example $(\rho + \pi)$.

The conclusion is that the average spin of the diffractively produced $(\pi^+\pi^+\pi^-)$ system increases with mass. This phenomenon is also present for $(\pi K\bar{K})$ and $(\bar{K}\pi\pi)$, and could therefore be valid for any 3-meson final state[42]. This is of great interest for the high-p_T phenomenon. In fact, high spins could be a source (less interesting and hopefully not true) of the striking high-p_T effects observed in many reactions.

14.16 $(\pi\pi)$ scattering length higher than the Weinberg expectations

This result was reported by Dr. A.D. Martin[43] (Geneva-Saclay). The analysis is based on 30,200 $(\pi\pi)$ events from $K^+ \to \pi^+\pi^-e^+\nu$, and the values of the $(\pi\pi)$ scattering length $a_S^{I=0}$ turns out to be somewhat bigger than the Weinberg expectations.

14.17 New results on $(\pi\pi)$ phases

These were also presented by Dr. Martin[43]. Let us mention the $(\pi\pi)$ phase shifts from the same large statistical sample quoted above:

$$\delta_S^0 - \delta_P^0 = 0.20 \pm 0.04 \;,$$

which is the best result obtained so far (as usual, S and P stand for S- and P-waves).

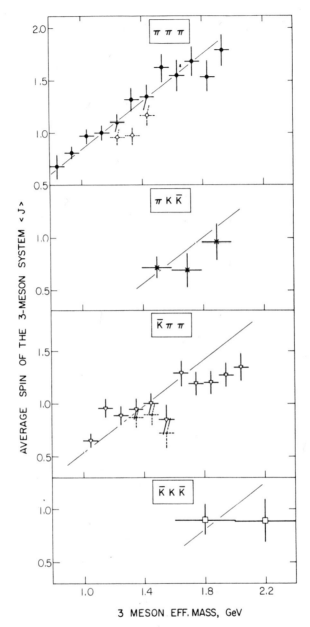

Fig. 34 Evidence that the average spin ⟨J⟩ increases with the 3-meson system. The top graph shows ⟨J⟩ versus the mass of the 3π system. The broken flags are obtained excluding 2^+ resonances. The solid lines show the parametrization ⟨J⟩ = 0.55 + 1.1 Q_{eff}.

15. THEORETICAL CONTRIBUTIONS

Here, there would be a lot to report on. Dr. P.H. Frampton[44] reviewed all dual and string models, and Dr. A. Neveu[45] the quark confinement theory, to which I am particularly allergic. Finally, Dr. G. 't Hooft[46] revived all our hopes of one day seeing weak, electromagnetic, and strong interactions unified.

As usual, many theoretical predictions were formulated up to the "highest energies" (higher than ISR), the beloved refuge of theoretical reluctance to explain lower-energy experimental results; again and again, the question is: Where does asymptopia lie?

Let me say a few words on some contributions related to the "whys" listed in Section 17 below.

15.1 Hope of understanding $\sigma(pp)_{el}/\sigma(pp)_{tot}$

A well-known experimental quantity which appears to have reached the asymptotic energy limit, is awaiting a theoretical explanation since a long time:

$$\frac{\sigma_{el}(pp)}{\sigma_{tot}(pp)} \simeq 0.20 .$$

Dr. M. Ciafaloni[47] has reported on a theoretical attempt to explain this. Theories with $U(N)$ internal symmetries have expansions according to topology, e.g. $G^2 = g^2 N = \text{const}$. Selecting planar diagrams gives $g^2 = O(1/N)$, which can, of course, be small. However, no definite predictions exist so far. The hope is that one day

$$\frac{\sigma_{el}}{\sigma_{tot}} = O(1/N^2) .$$

15.2 Hope of explaining the small Pomeron pole

Another mystery in high-energy physics is why the Pomeron pole is so small. Dr. R. Webber[48] has contributed a theoretical attempt, working on a relationship between multiparticle production and slopes of Regge trajectories. If the charge-exchange overlap function has the following rapidity densities,

$$\nu_\rho \simeq \tfrac{1}{2}\nu_{I\!P} \simeq 0.3 \quad \text{(as expected)} ,$$

then it follows that the slope of the ρ-trajectory α'_ρ is related to that of the Pomeron $\alpha'_\mathbb{P}$, as follows:

$$\alpha'_\rho \approx 3\alpha'_\mathbb{P} \approx 0.85 \text{ GeV}^{-2},$$

and this is very much desired. The Pomeron slope has important implications on the multiparticle production models. For example,

$$\alpha'_\mathbb{P} \simeq 0.3 \text{ GeV}^{-2}$$

gives the following value for the rapidity density ν:

$$\nu \simeq 0.6,$$

which corresponds to ~ 5 particles per cluster.

15.3 Hope of understanding the concavity of $\sigma_{tot}(pp)$

It is well known that the Reggeon calculus gives to $\sigma_{tot}(pp)$ a contribution which is roughly linear in ln s. As it is known that $\sigma_{tot}(pp)$ first decreases with energy, reaches a plateau, and then increases, one needs a large breaking of exchange degeneracy of the secondary Regge residues to account for the experimental data. The Gribov-Reggeon calculus can describe all elastic data near t = 0, with very few parameters, provided that i) the (bare) Pomeron intercept is above one: $\alpha_\mathbb{P}(0) \sim 1.3$; and ii) one gives up the two-component Harari-Freund duality.

This has been done and reported by Dr. A. Capella[49], who consequently predicts the ratio of

(Re/Im) for (pp) and ($\bar{p}p$),

up to the highest energies.

16. A NEW DRIFT CHAMBER

As you know, theoretical models often need inspiration from new experimental results, and these are bound to new tricks and discoveries on the technical side.

Dr. G. Charpak[50] has outdone himself. After having invented the so-called proportional chambers, he has announced a very important step in the field: a new drift chamber with light-emission instead of amplifying wires. The properties of this instrument are: 200 μ precision and 100 times the rate of the old Charpak-type chambers. The results presented by Dr. Charpak show that these chambers have a plateau from 10 particles per sec·mm² to 3×10^6 particles/sec·mm².

17. THE OUTSTANDING PROBLEMS

We now turn to those problems -- some very fundamental, others of detailed interest -- which are a constant challenge to all of us, especially to those who are youngest:

1) Why are S- and P-waves in non-leptonic decays correlated?

2) Why is the elastic scattering imaginary at high energy, or in Regge-language, Why do even-signatures dominate and odd signatures become negligible?

3) Why is the Pomeron slope so small?

4) Why is $\dfrac{\sigma(pp)_{elastic}}{\sigma(pp)_{total}} \simeq \dfrac{1}{5}$?

5) Why are total scattering cross-sections for strange particles smaller than those for non-strange particles? Why are meson cross-sections smaller than baryon cross-sections?

6) Why does $\dfrac{\nu W_2^n}{\nu W_2^p} \to \dfrac{1}{4}$ as $\omega \to 1$?

7) At relatively low energies (few GeV's) we observe that the inclusive electromagnetic coupling of the hadrons is point-like. Is this because the hadrons are made of super-elementary constituents? and if so, Why they do not show up?

8) Why is $\dfrac{\sigma(e^+e^- \to hadrons)}{\sigma(e^+e^- \to \mu^+\mu^-)}$ not 2, but most likely between 5 and 6 at the highest SLAC energy?

9) Why are the weak non-leptonic decay rates 3 orders of magnitude greater than the weak leptonic decays, i.e.:
$\dfrac{\Lambda \to \beta}{\Lambda \to all} \simeq 10^{-3}$?

10) Why is the inclusive ratio $e/\pi \sim 10^{-4}$?

11) Why does not $K_L^0 \to \mu^+ + \mu^-$ go as it should, while $\nu + p \to \nu +$ anything goes; or more generally, why are weak neutral non-strange currents there and weak neutral strange currents NOT there? Is the answer: charm, heavy leptons, or something else?

12) Why are the J and ψ particles there? Is the answer: charm, colour, or something totally different?

13) Why does $K_L \to 2\pi$ go? Or, more generally, why is CP violated? Is this because a new interaction -- superweak -- is at work?

14) Why is the Cabibbo angle $\sim 20°$? Can the answer be found via spontaneous symmetry breaking?

15) Why is SU(3) there? Is this because the so-called elementary particles are made of quarks?

16) Why is 3/2 the maximum value of isospin found in the baryon spectrum and 3 the maximum value of strangeness? Is this because baryons are made of three quarks?

17) We observe regularities which go beyond SU(3). Is this because unitary spin and Dirac spins are correlated?

18) Why do single quark transitions dominate? i.e. Why is it so easy to have spectator quarks?

19) Why have quarks so far not been found? Is it because the confinement theories predict the truth; or because the production process is peculiar, or because quarks do not exist in Nature at all?

20) Why is the muon there? Should other leptonic states (heavy leptons of the old type) exist?

21) We do not observe any of the processes predicted by theory to have infinite rates, such as all higher order weak interaction reactions and electromagnetic mass differences. Are we sure that something physically relevant is not missing, which causes all attempts to have a renormalizable theory of weak interactions to fail?

22) Why is the weak charge universal?

23) Why is the electric charge universal?

24) The bare electric charges of the electron and the proton are equal. Is this related to their $(1 + \gamma_5)$ coupling as particle states in weak interactions?

25) Why are the proton, the neutron, the Λ^0, as examples of baryonic states, and the e^-, μ^-, ν_e, as examples of leptonic states, all left-handed when they interact weakly? Is this due to the fact that they transform each other?

26) We observe six types of fundamental interactions. Are these all orthogonal to each other, or have they a common origin?

27) Why is the proton so stable? i.e. Why do we exist at all?

To be more complete let me point out that, according to the specialists, the last problem has an answer in the well-known field of religions.

I have mentioned this possible solution because sometimes theoretical models do produce religious feelings.

REFERENCES

Unless otherwise stated, page numbers refer to the Proceedings of the EPS Internat. Conf. on High-Energy Physics, Palermo, June 1975.

1) S.C.C. Ting, p. 15.

2) V. Lüth, p. 54.

3) P. Monacelli, p. 75.

4) G. Feldman, p. 233.

5) M. Bernardini et al., Proc. 1971 EPS Conf. on Meson Resonances and Related Electromagnetic Phenomena, Bologna (Editrice Compositori, Bologna, 1972), p. 489; and Phys. Letters 51B, 200 (1974).

6) V. Alles-Borelli et al., Nuovo Cimento 4, 1156 (1970); M. Bernardini et al., Nuovo Cimento 17A, 383 (1973).

7) G. Preparata, p. 422.

8) C. Rubbia, p. 111.

9) D. Cline, p. 335.

10) B.P. Roe, p. 497.

11) B.G. Pope, p. 201.

11a) M.L. Good, p. 220.

12) N. Cabibbo, p. 133.

13) P.T. Matthews, p. 178.

14) Y. Yamaguchi, p. 186.

15) K.W. Chen, p. 458.

16) D. Schildknecht and F. Steiner, Phys. Letters 56B, 36 (1975).

17) R.E. Taylor, p. 377.

18) G. Weber, p. 1183.

19) C.A. Heusch, p. 477.

20) D.O. Caldwell, p. 450.

21) J.H. Field, p. 247.

22) M. Basile et al., The present knowledge of $|F\pi|$ and the value of a low energy (e^+e^-) storage ring, to be published in Nuovo Cimento.

23) F. Palmonari, p. 261.

24) M. Rollier, p. 279.

25) H. Faissner, p. 291.

26) M. Gourdin, p. 308.

27) W.D. Apel, p. 971.

28) H.J. Lipkin, p. 1000.

29) K.R. Schubert, p. 1066.

30) B.G. Duff, p. 976.

31) C. De Marzo, p. 1070.

32) G.C. Mantovani, p. 961.

33) W.W. Neale, p. 1007.

34) J.R. Fry, p. 949.

35) P.V. Chliapnikov, p. 897.

36) V.P. Kenney, p. 815.

37) F. Schrempp, p. 682.

38) A. Babaev et al., Paper G2-02; See also F. Schrempp, p. 682.

39) P. Darriulat, p. 840.

40) R.A. Salmeron, p. 1047.

41) I.A. Radkevich, p. 1019.

42) P. Schmid, p. 553.

43) A.D. Martin, p. 545.

44) P.H. Frampton, p. 1250.

45) A. Neveu, p. 1273.

46) G. 't Hooft, p. 1225.

47) M. Ciafaloni, p. 1152.

48) B.R. Webber, p. 1111.

49) A. Capella, p. 1145.

50) G. Charpak, p. 1291.

A SURVEY OF THE THEORETICAL MODELS PROPOSED FOR THE NEW PARTICLES

G. Morpurgo

Istituto di Fisica dell'Università di Genova

Istituto Nazionale di Fisica Nucleare-Sezione di Genova

1-Introductory remarks

These lectures are only meant as an introductory survey of the present (July 1975) situation in this field. In spite of the fact that the discovery[1] of the narrow resonances $J(3.1)$ (or $\psi(3.1)$) and $\psi'(3.7)$ is recent, many other surveys have already appeared[2]; all of them were useful in preparing these notes which have no pretense at originality; the only difference with the previous surveys may reside in the emphasis attributed to this or that topic, in the order of presentation and in the attempt to include the main facts presented at the recent EPS Conference in Palermo. It should also be added that this survey is incomplete under many respects both because it was felt inappropriate to increase excessively its size and because in this School other lectures will be devoted to this "hot" subject, and we have tried to avoid excessive overlaps.

The number of theoretical papers published since the announcement of the discovery of the J (or ψ) and ψ' is huge; the main difficulty for a survey speaker is to select, among all these ideas, those which he considers most relevant and to give an order to his presentation. The selection and ordering adopted here are evident from the index.

The larger part of this survey is devoted to the charm scheme and to exemplify various colour schemes. As far as the charm picture is concerned our conclusion will be that the Okubo-Zweig rule, by itself, is not sufficient to inhibit the decays of the hidden charm mesons by the actual amount required. As far as the colour picture is concerned its many difficulties will also emerge clearly. Altogether the feeling I have, at the time of writing, is that a two body bound state seems suggested by the facts but that one should keep an open mind as to the nature of the two objects which are bound. Our procedure in what follows will be that of making, whenever possible, the simplest kind of evaluations without committing ourselves to a specific field theoretical scheme.

In fact many concepts to be used will be familiar concepts of the non relativistic quark model for which reference is made to our previous lectures at this School[27,21].

For what concerns the nomenclature we shall use almost always in what follows, for the sake of uniformity, the symbol $\psi(3.1)$ rather than $J(3.1)$: also we write 3100 instead of 3095 for the mass of the ψ, with a similar rounding for that of the ψ'.

2-A few experimental facts

These are listed in the following table I.

Table I

Particle	J^{PC}	M(GeV)	Γ_{tot}(KeV)	Produced in:
ψ	1^{--}	3.095	69 ± 15	P-Nucleon, e^+e^-, γ-Nucleon
ψ'	1^{--}	3.684	225 ± 56	e^+e^-, γ-Nucleon
ψ''		~ 4.2	Broad resonance ($\Gamma \cong 300$ MeV) or threshold phenomenon	e^+e^-

The more important comments, preliminary to any discussion of the situation, are:

1) The ψ, ψ' resonances are "narrow"; here narrow means that one has for the width Γ:

$$\Gamma(3.1 \to \text{hadrons}) \approx 10 \; \Gamma(3.1 \to \mu^+\mu^-)$$

whereas for a strong decay such as that of the ρ meson:

$$\Gamma(\rho \to \text{hadrons}) \approx 20.000 \; \Gamma(\rho \to \mu^+\mu^-)$$

Similar remarks apply to the 3.7 resonance.

2) The partial widths of the decays of the 3.1 and 3.7 states into e^+e^- are:

$$\Gamma(3.1 \to e^+e^-) = 4.8 \pm 0.6 \text{ KeV}$$

$$\Gamma(3.7 \to e^+e^-) = 2.2 \pm 0.3 \text{ KeV}$$

The $\mu^+\mu^-$ decay mode has (inside the experimental errors) the same partial widths. The partial widths for other specific decay modes are reported in the Appendix A. Note in particular that the partial width for the decay into hadrons of the ψ or ψ' is due, in part, to the usual one photon intermediate state mechanism which is operant outside the resonance, and in part to an effective additional real hadronic decay. Therefore[4] the total $\Gamma(3.1)=69\pm15$ KeV can be decomposed as follows: 9.6 KeV are accounted by the decays into e^+e^- and $\mu^+\mu^-$; 2.5 times 4.8 KeV $\approx$ 12 KeV are due to hadron decays due to the one photon intermediate state mechanism (the factor 2.5 is explained below); and the rest, that is $\approx$ 48 KeV are true hadronic decays, which however can include also one or more γ rays.

3) The traditional quantity

$$R = \frac{e^+e^- \to \text{hadrons}}{e^+e^- \to \mu^+\mu^-} \tag{1}$$

is $\approx$ 2.5 up to $\sim$ 3.6 GeV (except, of course, at the 3.1 resonance); it increases to 5.5 at and above 4.5 GeV (in the c.of m. system); the value 2.5 explains our previous use of this number.

4) To an accuracy 1/10 in intensity ($\sim$ 1/3 in the matrix element) no other narrow resonances in e^+e^- are detected in the region 1.9 to 5.9 GeV[5] (there is however still an unexplored region of $\sim$ 400 MeV centered around 2.7 GeV).

3 - The weak vector boson W^o

Immediately after the early data on the 3.1 resonance (and before the discovery of the 3.7) the possibility was considered[6] that the 3.1 resonance might be identified-in spite of the low mass-with the neutral vector meson W^o mediating the weak neutral interactions.

Write, in an obvious notation:

$$\mathcal{H}' = g\, \bar{\psi}(3.1)\left(\bar{e}\, \gamma_\mu (a+b\gamma_5) e\right) \qquad (a^2+b^2 =1)$$

Using the experimental value of $\Gamma(\psi \to e^+e^-)$ we get:

$$\Gamma(\psi \to e^+e^-) = 4.8 \text{ KeV} = \frac{g^2}{12\pi} M_\psi$$

that is:

$$g^2 = 5.7 \times 10^{-5}$$

We thus get for the Fermi constant $G^{(o)}$ corresponding to the exchange of $\psi(3.1)$:

$$\frac{G^{(o)}}{\sqrt{2}} \simeq \frac{g^2}{2M_\psi^2} \approx \frac{1}{2}\frac{10^{-5}}{M_P^2}$$

that is (neglecting factors) $O(\sqrt{2})$)

$$G^{(0)} \approx G_F$$

In spite of this striking coincidence in the orders of magnitude, there are however a number of facts which strongly appear to preclude this simple interpretation; they are listed below:

1) two (or more) W^0's.
2) the broad enhancement in the e^+e^- cross section for hadron production at 4.2 GeV.
3) The q^2 dependence of the neutrino form factor of the nucleon(such a low mass W should influence in a visible way the q^2 dependence.).
4) Definite G parity of ψ and ψ' (this point should be discussed in more detail; perhaps we may come back to it in the discussion session)
5) The $\gamma\psi$ couplings (they have the same general order of magnitude as the usual vector meson γ couplings).
6) Most important ! The photoproduction experiments (γ Nucleus $\rightarrow \psi$ Nucleus). From these one can deduce[7] by the usual argument summarized in the figure below:

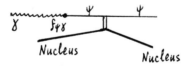

$\sigma_T(\psi N) \approx 1$ mb, typically strong (N means, here, nucleon). To this list we add, as indicative, the circumstance[8] that no evidence of parity non conservation appears from the angular distribution in the $e^+e^- \rightarrow \mu^+\mu^-$ reaction at the energy of the ψ state. A study of C conservation in the decays of the ψ has also been proposed[9][10].

4 - A digression on the $\chi\psi$ couplings

Both because of the importance of the photoproduction processes of the new particles for establishing the non weak nature of the decays of the new particles, and for its own interest, it is appropriate to derive the order of magnitude of the coupling constant $f_{\psi\gamma}$; this coupling, on the mass shell of the ψ, is derived experimentally from the graph

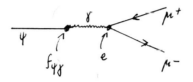

The interesting point is that the order of magnitude of $f_{\psi\gamma}$ is the same typical order of magnitude of the usual $\rho\gamma$ coupling constant.

From:

$$\Gamma_{\psi\to\mu^+\mu^-} = \frac{4\pi}{3}\alpha^2\left(\frac{f_{\psi\gamma}}{m_\psi^2}\right)^2 m_\psi + O\left(\frac{m_\mu^4}{m_\psi^4}\right)$$

it follows:

$$\left(f_{\psi\gamma}/m_\psi^2\right)^2 \simeq 7\ 10^{-3} \tag{2}$$

to be compared with

$$\left(f_{\rho\gamma}/m_\rho^2\right)^2 \simeq 3.6\ 10^{-2} \tag{3}$$

The order of magnitude is the same.

We can anticipate here that the charm model with $Q_{p'} = \frac{2}{3}$ gives, in its naive unbroken version:

$$\left(f_{\rho\gamma}/m_\rho^2\right)^2 / \left(f_{\psi\gamma}/m_\psi^2\right)^2 = \frac{9}{8} \tag{4}$$

In other words the ratios 9:1:2:8 are the natural extension to include "hidden charm" mesons of the "classical" 9:1:2 ratios in the case of the ρ, ω, ϕ mesons. From (2) and (3) the observed value for the quantity on the left hand side of (4) is ~5, but the usual mass-breaking argument can be invoked, if one so likes, to account for the discrepancy[11].

5-The charm interpretation

5.1-The motivation for charm

I will be very concise on this topic[12] because it will be dealt in the lecture by Professor Maiani. I only recall that in order to avoid the $\Delta S = 1$ neutral currents Glashow, Iliopoulos and Maiani[13] proposed that a fourth quark should exist-compare also Ref.[14] -and that it should be appropriately coupled to the weak charged vector boson. In the following we shall call the fourth quark p' or c, indifferently.

In the limit $m_{p'} = m_p$ the coupling can be chosen in such a way that the $K_L \to \mu^+ \mu^-$ decay rate vanishes. This is achieved by writing the (charged) weak interaction as: $W_\mu^+ J_\mu$ + h.c. with:

$$J_\mu = \bar{p} \Gamma_\mu (n \cos\theta + \lambda \sin\theta) + \bar{p}' \Gamma_\mu (-n \sin\theta + \lambda \cos\theta) \quad (5)$$

The vanishing of the $K_L \to \mu^+ \mu^-$ decay rate arises (in the lowest order) from the cancellation of the two graphs:

In the limit $m_p = m_{p'}$, it can be shown[15] that the above cancellation holds to all orders. In terms of the $m_{p'} - m_p$ mass difference (we are considering here effective masses) one can thus calculate the $K_L \to \mu^+ \mu^-$ decay rate as well as, for instance, the K_L, K_S mass difference. Such estimates, particularly the latter, thus allow to deduce an order of magnitude of $m_{p'} - m_p$. Although there are ambiguities[12] in these calculations, it is nevertheless true that an estimate $m_{p'} - m_p \approx 2\text{GeV}$ had been given on this basis before the discovery of the narrow resonances.

A final remark is this: due to the Cabibbo structure of the weak current (5), it is evident that $p' \to \lambda + W^+$ is preferred by a factor $\cotg \theta_c$ over $p' \to n + W^+$; this remark has an obvious relevance in connection with the decays of the charmed hadrons.

5.2 - The new states in the charm scheme

In the charm scheme the possible binding of the fourth quark implies a variety of new states; for simplicity we confine here to the mesonic ones; an analysis of the baryonic states would be valuable but would take too much time.

We recall that the fourth quark (p' or c) is assumed to have charge $Q_{p'} = 2/3$; it has also zero isospin and zero strangeness.

The new mesons which include one c quark and/or antiquark are listed in the table II.

Table II
Charmed and Hidden - charm mesons

Composition	name	charge	I	S	Note
$c\bar{p}$	D^0	0	$\frac{1}{2}$	0	$D^0, \bar{D}^0$ should give rise to a behaviour similar to $K^0, \bar{K}^0$ but the lifetimes of "D_L" and "D_S" should not be too different.
$c\bar{n}$	D^+	1	$\frac{1}{2}$	0	
$c\bar{\lambda}$	F^+	1	0	1	
$\bar{c}p$	$\bar{D}^0$	0	$\frac{1}{2}$	0	
$\bar{c}n$	D^-	-1	$\frac{1}{2}$	0	
$\bar{c}\lambda$	F^-	-1	0	-1	
$c\bar{c}$		0	0	0	

In the charm scheme the proposition[16] is that the new mesons are identified with the "hidden charm" mesons of the last row; precisely the assumption is that:

$$\psi(3.1) \equiv c\bar{c}\,(1^3S_1) \quad \text{and} \quad \psi'(3.7) \equiv c\bar{c}\,(2^3S_1)$$

Of course mixtures with the $^3D_1$ states (with amplitudes determined by the detailed dynamics) are possible in principle and should be taken into account, especially for the 3.7. The main point, however, at this stage, is that both states are "hidden" charm states and that the $\psi'(3.7)$ is interpreted as a radial excitation; this of course raises a first problem, namely that of finding the many radial excitations corresponding to the normal mesons (E, ρ' ...?) or to give reasons for their non detection.

5.3 – Hidden charm and the Okubo-Zweig[17] rule

Having identified, in the charm scheme, the new narrow resonances with the $c\bar{c}$ states, the main question—why they are so narrow— has still to be answered; the conventional answer, which will now be presented, is, however, in my opinion, not convincing or at least lacking in some respect. In other words I feel that in the hidden charm picture a really convincing understanding of the narrowness of the new states has still to be found.

The conventional explanation is based on the operation of the so called Okubo-Zweig rule. But two points are far from clear: 1) which is the explanation of the O.Z. rule?; Does the O.Z. rule provide an inhibition mechanism powerful enough to explain the huge depression in the observed rates? We shall try to answer these questions in the subsequent sections.

5.4 – The empirical evidence for the Okubo-Zweig rule

The Okubo-Zweig rule can be stated as follows: processes which must take place through non connected quark diagrams are inhibited. The text book illustration of this rule refers to the ϕ meson (compare the figure below):

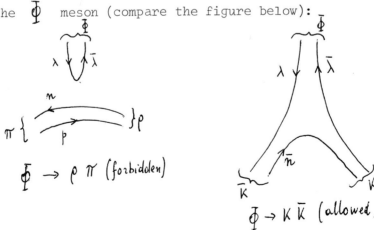

Another "classical" case is illustrated on comparing the
$\phi \to \eta \gamma$ decay with the $\phi \to \pi^0 \gamma$ decay; the first is O.Z.
allowed because the η contains a large $\lambda \bar{\lambda}$ component;
the second is O.Z. forbidden because the π^0 has no $\lambda \bar{\lambda}$
component.

Both from the above decays and from the reactions[18]:

1) $\pi^- P \to \phi N$ 2) $\pi^+ P \to \phi \Delta^{++}$ 3) $\pi^+ P \to \phi P \pi^+$
4) $PP \to \phi PP$ and others one gets for the typical inhibition in the <u>rate</u> produced by the O.Z. rule in the processes a factor from 50 to 100. The decay of the f'(1520)— which in the quark model description should be a rather pure 2^+ $\lambda \bar{\lambda}$ structure—does not contradict, but does not improve, this factor 100.

For the "hidden charm" mesons the assumption[16] is now that the same type of inhibition mechanism which is at work in the case of the ϕ also operates in the decay of the $c\bar{c}$ structures such as the ψ or the ψ'; the argument being that in the same way as the ϕ is an almost pure $\lambda \bar{\lambda}$ structure, the ψ and ψ' are almost pure $c\bar{c}$ structures; and that, therefore, if the low lying states (into which ψ and ψ' can decay) also contain very small $c\bar{c}$ amplitudes, the decay of ψ and ψ' is inhibited by the Zweig rule.

Before entering in a quantitative discussion of this, I like to mention, however, an experiment which is relevant to the O.Z. rule; this experiment, if complemented by others of a similar nature, appears to throw some doubts on the formulation of the O.Z. rule given at the beginning of this section.

Compare the process

$$\pi^+ P \to K^+ \phi \Lambda \qquad (A)$$

with the processes:

$$\pi^+ P \to \phi \Delta^{++} \quad \text{or} \quad \pi^+ P \to \phi P \pi^+ \qquad (B)$$

(A) should be (according to the previous formulation of the O.Z. rule) O.Z. allowed, while (B) should be O.Z. forbidden as the diagrams below indicate. At energies such that the differences in the thresholds can be neglected (A) should be highly favoured over (B) because the diagram A is connected, while B is not.

The experiments to which I referred above[19] (reported at the Palermo Conference) do not support this prediction. In these experiments use has been made of an incident proton beam rather than of an incident pion (above we used the pion for illustration only because it is somewhat simpler to draw the appropriate diagrams for the pion than for the proton) but this difference does not affect the argument.

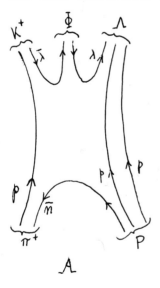
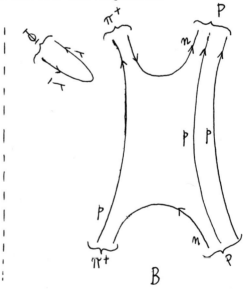

The incident proton beam has an energy of 24 GeV. The results are reported in the table III below; in the first row the table gives the cross section (in μb) for the reaction PP $\rightarrow$ PPV° + n pairs $(\pi^+\pi^-)$; in the second row the cross section for the reaction PP $\rightarrow$ V° + anything. Here V° is either ρ^0 or ω or ϕ.

Table III

The production of ρ^0, ω, ϕ in PP collisions at 24 GeV/c

	ρ_0	ω	ϕ
$\sigma(PP \rightarrow PP\, n\, (\pi^+\pi^-)V°)$ (μb)	355 ± 40	355 ± 50	6.6 ± 2.4
$\sigma(PP \rightarrow V°+ \text{anything})$ (μb)	3490 ± 420		158 ± 35

It appears from the table that, indeed, the production of the ϕ is suppressed with respect to that of ρ^0 or ω both in the process of the first row and in that of the second; but the point of interest is now different. If we compare the values for the ratio ϕ/ρ^0 in the first and in the second row of the table we should expect (if the Okubo-Zweig rule operates) a considerable larger number in the inclusive case—second row— than in the exclusive one —first row— because the inclusive case contains also diagrams (analogous to that of fig.4) in which the ϕ is produced, e.g., in association with a K and a Λ (O.Z. allowed), while the exclusive case of the first row does not. The table shows that the numbers are, instead, 1/53 and 1/22. There is no substantial factor between the two cases and the small factor (≈ 2) can be accounted probably by the different method used for recognizing the ϕ's in the two cases. The ϕ's of the first row are detected through the (typical) mode K_L, K_S while the $\bar{\phi}$'s of the second row are detected through the (less specific) mode K^+K^-.

Clearly other carefully thought experiments of the same kind in $P\bar{P}$, π^+P, π^-P etc. are necessary before a firm conclusion can be reached. Should they confirm the above result, the O.Z. would become even more misterious than it is; it might remain a limited dynamical accident typical of the ϕ (and f').

In the following, however, we shall <u>assume that the O.Z. rule holds; moreover we shall assume that if-to exemplify- the ϕ meson were a pure $\lambda\bar{\lambda}$ state it would not decay at all into $\rho\pi$</u>; and that likewise, if the ψ were a pure $c\bar{c}$ state it would not decay at all into hadrons containing <u>only</u> non charmed quarks; both these assumptions are, certainly, not correct because $\phi \to \rho\pi$ can take place through:

$$\phi \to K\bar{K} \to \rho\pi$$

where the two steps are both O.Z. allowed; and a similar process holds for the virtual process:

$$\psi \to D\bar{D} \to \text{normal hadrons}$$

even if the $D\bar{D}$ threshold is higher than the mass of the ψ[20].

Still we maintain the (underlined) assumption stated above because we intend to take at this point the most favourable attitude with respect to the effectiveness of the O.Z. rule. In spite of this we shall arrive to the conclusion already stated in the introduction: namely that it is not easy to account, using the O.Z. rule alone, for the extreme narrowness of the ψ and ψ'.

5.5 - The mixing and mass problem (phenomenological)

Following the approach indicated at the end of the previous section the $\phi \to \rho\pi$ decay is attributed to the fact that the ϕ is a mixture:

$$\phi = \lambda\bar{\lambda} + \alpha \frac{1}{\sqrt{2}}\left(p\bar{p} + n\bar{n}\right)$$

To get the correct rate of decay it is necessary that $\alpha \approx 3.8 \cdot 10^{-2}$ corresponding to a linear mixing angle between ω_0 and ω_ρ of 37.5° (the "ideal" mixing angle - ϕ pure $\lambda\bar{\lambda}$ state - is 35,3°). Assuming that a "normal" decay width for the $\psi(3.1)$ into hadrons should be of the order 200 MeV, the inhibition which is at work in the case of the $\psi(3.1)$ must be larger by a factor from 50 to 100 than the inhibition for the ϕ.

This implies: a) that the amplitudes of the normal states in the ψ wave function must be smaller by an order of magnitude than those of the non $\lambda\bar{\lambda}$ states in the ϕ. b) that the amplitudes of $c\bar{c}$ in the lower states (into which the ψ can decay) should also be very small. As we shall see the first requirement is fulfilled; the second is, most probably, not.

To show this we shall use the simple formalism of the mixing problem, familiar from the case of the ω, ϕ mixing. It is however appropriate to recall that the simple mixing formalism is in reality, as explained in more detail in ref.[21] an approximation-indeed a drastic approximation- to a complicated situation; although a discussion of this point here would be out of place, we have mentioned it because the limitations of the standard treatment of the mixing problem should, at least, be kept in mind.

Of course the mixing problem which we have to consider

is now a three channel mixing problem, that of determining the mixing matrix connecting the three channels (with I=0) $\frac{1}{\sqrt{2}}(p\bar{p}+n\bar{n})$, $\lambda\bar{\lambda}$, and $c\bar{c}$. In the following sections we shall give first an order of magnitude treatment and next a more complete one of this problem.

5.51 - Order of magnitude treatment of the mixing problem

To get a feeling for the orders of magnitude involved we assume[22], first, that the potential between quarks and/or antiquarks is the sum of two parts A and B represented, schematically, in the figure below:

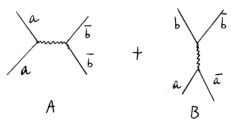

The part A of the potential is a "scattering potential" having matrix elements which connect the state a, $\bar{b}$ to the state a,$\bar{b}$; here a and $\bar{b}$ are a quark and an antiquark of the same or of different type; the part B of the potential is an "annihilation potential" which has only matrix elemnts connecting states $a\bar{a}$ to $b\bar{b}$ (again a and b can be the same or different). It is assumed that all the matrix elemnts of the part A of the potential are equal-we call A their value- and a similar assumption is made for the part B. With these simplifying assumptions the 3x3 mixing matrix is:

	$\frac{1}{\sqrt{2}}(p\bar{p}+n\bar{n})$	$\lambda\bar{\lambda}$	$c\bar{c}$
$\frac{1}{\sqrt{2}}(p\bar{p}+n\bar{n})$	$m_p+m_n + A + 2B$	$\sqrt{2}\,B$	$\sqrt{2}\,B$
$\lambda\bar{\lambda}$	$\sqrt{2}\,B$	$2m_\lambda + A + B$	B
$c\bar{c}$	$\sqrt{2}\,B$	B	$2m_c + A + B$

THEORETICAL MODELS PROPOSED FOR NEW PARTICLES

where $m_p \simeq m_n$, m_λ and m_c are the effective masses of the p, n, λ and c quark. In this approximation it is easy to see that:

$$M_{K^*} = m_\lambda + m_p + A$$
$$M_\rho = 2m_p + A$$
$$M_\Phi = \text{(to zero order)} = 2m_\lambda + A + B$$

and, therefore:

$$\Phi + \rho - 2K^* = B_V = \frac{\omega - \rho}{2} \simeq 6 \text{ MeV} \tag{6}$$

where now Φ stays for M_Φ, ρ stays for M_ρ etc.; and the suffix V appended to B reminds that we are referring to the vector mesons. Of course the value B_V=6MeV should not be taken too seriously; it indicates, however, that the non diagonal matrix elements in the mixing matrix are small.

Using the first order perturbation formula:

$$\psi_i^{(1)} = \psi_i^{(0)} - \sum_K \frac{M_{iK}}{E_i^{(0)} - E_K^{(0)}} \psi_K^{(0)}$$

it follows from the previous value of B_V:

$$\psi = c\bar{c} - 3.5 \cdot 10^{-3} \frac{1}{\sqrt{2}} (p\bar{p} + n\bar{n}) - 3 \cdot 10^{-3} \lambda\bar{\lambda}$$

where ψ is the wave function of the $\psi(3.1)$.

It is apparent that $\psi(3.1)$ is, <u>to a few parts per thousand in amplitude</u>, a pure $c\bar{c}$ state. This is a good starting point for using the O.Z. rule. Note incidentally that the same calculation gives:

$$\Phi = \lambda\bar{\lambda} - 3.5 \cdot 10^{-2} \frac{1}{\sqrt{2}} (p\bar{p} + n\bar{n})$$

The situation for the <u>pseudoscalar mesons</u> is, however, entirely different. Even if, as shown in ref.[23], the 2-parameter potential used cannot describe the mass situation of the pseudoscalar mesons and, moreover, a first order perturbative solution is now no more legitimate, such a solution is still useful to show that the order of magnitude of B in the pseudoscalar case is much larger than in the vector case. One finds, instead of (6):

$$B_{ps} \approx \eta' + \pi - 2K \approx 160 \text{ MeV} \tag{7}$$

Two consequences arise from this large value of B_{ps}:
1) The η' meson contains with a large (15 to 20%) amplitude a $c\bar{c}$ component.
2) The pseudoscalar analogous of the ψ(3.1) (let us call it η_c) contains with large values of the amplitudes (again 15 to 20%) both a $\frac{1}{\sqrt{2}}(p\bar{p} + n\bar{n})$ and a $\lambda\bar{\lambda}$ component.

As we shall see (Sect.5.9, 5.10) these large amplitudes (in particular that of the η'), if correct, prevent an explanation of the narrowness of ψ and ψ' based exclusively on the O.Z. rule.

5.52-A more exact solution of the mixing problem

Although we have not the time for discussing this problem in detail, a more accurate study of the mixing matrix can be and has been performed in several ways. Okubo et al.[24] as well as Gourdin[25] have taken a "group theoretical" approach introducing violations of SU_4 and determining its effects on the masses and on the admixtures.

The Helsinki group[26], on the other hand, has used a quark model approach, entirely similar to that described, e.g., in ref.[27] for the analogous problem in the absence of the

THEORETICAL MODELS PROPOSED FOR NEW PARTICLES

$c\bar{c}$ states. The problem, as treated by the Helsinki group, is a five parameter problem, in the sense that the mixing matrix contains five parameters after some appropriate simplifying assumptions are introduced. For the vector case these five parameters are determined using the knowledge of the masses of the K*, ρ, ω, ϕ, ψ(3.1); and, doing this, one again finds, as in the previous section, that the admixtures of $c\bar{c}$ states with normal states and viceversa are of the order of a few parts per thousand.

In the pseudoscalar case no data on the mass of the η_c are available; the fifth necessary input (in addition to the masses of the K, π, η and η' mesons) is then introduced—in ref.[26]—by assuming that the mass of the D meson is in between 3.8/2 GeV and 4.2/2 GeV. This supposition is quite natural in the charm model, because it amounts to the assumption that the rise in R starting at $\sim$ 3.8 GeV and ending near to 4.2 GeV is attributed to the production of a $D\bar{D}$ pair (and, more generally to associated production of a pair of charmed mesons); this assumption is we repeat, part of the charm picture and therefore quite acceptable in terms of logical selfconsistency of the picture. Changing the mass of the D meson inside the interval mentioned above does not introduce much difference in the admixtures of interest. The table IV below refers to a value M_D = 2.03 GeV and the numbers represent the mixing amplitudes.

Table IV
Mixing amplitudes in the pseudoscalar case
(linear mass formulas)

	$\frac{1}{\sqrt{2}}(p\bar{p} + n\bar{n})$	$\lambda\bar{\lambda}$	$c\bar{c}$
η	0.849	− 0.526	0.057 (*)
η'	0.494	0.827	0.268 (*)
η_c	−0.188 (*)	− 0.199 (*)	0.962

The attention should be focussed, in particular, to the starred entries which represent the admixtures of $c\bar{c}$ in the normal mesons and of normal states in η_c. We see that, for instance, η' contains a 25% amplitude of $c\bar{c}$ state and η_c contains with amplitude $\approx$ 20% states different from $c\bar{c}$.

Note that qualitatively similar conclusions for the magnitude of these admixtures are also reached by the calculations of ref.[24] and [25]. Only if the ninth pseudoscalar meson is to be identified with the E meson rather than with the η' the admixtures become much smaller. However this identification, which would imply an assignment 2^- to the η', would create serious problems to the quark model; because all this discussion has been made in the frame of the quark model, we shall continue to assume here that the η' and not the E is the ninth pseudoscalar meson.

It has already been anticipated and it will be shown in the following sections that the large admixture of $c\bar{c}$ in the η' creates serious problems to an explanation of the narrowness of the ψ and ψ' based exclusively or mainly on the operation of the Okubo-Zweig rule.

We now proceed to treat in the "hidden charm" picture three problems:
1) The expected position and the decay of the η_c (Sect.5.6-5.7)
2) The expected position of the P states (Sect.5.8)
3) The decays (γ and "strong") of the ψ and ψ' (Sect. 5.9, 5.10, 5.11)

<u>5.6-The position of the η_c with respect to the $\psi(3.1)$</u>

Of course a state $\eta_c = c\bar{c}$ ($^1S_0$) -a pseudoscalar hidden charm meson- and its radially excited states are expected in this picture. Here we shall confine to the lowest state $1\,^1S_0$.

We intend to show[28] that, if η_c is below the 3.1 state, it should not be too much below; indeed if $M_{3.1} - M_{\eta_c}$ is too large the decay:

$$\psi(3.1) \to \eta_c + \gamma \qquad (8)$$

would have such a rate to produce-alone- a too large width for the $\psi(3.1)$.

Note in fact:
1) that the O.Z. rule has no part in this decay.
2) that the decay (8), being a pure M1 transition is independent of the specific and uncertain-to some extent- assumptions for the radius of the wave function; it only depends on the magnetic moment of the charmed quark.
3) The calculation, once the magnetic moment of the charmed quark is known, is entirely similar to the old one (Becchi and Morpurgo[29], 1965) for the $\omega \to \pi\gamma$ rate.

Considering the result of that calculation:

$$\Gamma(\omega \to \pi\gamma) = \frac{\alpha}{3}(2\mu_p)^2 k_\gamma^3 \qquad (9)$$

where μ_p is the magnetic moment of the proton ($\mu_p = 2.79 \frac{e\hbar}{2M_p c}$) only one modification is necessary: the factor $2\mu_p$ in the equation (9) has to be replaced by $(4/3) \cdot 2\mu_{p'}$, where the factor 4/3 comes from a Clebsch-Gordan coefficient and:

$$\mu_{p'} = x \mu_p \qquad (10)$$

Here the numerical factor x takes into account the fact that the magnetic moment of the charmed quark is presumably different—and smaller—than that of the p and n quarks. The value of x in (10) will be considered below. Instead of equation (9) we therefore get for the decay (8) the rate:

$$\Gamma(3.1 \to \eta_c \gamma) = \frac{\alpha}{3}\left(\frac{4}{3} \cdot 2\mu_{p'}\right)^2 k_\gamma^3 =$$

$$= \frac{16}{9} \Gamma(\omega \to \pi\gamma) \frac{(M_{3.1} - M_{\eta_c})^3}{(380 \text{ MeV})^3} x^2 \qquad (11)$$

It is difficult to form an idea of the value of x; if the ratio between the magnetic moments of the normal and charmed quarks is the same ratio as their "effective" masses—an entirely arbitrary assumption—then x=1/6; we must then have:

$$|M_{3.1} - M_{\eta_c}| < 200 \text{ MeV}$$

to get, correspondingly:

$$\Gamma(3.1 \to \eta_c \gamma) < 50 \text{ KeV}$$

THEORETICAL MODELS PROPOSED FOR NEW PARTICLES 825

If the observed branching ratio of the $\psi(3.1)$ into $\eta_c \gamma$ should turn out to be say, less than 1%, then, with x=1/6, the mass difference between 3.1 and η_c should not be greater than 20 MeV. At the moment little can be said because there is no experimental indication for the η_c [58].

5.7 — The decay of η_c.

In spite of the lack of evidence for the η_c [58] (or, if one prefers, as an attempt to understand this lack of evidence) one can try to give an estimate for the decay rate of the η_c. If the η_c contains, as discussed previously in connection with the mixing formulas, a 20% amplitude of normal hadrons, one should have:

$$\Gamma_{\eta_c}(\text{total}) \approx (20\%)^2 \, \Gamma(\text{typical strong}) \approx 4 \div 8 \text{ MeV}$$

if the decay is only inhibited by the Okubo-Zweig rule.

On estimating:

$$\Gamma(\eta_c \to 2\gamma) \approx \frac{m_{\eta_c}^3}{m_\eta^3} \Gamma_\eta \approx 10^5 \text{ eV}$$

we would then conclude that:

$$\Gamma_{\eta_c}(\gamma\gamma) / \Gamma_{\eta_c}(\text{total}) \approx \frac{1}{40} \div \frac{1}{80}$$

In particular:

$$\psi(3.1) \to \eta_c + \gamma \to 2\gamma + \gamma$$

would be expected to have a very small branching ratio. Similarly the Primakoff photoproduction of η_c would not be easy to detect because of the small branching ratio of the 2γ decay channel of the η_c.

If there are additional inhibitions in the strong decays of the η_c with respect to those provided by the O.Z. rule, then the Primakoff detection of the η_c (by the observation of its 2γ decay) becomes more favourable; for this reason we record the total Primakoff production cross section of the η_c on Pb, as given in ref.[30].

$$\sigma_{Pb}(\gamma \to \eta_c)_{coherent} \approx 380 \text{ nb} \quad E_\gamma = 100 \text{ GeV}$$
$$\approx 1.6 \,\mu b \quad E_\gamma = 200 \text{ GeV}$$

5.8 - The P states

If $\psi(3.1)$ and $\psi'(3.7)$ are interpreted as $1\,^3S_1$, and, respectively, $2\,^3S_1$ two body bound states, then, in general, additional states are expected in any bound state model. Except for the $^1S_0$ states, just discussed, the lowest among these new states are presumably P states; such states cannot, due to their quantum numebrs, appear as resonances in e^+e^- collisions; if one of these P states is below the 3.7 state, it can, however be feeded by a γ decay:

$$\psi'(3.7) \xrightarrow{E1} P + \gamma \tag{12}$$

The rate of this decay depends on several factors: the radius of the initial and final states (it is an electric dipole transition) and the difference in energy between the 3.7 and the P state.

As a matter of fact we are once more confronted with a phenomenon already met in the course of this survey; namely we shall see that, for a reasonable position of the P states, the rate of the process (12) is expected to be too large to be compatible both with the measured total width of the 3.7 state and with the experimental upper limit on

THEORETICAL MODELS PROPOSED FOR NEW PARTICLES

the intensity of monochromatic γ rays from that state.

How can the presumed position of the P states be determined? There are, essentially, two requirements which have to be fulfilled: a) Using the formula:

$$\Gamma_{V \to e^+e^-} = \text{const.} \frac{|f_V(0)|^2}{m_V^2} \tag{13}$$

where m_V is the mass of the decaying Vector meson and $f_V(0)$ its wave function at the origin, the ratio:

$$\chi \equiv \frac{|f_{1S}(0)|^2}{|f_{2S}(0)|^2} = \left(\frac{3.1}{3.7}\right)^2 \frac{\Gamma_{3.1}(e^+e^-)}{\Gamma_{3.7}(e^+e^-)} \cong 1.53 \tag{14}$$

has, experimentally, the value indicated (~1.53) on the right hand side of the equation (14). b) The energy difference between the 2S and 1S states is known (~ 600 MeV).

Now the requirement a) with the value 1.53 implies that the potential which binds the two objects (say c and $\bar{c}$ if they are charmed quarks, but the argument is general) must be rather different from a Coulomb potential which would give χ = 8. Even suppressing the factor $(3.1)^2/(3.7)^2$ in (14) which is due to the extremely uncertain mass factors m_V^2 in (13), the number 1.53 only increases to 2.17, still far from 8.

Note that a Coulomb potential would give 2S and 2P degenerate and thus create no problems for the γ decay rate 3.7 → P + γ. Take, instead an harmonic oscillator potential, an extreme opposite case. This produces χ = 0.66 and P midway between 3.1 and 3.7; in this case, due to the large energy difference between the 3.7 and the P state, the corresponding rate of transition is expected to be large.

The values of χ and of the position of the 2P level are listed below for a variety of potentials (the use of a Schrödinger non relativistic equation is implied).

Potential	χ	$E_{2S} - E_P$
Coulomb	8	0 (degenerate)
Hülthen	>8	$E_P > E_{2S}$
Harm. Oscillator	0.66	$E_{2S} - E_P = E_P - E_{1S}$
Linear	1	E_P somewhat higher than harm. oscillator

Eichten et al.[31] have chosen a combination of Coulomb+linear potential:

$$V(r) = -\frac{\alpha_s}{r}\left(1 - \frac{r^2}{a^2}\right)$$

They have determined the parameters α_s, a and m_c (the "effective" mass of the charmed quark) using as inputs the experimental values of $E_{2S} - E_{1S}$, of $\Gamma(1\,^3S_1)$ (they use 5.5 KeV) and selecting m_c (1.6 GeV) so as to be in agreement with the very "naive" condition $2m_c + 0.75 \cong 2M_D$ where M_D is assumed to be $\approx$ 2 GeV[32].

They get in this way $\chi = 1.15$, $r_{Av} \cong 0.7$ Fermi and the situation of the levels reproduced below. (In the figure the splitting of the P levels is indicated, but, to get it the introduction of an appropriate spin-orbit potential is necessary). With the above choice of a, α_s and m_c the center of the P levels is calculated (in ref.[31]) to be 230MeV below the 3.7 level. Because everything is known, and in particular the wave functions, the γ widths for the transitions from the 3.7 state to the various P levels as well as

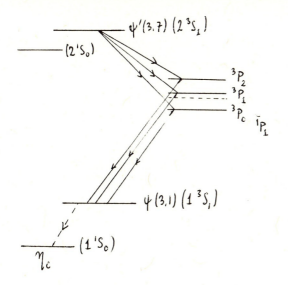

The theoretical situation of the S and P levels[31]; the ^1P state cannot, due to C conservation, be fed from the 3.7 state by a γ ray.

to the η_c can be easily evaluated. Calling (as in ref.[31]) I_1 and I_4 two radial integrals, the following expressions and values for the γ widths are obtained (table V)

Table V

Calculated widths of the $3.7 \to P + \gamma$ and of the $3.7 \to \eta_c + \gamma$ transitions.

Transition	Γ_γ	(in KeV)	Remarks
$2^3S_1 \to {}^3P_2$	$5 I_1 \propto k^3$	120	(*) The orthogonality of the 2S and 1S radial wave functions is responsible of the small value of $\Gamma_\gamma(\psi' \to \eta_c)$.
$\to {}^3P_1$	$3 I_1 \propto k^3$	70	
$\to {}^3P_0$	$1 I_1 \propto k^3$	25	
$2^3S_1 \to 1{}^1S_0$	$I_4 \propto k^7$	~1 (*)	

At present the data give:

$$\psi'(3.7) \to \psi(3.1) + 2\gamma < 6.6\% = 14 \text{ KeV}$$

Note also that the sum of the Γ_γ to the three P levels from the $\psi'(3.7)$ entirely exhausts the ψ' width; again there seems to be in the calculated figures a systematic overestimate of the rates. If, in this case, this is serious or not is difficult to say at present[58].

5.9 - The decay of $\psi(3.1)$ into γ + normal hadrons

We confine here to consider the decay:

$$\psi(3.1) \to \eta' + \gamma \tag{15}$$

because it is important in showing that the O.Z. rule alone is not sufficient to reduce the widths of some decays to acceptable values; in fact due to the 20% impurity in $c\bar{c}$ of the η', the width of the process (15) is expected to be very large.
It is:

$$\Gamma(\psi(3.1) \to \eta'\gamma) = \frac{16}{9} \Gamma(\omega \to \pi^\circ \gamma) \frac{k^3 x^2 y^2}{(380 \text{ MeV})^3}$$

where x is the reduction factor of the magnetic moment introduced previously and y=20% is the $c\bar{c}$ amplitude impurity in the η'.

With k=1.4 GeV and taking x=1/6 we get:

$$\Gamma \simeq 0.10 \text{ MeV}$$

Much too large ! To "repair" this one can ask for the intervention of a form factor such as:

$$F^2 = \exp - k^2 \langle r^2 \rangle / 3$$

THEORETICAL MODELS PROPOSED FOR NEW PARTICLES

For $\langle z^2 \rangle^{\frac{1}{2}} \sim 1/2m_\pi$ it is $F^2 \sim 2 \cdot 10^{-4}$. This intervention of F^2 is certainly a possibility and, perhaps, is in fact true. But, if so, it is this factor (more than the O.Z. rule) to play a dominant role in reducing the excessive "exclusive" widths. Note, incidentally, that F^2 has here this large reducing effect because k is large, but has, of course small effects for channels with k small.

5.10 – The strong decays of the $\psi(3.1)$

Once more we exemplify by considering:

$$\psi(3.1) \rightarrow \eta' + \omega \tag{16}$$

Here η' is very impure (20% in amplitude), whereas ω is rather pure (say 2% in amplitude). According the O.Z. rule the decay takes place through these impurities (compare the diagram below):

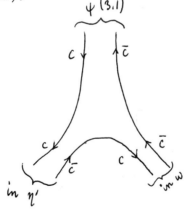

Except for a small correction in the phase space, we can estimate the order of magnitude of the rate of this process from the previous one $\psi \rightarrow \eta' + \gamma$ by the following replacements:
1 instead of $\alpha = 1/137$ (there are no γ's)
1 instead of 1/36 (no reduction factor for the magnetic moment)

$4 \cdot 10^{-4}$ instead of 1 (this comes from the square of the $c\bar{c}$ impurity in the ω).

Altogether the previous amplitude for the $\psi(3.1) \to \eta'\gamma$ (Sect.5.9) is multiplied by ≈ 2. It is, again, large: $\Gamma \approx 0.2$ MeV; the factor F^2 is still needed.

5.11 - The "strong" decays of $\psi'(3.7)$

Two important features of the experimental situation here[33] are the following: we have reasonably good data on these reactions:

a) $\psi'(3.7) \to \psi(3.1) + \pi^+ + \pi^-$ Br.ratio $32 \pm 4\%$
b) $\psi'(3.7) \to \psi(3.1) + \eta$ Br.ratio $4 \pm 2\%$

The spectrum of $M_{\pi\pi}$ in the reaction a) is also known; schematically it has a shape of the form indicated below (peaked towards the high end), very different from the pure phase space spectrum:

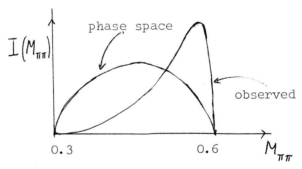

One of the most debated questions has been, for some time, if the reaction a) is or is not inhibited (it should be, according to the O.Z. rule); also there has been some discussion on how to explain the $M_{\pi\pi}$ distribution.

We follow, in discussing these questions, the simple calculations of ref.[34]. Writing:

$$f_{\psi'\psi\pi^+\pi^-}$$

THEORETICAL MODELS PROPOSED FOR NEW PARTICLES

and
$$f_{\rho'} \, \rho' \rho \, \pi^+ \pi^-$$

for the effective Lagrangians giving rise respectively to the process a) and to the decay of the ρ' into 4 pions. Here $f_{\psi'}$ and $f_{\rho'}$ are adimensional constants. Using $\Gamma(\psi' \to \psi \pi^+ \pi^-) \approx 100$ KeV., and $\Gamma(\rho' \to \rho \pi^+ \pi^-) = 150$ MeV we get:

$$f_{\rho'} = 87 \qquad f_{\psi'} = 9.2$$

This can be tentatively interpreted by stating that the rate of a) is inhibited with respect to a strong decay by a factor 10^2.

To the same general conclusion we arrive if, to explain the shape indicated in the previous diagram, we imagine the decay process to take place through an intermediate state, in which a broad spin zero meson ε (decaying into two pions) is emitted. Assuming $\Gamma(\varepsilon \to 2\pi) = 0.4$ GeV, $M_\varepsilon = 750$ MeV and the interactions:

$$g_{\psi'\psi\varepsilon} \, \psi' \psi \varepsilon + g_{\pi\pi\varepsilon} \, \pi \pi \varepsilon$$

or

$$g_{\rho'\rho\varepsilon} \, \rho' \rho \varepsilon + g_{\pi\pi\varepsilon} \, \pi \pi \varepsilon$$

one gets for the (dimensional) couplings $g_{\psi'\psi\varepsilon}$ and $g_{\pi\pi\varepsilon}$

$$\left(g_{\psi'\psi\varepsilon} / g_{\rho'\rho\varepsilon} \right)^2 \simeq \frac{1}{50}$$

We end this section with one remark concerning the process b) $\psi' \to \psi \eta$. If SU_3 were good and in the ideal case in which $\eta \equiv \eta_8$ the decay:

$$\underset{SU_3 \text{ singlet}}{\psi'(3.7)} \to \underset{\text{singlet}}{\psi(3.1)} + \underset{\text{octet}}{\eta}$$

should be forbidden. It is therefore somewhat puzzling that the decay $\psi' \to \psi \eta$ takes place at an observable rate ($4 \pm 2\%$) in spite of this inhibition, that due to the O.Z. rule, and the extremely small phase space available. This circumstance should be understood.

5.12 - A few scattered comments on the detection of charmed mesons

Because this subject will be treated subsequently and repeatedly in this School I will confine, in these introductory lectures, to very few and superficial comments[35].

1. Let us begin with the production of charmed mesons (as distinct from hidden charm mesons) at SPEAR.

Clearly the reaction:

$$e^+ + e^- \to D + \bar{D} \qquad (17)$$

becomes possible above $2 M_D$. In the charm model the increase in the quantity R (equation (1)) in the region 3.8 to 4.3 GeV. (c.of m.) is attributed to this real production of D mesons (possibly in association with additional pions). One should then observe the non leptonic decays of the D's (which will be discussed by professor Cabibbo) and one should also observe the leptonic decays. Consider, for instance, a D^+ decaying into $e^+ \nu$ and a D^- decaying into $\mu^- \bar{\nu}$. By looking to the $e^+ \mu^-$ or $\mu^+ e^-$ pairs one could thus get information on the associated production of $D^+ D^-$. According to preliminary information[36] from SPEAR one has noted indeed an increase in the number of $e^+ \mu^-$ and $e^- \mu^+$ pairs when the total energy increases in the region of energies stated above. If these observations, and in particular, this sort of threshold, is going to be confirmed (and if it will be possible to be sure that these leptons do not come from the decays of normal mesons) this will constitute a point in favour of charm.

However it should be kept in mind that similar effects, qualitatively, can arise also in other ways; for instance from the associated production of heavy leptons, or of Weak Vector bosons.

In these conditions the first question to be asked is, I feel, the following: assuming that the existence of this threshold around 4 GeV. at SPEAR in the production of $e^{\pm}\mu^{\mp}$ pairs is confirmed, is it possible to decide (by the study, for instance of the energy spectrum) if these $e^{\pm}\mu^{\mp}$ pairs come from the decays of fermions or of bosons?[58].

2. Another class of events typical of the existence of charm consists, due to the properties of the weak current (5), of the events with apparent $\Delta S = -\Delta Q$.

Consider, e.g. a ν - neutron collision; the process illustrated below can then take place.

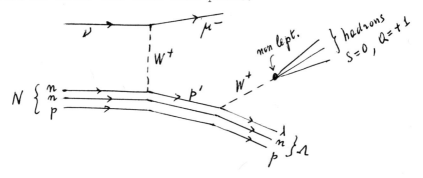

The figure is self explanatory. Initially we have a neutron (Q=0, S=0) and finally a Λ (S=-1) + non strange hadrons with Q=+1.

One event of this kind has been observed (Ref.[37]) but in spite of further search and a large increase in statistics no other events were found[38].

3. $\mu^+\mu^-$ pairs with "flat" energy spectrum. If the decaying W^+ in the figure above goes into $\mu^+\nu$ rather than into hadrons, we have $\mu^+\mu^-$ production by neutrinos; the invariant mass of the $\mu^+\mu^-$ pair is clearly not peaked at any particular value, that is has a rather flat distribution. Events of this type have been observed[39] at NAL; certainly charm is a possible interpretation of these events; but this interpretation is probably not unique; and an additional difficulty is created by the detection of few events where both muons have the same charge[38].

4. Events in emulsion. Here we give only two references; the interested reader should personally look at the original papers to form an opinion. The first reference is to the Niu event[40], detected in very high energy cosmic rays (incidentally much theoretical work in Japan on SU_4 was stimulated by this event). The second reference is to an event -due to a group from the University of Buffalo- produced by 300 GeV protons at NAL[41].

Note that if the lifetime of the D meson is of the order 10^{-13} sec. (compare the Appendix B) nuclear emulsions have many good features for the detection of these particles.

The reason is that the spatial resolving power of nuclear emulsions (few μ) is much better than that of the bubble chamber (0.5 mm). Independently of any specific theory systematic work with nuclear emulsions in the 100 GeV region can be rewarding.

5. We end this section by noting two points which are often mentioned as indicating qualitative evidence against the production of charmed mesons:

a) there is no substantial increase in the K production at SPEAR in the region above the presumed $D\bar{D}$ threshold. Because of the structure (5) of the weak current the D meson should give rise to a rather abundant production of K mesons in its decay[42].

b) Fixing the attention on a particular decay channel of the D (such as, for instance K𝛾) a peak in the mass distribution in that channel has been searched but not observed. The same work has been done for a variety of different channels. Of course to this objection the charm lovers may answer that the branching ratio for the decay of the D meson into any particular mode can be rather small.

We conclude this short survey of the charm interpretation of the new particles by the following trivial remark; the only decisive test of the model will consist in the detection of real charmed mesons (or baryons) having the properties which these particles should have. Even if(as discussed above) this is difficult, it is likely that an experimental answer to this question may not take too much time. If the charm interpretation will thus be confirmed, the difficulties in explaining the narrowness of the hidden charm- discussed here in some detail- will have to be reexamined, and will constitute an interesting problem.

6-The colour scheme(s). General remarks

So far we have discussed the new resonances as possible manifestations of the charm degree of freedom. Another popular interpretation (judging from the number of papers[43])consists in assuming that they are instead manifestations of the colour degrees of freedom. It should be noted that the lovers of charm do not necessarily dislike colour; nor, perhaps, viceversa. However for the lovers of charm the colour is frozen up to a high c.of m. energy and for the lovers of colour the reverse is true.

I am going to list below the arguments which are given for the introduction of colour[44]. I will then add a few comments and next proceed to exemplify one or two cases of colour models for the new resonances.

6.1 - The (standard) motivations for colour

The list of these motivations follows:

1) Quarks should have integer charge but behave with fractional charge in "low energy" phenomena.

2) One can have nodeless wave functions for baryons compatible with Fermi statistics.

3) $\pi^0 \to \gamma\gamma$: Adler anomaly calculations. The relevant formulas are:

$$\tau^{-1}_{\pi^0} = \frac{\mu^3}{64\pi} F^2 \quad \text{with} \quad F(0) = -\frac{\alpha}{2\pi} \frac{2\sqrt{2}\,\mu^2}{f_\pi} S$$

The experiment gives: $S \simeq \frac{1}{2}$; one triplet of fractionally charged quarks gives $S = \frac{1}{6}$; with three triplets (either of Han-Nambu[45] type or fractionally charged) we get $S = \frac{1}{2}$.

4) The asymptotic behaviour of the ratio:

$$R_\infty = \frac{e^+ e^- \to \text{hadrons}}{e^+ e^- \to \mu^+ \mu^-} = \sum_i Q_i^2 \text{(for fermions)} \tag{18}$$

is difficult to achieve (if the value 5.5 is taken as the asymptotic value). One triplet of fractionally charged quarks gives 2/3; to get 5.5 is a long way; colour helps; three triplets of the Han-Nambu type give 4.

5) Saturation

6) Confinement (?)

Before commenting briefly on the above motivations, I would like to make the following general remark: also if we have only three (or four) quarks it is not obvious that they are particles in the usual sense of the word (with asymptotic fields); it is quite possible that quarks (also if only three) should be conceived-in a way as yet obscure-as isomorphic to some collective degrees of freedom of hadronic matter (as we did remark long time ago when commenting[46] on some possible analogies with the Goldhaber-Teller excitations in nuclei). What is difficult to achieve is to find the proper way to assign to such degrees of freedom the quantum numbers typical of quarks (in particular half integer spin).

However if the number of quarks is only three -or four-[47] there is also perhaps some moderate chance they can be particles in the usual sense: particles which the quark hunters may one day find.

But if the number of necessary quarks increases, through the addition of colour[48], to, say, twelfe, it looks somewhat unpleasant to conceive all these twelfe quark fields as independent asymptotic fields; practically one would then be compelled to think of them in a new way. Therefore to be able to decide, on the basis of the empirical evidence, if the introduction of colour is strictly necessary is, in this respect, extremely important. The question is therefore: how compelling are the arguments 1 to 6 in the list given above. I will briefly express my point of view on each of them; note that especially for the point 2) and especially 4) I do not share the confidence held by the majority of theorists.

1) The requirement in question has been the motivation of the Han-Nambu three triplet scheme. It is not necessary

but can be aesthetically appealing.

2) It is well known that Fermi statistics (for baryons) can be satisfied also with one triplet of quarks only; the space part of the wave function must be antisymmetric. It is possible to lower such states in energy by the assumption of Majorana exchange forces between quarks attractive in antisymmetric states, repulsive in symmetric; but it is an open dynamical problem (on which we have commented at some length elsewhere[49]) to understand if this is compatible or not with the observed baryon spectrum; an antisymmetric space wave function is not necessarily incompatible with the electromagnetic form factor of the nucleon.

3) I have no experience on the $\pi^0 \to \gamma\gamma$ problem. I may refer for a discussion to a paper[50] by Drell. To understand the connection between a treatment of the pion field as a bound state and PCAC is still an open problem.

4) R_∞. The equation (18) is only true in some -asymptotically free- theories; moreover it looks to me very naive to believe (each year) that we are asymptotic.

5) The problem of saturation is serious; it has always been. The introduction of colour can be a solution; note that the problem of saturation might disappear at the origin if quarks could be conceived, in some sense, as collective degress of freedom[46].

6) In the class of colour models of the new resonances to be discussed below, confinement is not achieved through the restriction (which appears in some models considered e.g. by Gell Mann) that the only states to be allowed are colour singlets; confinement remains a dynamical problem; the models to be considered below are based precisely on the possibility of colour excited states.

6.2 – Three triplet models

One can invent, presumably, as many varieties of colour as one likes. Here we shall confine our attention to the three triplet schemes. We assume that to each triplet p, n, λ an index i is appended, the colour index, capable of three values. We therefore have three triplets:

$$SU_3' \downarrow \begin{matrix} p_1 & n_1 & \lambda_1 \\ p_2 & n_2 & \lambda_2 \\ p_3 & n_3 & \lambda_3 \end{matrix}$$
$$SU_3 \longrightarrow$$

We call, as indicated above, SU_3 the unitary transformation acting on the horizontal (that is mixing p, n, λ with given index i) and SU_3' the unitary transformation which acts along the vertical, that is mixes states of equal quantum numbers but different colour index.

It is assumed that the zero order hamiltonian is invariant both with respect to SU_3 and SU_3'.

We introduce the usual SU_3 matrices λ_i; and we complement them with SU_3' matrices λ_i'; and classify each state by giving both its SU_3 and SU_3' quantum numbers; we use for this a symbol of the type ($SU_3 \mid SU_3'$). Thus[51]:

$$\text{quarks} = (3 \mid \bar{3}) \qquad (19)$$

$$\text{Mesons} = (3 \mid \bar{3}) \times (\bar{3} \mid 3) = \underbrace{(1 \mid 1)}_{\text{normal mesons}} + \underbrace{(8 \mid 1) + (1 \mid 8) + (8 \mid 8)}_{\text{coloured mesons}}$$
$$(20)$$

The normal quark model states are colour singlets; the new mesons, and in particular the new narrow resonances, are to be identified, in the colour schemes, as non singlets of SU_3' (coloured mesons in equation (20)).

If SU_3' is a good group under all the strong interactions clearly a first important consequence typical of all these schemes is: production of coloured hadrons in strong interactions must be associated.

To be more complete we can, in fact, consider two cases for the properties of the strong interactions in the colour space (both imply associated production of colour particles). Case α): SU_3' is exact as far as strong interactions are concerned; this is the case mentioned above.

Case β): SU_3' is not exact for the strong interactions but a subgroup SU_2' of SU_3' is. Call I_c the colour isospin associated to such a subgroup. The normal states have then $I_c=0$, the coloured states $I_c=1$. Again the conservation of I_c implies associated production.

6.3 - Electromagnetic interactions

We can reproduce the old results (as far as the <u>first order</u> electromagnetic matrix elements are concerned) also if to the usual current J_μ (usual) behaving as $(8 \mid 1)$ we add a new piece J_μ' having the structure $(1 \mid 8)$. Thus:

$$J_\mu = J_\mu \text{(usual)} + J_\mu' = (8 \mid 1) + (1 \mid 8) \qquad (21)$$

Indeed because the normal old states are colour singlet, it is:

$$\langle 1(SU_3'), \ldots \mid J_\mu' \mid \ldots, 1(SU_3') \rangle = 0 \qquad (22)$$

that is the part J_μ' added to the current does not produce transitions between colour singlets.

The added part of the current does, however, produce transitions between the normal and the coloured states; and therefore the absorption of one real or virtual photon from a normal state can produce a coloured state. Also a coloured

state can easily decay into a normal state plus one photon
through the part J' of the current. (As we shall mention in
a moment this is in fact one of the greatest difficulties
for the colour interpretation of the new particles).

With the general structure (21) for the electromagnetic current there is still a variety of possible models.

Consider the expression of the charge Q and fix the attention on its $SU_3 \times SU_3'$ structure. Schematically it is:

$$Q = \frac{1}{2}\left(\lambda_3 + \frac{1}{2\sqrt{3}}\lambda_8\right) \times 1 + 1 \times \left(\frac{a}{2}\lambda_3' + \frac{b}{\sqrt{3}}\lambda_8'\right) \quad (23)$$

where the first part is the "old" charge (octet in SU_3 and singlet in SU_3') and the second part is the new charge (singlet in SU_3 and octet in SU_3'). In the expression (23) a and b are arbitrary parameters (in principle the total charge can have arbitrary eigenvalues; note that the equation (22) is satisfied no matter what a and b are).

In the expression for Q given above λ_3 and λ_8 (dashed or non dashed) are the usual matrices:

$$\lambda_3 = \begin{pmatrix} 1 & & \\ & -1 & \\ & & 0 \end{pmatrix} \qquad \lambda_8 = \frac{1}{\sqrt{3}}\begin{pmatrix} 1 & & \\ & 1 & \\ & & -2 \end{pmatrix}$$

If we now require that Q is an integer and <u>arbitrarily</u> decide that

$$|Q| \leq 1 \quad (24)$$

it is straightforward to check that only three possibilities exist:

A) a=1, b=-1/2 ; B) a=0, b=1 ; C) a=-1, b=1/2

The corresponding Q matrices are:

Choice A

	p	n	λ
1	1	0	0
2	0	-1	-1
3	1	0	0

Choice B

	p	n	λ
1	1	0	0
2	1	0	0
3	0	-1	-1

Choice C

	p	n	λ
1	0	-1	-1
2	1	0	0
3	1	0	0

Are A, B, C equivalent? It depends on the interactions which violate SU_3' in addition to the electromagnetic ones. If SU_3' were exact, clearly A, B, C would be identical by a permutation of 1, 2, 3. On the other hand, by definition, SU_3' violating interactions are not invariant with respect to a permutation of 1, 2, 3 (this applies in particular to the weak interactions). However we do not know the specific form of the colour dependence of the SU_3' violating interactions (except for the electromagnetic ones which were discussed above). There is then no lack in generality in fixing the attention only on A <u>or</u> B <u>or</u> C.

We consider the choice C which is the conventional Han-Nambu choice; and we examine, in the next section, two possible schemes, among many.

6.3 — Two possible schemes (among many)

Having in mind the distinction α) and β) introduced in Section 6.2 for the SU_3' properties of the strong interactions we can discuss two broad classes of models[52]:

Models of type α): SU_3' destroyed only by the electromagnetic and weak interactions.

Models of type β): SU_3' invariance destroyed also by medium strong interactions which however do conserve I_c; I_c destroyed only by the electromagnetic and weak interactions.

The two types of models are characterized by the following properties:

Models of type α): 1) all the different colour states belonging to the same colour representation are almost degenerate except for the Coulomb interactions; 2) Defining the U_C isospin as associated to rotations in the colour 2,3 subspace- recall the form of the matrix C, symmetric in 2,3- a photon carries $U_C=0$:

$$(1 \mid 8) + (8 \mid 1)\Big|_{U_C = 0}$$

Therefore a photon, acting on a colour singlet can only create mesons with $U_C=0$. Thus the coloured states created in e^+e^- collisions-a virtual photon acting on the vacuum- have the form:

$$(1 \mid 8 (U_C = 0)) \qquad (25)$$

Because the ordinary SU_3 is broken, this type (25) of states splits into two types of states, which we can call

$$(\tilde{\omega} \mid 8(U_C=0)) \qquad \text{and} \qquad (\tilde{\phi} \mid 8(U_C=0)) \qquad (26)$$

(the upperscript $\sim$ is there only to recall that now the ω, ϕ mixing is not necessary the same as for the usual vector mesons). In this type of scheme the 3.1 and 3.7 states are identified with the two states (26) above.

Models of type β). Because in these models SU_3' is strongly broken (with I_C consreved) the stable states (stable except for e.m. and weak interactions) are now characterized by the eigenvalue $I_C=1$. A photon, acting on an SU_3' singlet state again produces an $U_C=0$ state which now, however, has to be decomposed into states with $I_C=1$ and states with $I_C=0$. States with $I_C=1$ (and charge zero) are indicated by the symbol $(\mid \rho_0)$, and the following identifications are suggested:

$$\psi(3.1) \equiv (\tilde{\omega} | \rho_0) \qquad \psi'(3.7) \equiv (\tilde{\phi} | \rho_0) \qquad (27)$$

(again the first symbol in (27) refers to SU_3 the second to SU_3'). As we repeat, except for weak and e.m. interactions the states (27) are stable.

States with $I_c=0$ are naturally indicated as ($|\omega_8\rangle$; we can have:

$$(\tilde{\omega}|\omega_8) \qquad \text{and} \qquad (\tilde{\phi}|\omega_8) \qquad (28)$$

As stated above they are also produced in e^+e^- collisions, but having $I_c=0$, the states (28) are expected to be broad. The colour lovers identify them with the 4.15 phenomenon and with another possible resonance at $\sim$ 4.8 GeV.

6.4 – Photon decays

The colour stable states can be produced by photons, but also decay by photon emission; in fact, as it has already been stated, these photon decays, if not inhibited in some way, are very dangerous for the model. There is no need of special calculations here to understand this; reference can be made to the same calculations sketched when dealing with the charmed mesons.

Most of the colour lovers prefer to avoid the use of the O.Z. rule for inhibiting the decays; then only form factor effects can be considered such as those mentioned in connection with the $\psi \to \eta' \gamma$ decay. These form factors can well be there, and probably are there; but while it is clear that they can depress individual channels with large energy release, it is not so clear that they can be of help for the sum over all possible channels.

Another remark which is often heard in connection with the problem of reducing the γ decay widths is: look at the η decay. In the η decay one has (compare the figure below):

$\eta \to \pi^+ \pi^- \gamma$ Br.ratio 5% (process of order α)
$\eta \to \pi^+ \pi^- \pi^0$ Br.ratio 53% (process of order α^2)

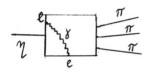
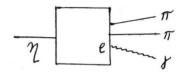

In other words the η decay into a real γ plus two pions is depressed 10 times with respect to the η decay into 3π which is of order α^2. This argument is invoked to indicate that we do not really understand all the reasons for inhibitions. This is certainly true (and, in fact, this η problem is an interesting one) but it is not very convincing to explain something using the lack of understanding of something else.

6.5 - Tests of colour schemes

We give below a list of qualitative predictions which, if not fulfilled, are serious problems to the colour models:
1) Importance of electromagnetic decays; this has been just discussed.
2) 72 coloured mesons are needed (leaving aside radial excitations, exotics, etc.). Although the masses can be distributed over some interval, many of them should lie in the 3 to 4 GeV. mass region. A coloured meson at mass $\sim$ 2 GeV. is somewhat unexpected, whereas a charmed meson at mass $\approx$ 2 GeV is expected; the 4.2 phenomenon has quite diffe-

rent interpretations in the charm and in the colour scheme; in the latter it is explained, e.g. in terms of the states (28); in the charm case it is due to the real production of a $D\bar{D}$ pair.

3) There should exist, in particular ($\rho | \rho$) mesons. For instance[53]

$$(\tilde{\tilde{\Phi}} | \rho_0)_{3.7} \rightarrow \pi^{\pm}_0 + (\rho^{\mp}_0 | \rho_0)$$

should have a stronger rate than:

$$(\tilde{\tilde{\Phi}} | \rho_0)_{3.7} \rightarrow \eta + (\tilde{\omega} | \rho_0)_{3.1}$$

There is no evidence, so far, for monochromatic pions from the 3.7 decay.

4) Associated production of ψ together with coloured baryons in PP collisions; it is compulsory in the colour scheme.

5) There should exist, among the 72 states mentioned above, several doubly charged states.

6.7 - Conclusion

We have expressed our point of view on the charm model at the very end of Sect.5.12. As to the colour models our feeling is that their flexibility is large; their outlook, in their present version, is gray (to use an expression from reference[43]). The reasons can be deduced from Sect.6.6.

We add that the discovery[58] of the intermediate states (at 3.5 and 3.4 GeV), points towards a two body bound state picture, in agreement with a charm-anticharm bound state picture. However only the future will tell us if the constituents are, or are not charmed quarks, endowed with the properties which these objects should have.

THEORETICAL MODELS PROPOSED FOR NEW PARTICLES 849

Of course, as always, doubts and uncertainties strongly arise as soon as one thinks to the nature of the quarks, not only the charmed one, but possibly also the others; these doubts have been expressed in the footnote[47], in the reference[46] and in the discussion in the text of Sect.6.1. But it would be pointless to insist on them until one has not a solution.

Before ending we must apologize for having omitted a number of important topics; not only (intentionally) the weak decays of the charmed mesons (as I repeat these will be treated by Professor Cabibbo in the School) but also -for reasons of time - other important aspects as for instance: a) the problem of the steep energy dependence of the production cross section of ψ by hadrons[54], b) a description of the attempts to calculations[55] of the production cross section of charmed particles by hadrons and especially c) the "ordinary" resonances or structures in $P\bar{P}$ and baryon antibaryon channels[56][57].

6.8-A short added note September 1975)

Already during the School in Erice (after these introductory lectures) evidence was presented in the talk by dr. Wiik (Desy) of a narrow state, at an energy of $\sim$ 3.5 GeV; the evidence had been obtained at Desy from a study of the reaction 3.7 $\rightarrow$ $\gamma\gamma$ + 3.1.

The above group reported this evidence also at the SLAC Conference (mid August 1975); calling P_c the state in question they give a value $\approx 4.10^{-2}$ for the product $(3.7 \rightarrow P_c + \gamma) / (3.7 \rightarrow \text{total}) \times (P_c \rightarrow 3.1 + \gamma)/(P_c \rightarrow \text{all})$. In addition <u>very</u> scanty evidence for a second state at $\sim$ 3.4 GeV was given by the same group. Also the decays 3.7 $\rightarrow P_c + \gamma$

followed by $P_c \to \pi^+\pi^- + 3.1$ or $K^+K^- + 3.1$ (and similar ones through the 3.4 state) were searched and one possible event for each case reported.

A study of the same nature, but with much more detailed information on the hadronic decays of the intermediate states and of their branching ratios) has been presented, always at the SLAC conference, by the SLAC group (talk by G. Feldman). From the SLAC and Desy results our conclusion is that two intermediate states (at masses ~ 3.5 and ~ 3.4 GeV) exist and possibly also a third nearby.

2) The SLAC group also presented evidence of a structure of the $e^+ e^-$ cross section for hadronic production in the region from 4 to 4.5 GeV. There appear to be two broad (50 or 100 MeV) bumps: one in the vicinity of 4.1 and the second centered around 4.4 GeV.

3) The SLAC group extended the search of narrow resonances sweeping the region from 5.9 to 7.6 GeV without finding any new phenomenon.

4) Evidence for a decay $3.1 \to 3\gamma$ was presented by the Desy group; they give a value of: Br.ratio $\psi(3.1) \to \eta\gamma$ $0.13 \pm 004\%$; Br.ratio $\psi(3.1) \to \eta'\gamma < 2\%$ - Moreover they interpret the data as indicating the sequence $3.1 \to X + \gamma$ with X centered around 2.8 GeV and decaying into two gammas (note: if X is the η_c of Sect.5.6 such a low η_c mass creates all the problems mentioned there). Two cases of possible decay of X into $P\bar{P}$ were also given.

5) The $e^{\pm} \mu^{\mp}$ statistics by the SLAC group has been increased by a factor ~ 3 with respect to that presented in Sect.5.12. The detailed analysis of the angular and energy distributions of the events tends to favour (with still gigantic errors) a fermion three body decay (heavy lepton?) with respect to a

THEORETICAL MODELS PROPOSED FOR NEW PARTICLES

boson two body decay. We feel, however, that a wait and see attitude should be taken with respect to this conclusion.

I thank dr. M. Bozzo of this department for a detailed discussion of the results presented at the SLAC Conference.

Appendix A

Branching ratios for several decay modes of ψ(3.1) and ψ'(3.7) (this is only part of the data shown by V. Lüth (SLAC) at the Palermo Conference).

a) ψ(3.1)

Mode	Br. ratio %
$e^+ e^-$	6.9 ± 0.9
$\mu^+ \mu^-$	6.9 ± 0.9
$\rho \pi$	1.3 ± 0.3
$2\pi^+ 2\pi^-$	0.4 ± 0.1
$2\pi^+ 2\pi^- \pi^0$	4 ± 1
$3\pi^+ 3\pi^-$	0.4 ± 0.2
$3\pi^+ 3\pi^- \pi^0$	2.9 ± 0.7
$\pi^+ \pi^- K^+ K^-$	0.4 ± 0.2
$K^0 K^{*0}(892)$	0.24 ± 0.05
$K^\pm K^{\mp *}(892)$	0.31 ± 0.07
$p \bar{p}$	0.21 ± 0.04
$\Lambda \bar{\Lambda}^0$	0.16 ± 0.08
$\bar{p}p \pi^0$ $\bar{N}P \pi^-$ $\bar{P}N \pi^+$	0.37 ± 0.19
$K K_S \mu$	< 0.02

b) ψ'(3.7)

$\psi'(3.7) \to \mu^+ \mu^- = \psi'(3.7) \to$
$\to e^+ e^- = 0.97 \pm 0.16 \%$

$\psi'(3.7) \to \psi(3.1) +$ anything
$= 57 \pm 8 \%$

In particular

$\psi'(3.7) \to 3.1 + \pi^+ \pi^-$ $32 \pm 4\%$

         $3.1 + \eta$ $4 \pm 2\%$

         $3.1 + \gamma\gamma$ $< 6.6\%$

$\psi'(3.7) \to$ hadrons and/or γ without passing through ψ(3.1): there seems to be no dominant channel.

Appendix B

Decays of charmed mesons. We confine to give some numbers which are obtained through simple dimensional considerations (compare ref.[2] especially M.K.Gaillard et al. and J.Ellis).

a) from dimensional arguments one can guess:

$$\Gamma_{total} (\text{charmed} \to \text{lepton} + \nu + \text{hadrons}) = \frac{G^2 M^5}{196 \pi^3}$$

where G is the Fermi constant and the M^5 dependence is the typical dimensional dependence which arises whenever the masses of the decay products can be neglected. The numerical factor is the same present for the μ decay. It follows that the order of magnitude for the total semileptonic decay of a D or F meson ($M \approx 2$ GeV) is expected to be 10^{12} sec.$^{-1}$.

b) If the non leptonic decay is enhanced above the semileptonic one by a factor 10 (similarly to the enhancement existing in the case of the strangeness changing decays) the total rate of decay of D or F mesons is expected to be around 10^{13} sec.$^{-1}$.

c) The two body leptonic decays are linear in M (the rate vanishes with vanishing lepton masses); the rates are expected to be:

$$\Gamma(D^+ \to \mu^+ \nu) \approx 2 \cdot 10^8 \text{ sec}^{-1} \ ; \ \Gamma(F^+ \to \mu^+ \nu) \approx 4 \cdot 10^9 \text{ sec}^{-1}$$

The factor 20 between the rates of F^+ and D^+ is due to the specific form (5) of the current which introduces a factor $\cos \theta_c$ and respectively $\sin \theta_c$ in the matrix elements.

Footnotes and References

(1) J. Aubert et al. Phys. Rev. Letters 33, 1404 (1974); S.C. Ting, Hadron and photon production of J particles (A Rapporteur Summary at the EPS Conference in Palermo, June 1975)
J. E. Augustin et al. Phys. Rev. Lett. 33, 1406 (1974)
C. Bacci et al. Phys. Rev. Lett. 33, 1408 (1974)
G. S. Abrams et al. Phys. Rev. Lett. 33, 1453 (1975)
L. Criegee et al. Phys. Letters 53B, 480 (1975)

(2) M. K. Gaillard, B. W. Lee and J. L. Rosner, Fermi Lab. Pub. 74/86 and 75/14 (addendum) – to be published in Revs. Mod. Phys.; H. Harari, ψ- chology, SLAC informal notes, November 1974; The Cern theory Boson Workshop, TH 1964 Cern preprint (Dec. 1974); J. Ellis, $e^+ e^-$ annihilation, the new particles and charm, (Cern preprint TH 1996 (March 1975); F. Gilman, SLAC Pub. 1537, Feb. 1975.

(3) The numbers in the table are taken from the rapporteur talk of V. Lüth at the EPS Conference in Palermo. Compare also: A. M. Boyarsky et al. Phys. Rev. Lett. 21, 1357 (1975) (data on the ψ); note that the total widths of the resonances are calculated from the data under the assumption that purely neutral decays are negligible.

(4) Compare the paper by Boyarsky et al. quoted in ref.[3]

(5) from 3.2 to 5.9 GeV: A. Boyarsky et al. Phys. Rev. Lett. 34, 762 (1975); in the lower intervals: C. Bacci et al. LNF preprint 75/38 (1975) and B. Bartoli et al. LNF preprint 75/40 (1975)

(6) G. Altarelli et al., Lettere al Nuovo Cimento 11, 609(1974); S. Borchardt et al., Phys. Rev. Lett. 34, 38 (1975).

(7) B. Knapp et al., Phys. Rev. Lett. 34, 1040 (1975); U. Camerini et al. SLAC pub. 1591, June 1975. Compare also D. Andrews

et al., Phys.Rev.Lett.34, 231 (1975); id.1134 (1975); J.Martin et al., Phys.Rev.Lett. 34, 288 (1975).

(8) Compare e.g. E.Paschos - Phys.Rev.Lett. 34, 358 (1975)

(9) C.Becchi and G.Morpurgo, Phys.Letters 56B, 85 (1975)

(10) For an alternative scheme involving weak vector bosons- also proposed in the early period after the discovery of the new narrow resonances -compare R.Marshak and R. Mohapatra,Phys.Rev.Lett. 34, 426 (1975)

(11) D.Yennie, Phys.Rev.Lett. 34, 239 (1975)

(12) Compare e.g. the discussion in the paper by M.K.Gaillard et al. under ref.$^{2)}$ as well as the paper by J.Primack in the Proc.of the XVI Int.Conf.on high energy physics (Batavia), Vol.2, page 307.

(13) S.Glashow, J.Iliopoulos and L.Maiani, Phys.Rev.D2, 1285 (1970)

(14) J.Bjorken and S.Glashow, Phys.Rev.Lett. 11, 255 (1964)

(15) G.Feldman and P.T.Matthews, Physics Lett.57B, 69 (1975)

(16) This proposition is already considered in ref.(2) (paper by M.K.Gaillard et al. -formula 4.16). We do not know if it had been considered before. After the discovery of ψ the identification of the ψ with a $c\bar{c}$ structure is discussed by Applequist and Politzer and by Glashow and de Rujula (papers of ref.$^{22)}$).

(17) The often called Zweig rule has a long story: it was first expressed in a purely algebraic fashion by S.Okubo, Phys.Letters 5, 165 (1963) and then formulated in the quark language by Zweig (Cern preprint TH 412, 1964); G.Morpurgo (Galilei Meeting in Pisa, September 1964 - Supp.Nuovo Cimento IV, p.448 (1966)) stressed the importance of this inhibition; compare also C.Becchi and G. Morpurgo, Phys.Rev.Lett.13, 110 (1964); J.Izuka also insisted on its importance and consequences (Suppl.Progr.

Theor.Phys.(Japan) 37-38, p.21 (1966)).

(18) A convenient way to obtain data and references on these reactions is to consult the compilations of data CERN-HERA 72/1, id.73/1 and LBL 53 parts 1 and 2 (1973); of course it is then necessary to look at the original papers to get a feeling on how the $\bar{\phi}$'s are detected.

(19) V.Blöbel et al.(Bonn-Hamburg-München collaboration): contribution to session A1 at the EPS Conference, June 1975, in Palermo.

(20) Several papers have considered the above processes: 1) N.Törnqvist -Cern preprint TH 2002 (March 1975); 2) J. Pasupathy, Tata Institute preprints ("Unitarity and partial width for $\phi \to 3\pi$ "; and "Unitarity corrections to Okubo Ansatz or Zweig's rule", May 1975).

(21) G.Morpurgo, Proc.of the IX Course of the Int.School of Subnuclear Physics "E.Majorana", (Erice 1971), Properties of the Fundamental Interactions, p.432 (Edited by A.Zichichi, Editrice Compositori - Bologna).

(22) A.de Rujula and S.Glashow, Phys.Rev.Lett. 34, 46 (1975). We follow here de Rujula and Glashow only in using a simplified potential between c and $\bar{c}$ like that described below ;we do not follow them in the Coulomb picture of charmonium. We feel that it is preferable to use, at this stage, a purely phenomenological approach. For this reason we shall not expand on the results from the asymptotically free gauge theories; these theories may, or may not, turn out to be a good description of nature. This also explains why we shall not discuss in these lectures the pioneer paper by Appelquist and Politzer on charmonium (Phys.Rev.Lett. 34, 43 (1975)).

(23) J.Pasupathy and G.Rajasekaran, Phys.Rev.Lett.34, 1250 (1975).

(24) V.S.Mathur, S.Okubo, S.Borchardt, Phys.Rev. D11,2572 (1975)

(25) M.Gourdin (preprint) PAR/1PTHE 75.5 (April 1975)

(26) K.Kajantie et al., Helsinki preprint 3-75, January 1975

(27) G.Morpurgo, Lectures on the quark model; from "Theory and phenomenology in particle physics" part A p.84-217 edited by A.Zichichi (Erice lectures 1968, Academic Press 1969)

(28) This point has been considered by several authors; we treat it here by the non relativistic quark model.

(29) C.Becchi and G.Morpurgo, Phys.Rev.140B, 179 (1965); also G.Morpurgo, ref.$^{21)}$.

(30) R.F.Dashen et al., Fermi Lab.pub. 75/18/THY, January 1975

(31) E.Eichten et al., Phys.Rev.Lett. 34, 369 (1975); T. Appelquist et al., Phys.Rev.Lett.34, 365 (1975). We follow here the treatment of Eichten et al.

(32) The argument underlying this value has been given in Sec.5.52

(33) From the report by V.Lüth at Palermo. Compare also G. Abrams et al., Phys.Rev.Lett. 34, 1181 (1975)

(34) F.M.Renard, Montpéllier preprint (1975); compare also J.Schwinger et al. UCLA preprint (April 1975); for an early discussion J.D.Jackson, LBL Physics notes (December 5, 1974)

(35) An early clear discussion on how to detect charm (as well as other "missing" particles) was given by R.K. Adair in his 1974 lectures at this School.

(36) M.Perl, SLAC pub. 1592 (June 1975)

(37) E.Cazzoli et al. Phys.Rev.Lett. 34, 1125 (1975)

(38) C.Rubbia-Rapporteur talk at the Palermo EPS Conference (June 1975)

(39) A. Benvenuti et al. Phys.Rev.Lett.34, 419 (1975); compare also ref.(38)

(40) K. Niu et al. -Progr.Theor.Phys. 46, 1644 (1975)

(41) P. Jain and L. Girard, Phys.Rev.Lett.34, 1238 (1975); compare also K. Hoshino et al. Progr.Theor.Phys.53, 1859 (1975)

(42) Compare however M.B. Einhorn and C. Quigg, Fermi Lab.pub. 75/21 THY Feb.1975

(43) A complete survey of the many varieties of colour schemes can be found in O.W. Greenberg, Dept.of Physics, University of Maryland Technical report 75-064 (January 1975) to be published.

(44) A detailed presentation of some of these arguments as well as a discussion of the various ways to introduce a "colour" degree of freedom can be found in Y. Nambu and M. Han, Phys.Rev. D 10, 674 (1974)

(45) M. Han and Y. Nambu, Phys.Rev.139B, 1006 (1965)

(46) Compare ref.(27), p.210

(47) But what should be the attitude if the data at, say, 20 GeV c.of m. should show, one day, the necessity of another quantum number(should one call it beauty, ugliness or perhaps, rotational internal number K?)

(48) Or if the conjecture of ref.(46) above should verify

(49) Compare ref.(27) sections 3.5 and 6.4

(50) S.D. Drell, Phys.Rev. D7, 2190 (1973)

(51) We identify the quarks as objects transforming as $(3|\bar{3})$, the original Han-Nambu choice. We might as well have identified them as $(3|3)$; some additional possibilities which arise from the latter choice have been examined by Greenberg in ref.^{43})

(52) In the notation of ref.[43] these are the models of classes III and IV respectively. To the list of references given in Greenberg's survey we add: for models of type β): N.Marinescu and B.Stech, Heidelberg preprint, May 1975, G.Feldman and P.Matthews, Imperial College preprint 1975. For more detailed calculations of the mass spectrum: F.Steiner, Universität of Hamburg preprint.
For a modified type β) model, F.Close and J.Weyers, preprint. For models, essentially, of type α) M.Kuroda and Y.Yamaguchi, University of Tokyo preprint 248, April 1975.

(53) This point is emphasized in a paper by H.J.Lipkin "Some general comments on the new particles "Rehovot preprint 1975 which also discusses other predictions shared by general classes of models.

(54) S.Ting - Ref.[1]

(55) V.Barger and R.Phillips - Rutherford Lab. preprint 75-047 (March 1975)

(56) Compare e.g. T.E. Kalogeropoulos in Experimental Meson Spectroscopy (Boston Conf.1974) AIP Conf.Proc.N.21 p.97

(57) Compare e.g. D.Weingarten and S.Okubo - Phys.Rev.Lett. 34, 1201 (1975) and the references quoted there.

(58) Compare, for, additional information, the added note in Section 6.8.

DISCUSSION

CHAIRMAN: G. Morpurgo

Scientific Secretaries: M. Günaydin, P. Marcolungo, H.D. Politzer

DISCUSSION No. 1

- FRAMPTON:

You said that there exist only a couple of isolated examples of the Zweig rule suppression -- you mentioned $\phi \to \rho\pi$ and $f' \to \pi\pi$. Is it not true that one may more fruitfully view the rule in terms of a particular value of octet-singlet mixing, and that one can thereby use SU(3) to relate the ϕ and f' suppression to many other observed coupling strengths? This may not enable us to explain the Zweig rule, but it replaces what needs to be explained by a general octet-singlet mixing rather than the successful prediction of a couple of very specific (and mysterious) disconnected quark diagrams.

- MORPURGO:

In addition to those decays (and to $\phi \to \eta\gamma$, $\phi \to \pi^0\gamma$) I mentioned a large number of production processes, suppressed by the Zweig rule. In fact, it was one of the points of the lecture to illustrate the rule by a large variety of cases.

- FRAMPTON:

I was really referring to decays of the other members of the ϕ and f' multiplets.

- MORPURGO:

You ask about the description of these decays in terms of a mixing angle; but then the Okubo-Zweig rule simply uses the fact that the mixing is near to its ideal value of 35.3° for vector mesons. Why it is so is part of the mystery, and the fact that for vector and 2^+ mesons the angle is so near to the "no mixing case" characterizes the Zweig rule, as I stated this morning.

- PETRONZIO:

Suppose that the lowest lying charmed particles were vectors instead of pseudoscalars; in this case, how would the mixing pattern that you presented this morning be changed?

- *MORPURGO:*

It would be only a slight change in the pseudoscalar matrix, which needs the value of the D mass as an input parameter, but is not particularly sensitive to that value. In fact, in this case I think that pseudoscalar masses would be very near the vector masses. I should add that there is another approach of Okubo et al. where this mixing problem is treated under general assumptions on SU(4) breaking and where the pseudoscalar mixing matrix is correlated with that of vectors by assuming that one breaking parameter of SU(4) is the same in both cases. For the mixing amplitudes in which we are interested here, there is not much difference between the two treatments.

- *MILLER:*

Please clarify your statement about inclusive pp $\to \phi$, etc., *vis-à-vis* exclusive processes.

- *MORPURGO:*

This morning I simply gave an example: I compared the process pp $\to \phi + n(\pi^+\pi^-) + $ pp which should be inhibited by the Zweig rule (and it is experimentally) with the process pp $\to \phi + $ anything, which at 24 GeV should not be inhibited by the Okubo-Zweig rule, but apparently is inhibited in the same way (within a factor of two) as is the exclusive process.

- *FERBEL:*

Regarding inclusive ϕ production, the so-called allowed process (where ϕ may be accompanied by two K's), may still be suppressed as it involves double strangeness exchange... It is just not a good reaction.

- *MORPURGO:*

You may remember that this morning I stated that a single reaction cannot be alone very indicative, but that one should consider many appropriate cases.

- *FERBEL:*

At what level would G-parity violations be of interest in ψ decay?

- *MORPURGO:*

This is difficult to answer, as the example of the η decay shows: there, α^2 decays are favoured over decays linear in α. I would like to add that at present, the best test of G-conservation seems to be not the 5π versus 4π production (because there is apparently a 20% chance of misidentifying π^0's and low energy γ's), but the absence of $\psi' \to \psi + \pi^0$.

- *FERBEL:*

I would like to ask Breidenbach: what about the apparent violations of G-parity at SPEAR, e.g. the η peak in the 5π decays of ψ; $\pi\pi\eta$ appears to be comparable to $\pi\pi\omega$. Is thus regarded as just noise, or can you say something definite?

DISCUSSION

- *BREIDENBACH:*

We are not prepared to say anything about the η.

- *t'HOOFT:*

I want to make a comment, calling attention to a theory that explains, though not quantitatively but still qualitatively, the way in which heavy quarks obey the Okubo-Zweig rule. This is the theory in which the quarks are kept together by gauge-vector gluons through the infrared-slavery mechanism. If you assume that the total object called ψ is much smaller in size than ordinary hadrons, then you can do a perturbation expansion with respect to an effective coupling constant $\alpha_{\text{effective}}$ is about $1/\log M/M_0$. (M is the bound state mass, equal to 3.1 GeV; a reasonable value for M_0 is 300 MeV.) You get for the width Γ/M = coefficient $(1/\log M/M_0)^6$ + higher orders, where you can calculate the coefficient. This theory gives precise predictions and should therefore be easy to kill.

- *MORPURGO:*

I stated this morning that I did not intend to commit myself to a particular field theoretical scheme; as to the theory you mention I do not think that it is very good because it gives a wrong value for $\Gamma_{3.1}(e^+e^-)/\Gamma_{3.7}(e^+e^-)$ which would be eight, if $\Gamma \propto |\psi(0)|^2$, and it is much less.

- *POLITZER:*

That prediction relies explicitly on Coulombic wave functions. But, it is clear from several considerations that the pure Coulombic picture cannot be right. For instance, typical masses and coupling constants give a radius of 3 fermis; but at that large a distance, quark forces are certainly strong. So, the Coulomb wave function must be modified and rather be confined to a smaller volume. There do remain several estimates and detailed calculations that can be made, using the coupling constant where it is justifiably small, or making more reasonable models of long distance behaviour.

- *MORPURGO:*

I certainly cannot exclude the role of asymptotic freedom in all this, but I feel that it is desirable to take, at the moment, a more phenomenological attitude in order to see what the data tell us. This is the attitude I took in my lectures; I feel that I stated this quite explicitly, although I agree with you that a presentation of the specific field theoretical scheme you are considering might have been both interesting and useful.

- *JOSHI:*

Zweig's rule operating close to a threshold seems inconsistent with linear trajectories. Consider the ϕ trajectory: $\text{Im}\alpha$ is small at the ϕ position. However, the Regge recurrence of ϕ will have a large $\text{Im}\alpha$. This sudden change in the $\text{Im}\alpha$ will make the ϕ trajectory highly non-linear. This is important for the ψ problem. If the 4.1 bump is broad, then the ψ trajectory will be highly non-linear and will play an important role in large angle pp scattering.

– *MORPURGO:*

I am not in a position to comment on this problem; you may be right.

DISCUSSION No. 2

– *CALDI:*

Would you comment on the mixing of colour-octet states with colour-octet singlet states, in particular with the ϕ; since if this mixing occurs, would not the dominant decays be through the ϕ?

– *MORPURGO:*

With reference to the models I described this morning, your problem arises only in those models where the states are classified on the basis of all the quantum numbers of $SU(3)'$ (not in the models where they are classified simply on the basis of I_c). The solution to the problem is very much dependent on the assumptions on the mixing forces, and I feel that it is premature to enter into this problem at the moment. Switching to another subject, may I ask a question of Professor Ting. I would like to ask if the branching ratio of something for the $K\pi$ decay is 5%, would you see a bump if the cross-sections involved for production have the same order of magnitude as that of the J?

– *TING:*

Yes.

– *PESKIN:*

You stated this morning that the decays of the η gave an example of the irrationality of photon decays, with an order α^2 decay dominating over an order α decay. But this case seems very special for two reasons. First, the decay $\eta \to \pi\pi\gamma$ should be suppressed by considerations of current algebra, while $\eta \to 3\pi$ can evade the constraint of current algebra. Secondly, the photon in $\eta \to \pi\pi\gamma$ occurs only in a radiative correction. Naively calculated, the contribution is logarithmically divergent, though still formally of order α^2.

– *MORPURGO:*

I would ask Prof. 't Hooft to answer this comment.

– *'t HOOFT:*

$\eta \to 3\pi$ decay is indeed forbidden if you assume that it is purely electromagnetic, and you apply PCAC current algebra. The point one has to make is that in a gauge theory, such an approach is inconsistent. As soon as you have breaking of a symmetry due to electromagnetism, then you get mass differences described by free parameters. Thus $\eta \to 3\pi$ comes from the difference between m_p and m_n, and no virtual photon is involved.

DISCUSSION

If you calculate the amplitude this way, you get no contradiction with experiment. Similar arguments should apply to the colour scheme, and consequently, you would have many colour decays without photons.

— *POLITZER:*

I wish to comment on two of your last remarks concerning the charm scheme. The quark-gluon gauge theories do provide a criterion of what is asymptotic in e^+e^- annihilation. By such estimates, 5-6 GeV is asymptotic. However, the mechanism which accounts for $R \sim 2\frac{1}{2}$ (instead of 2) below 3.5 GeV, i.e. first gluon corrections, suggests $R \sim 4.2$ at or above 5 GeV. With R remaining $\sim 5\frac{1}{2}$, I am compelled to believe that there may be both heavy leptons and D^+D^- contributing to the $e^{\pm}\mu$ pairs, or there must be more quarks, or at least more quark charge than the canonical c with $Q = 2/3$.

— *FERBEL:*

The SPEAR data are consistent with a constancy as well as with simply a slower rise of R above 4.15 GeV than observed for the rise below 4.15 GeV. Could Mr. Politzer comment on what would happen if R would continue rising slowly?

— *POLITZER:*

Trying to understand the earliest SPEAR data, I concluded that R could not continue rising. It did not. In the standard approach, the only possible smooth or regular behaviour is a slow approach to flat from above; anything else must come from thresholds and be transitory. I already mentioned that the high value of R suggests that there must be some more charged stuff in addition to charmed quarks. Of course, this picture may be totally wrong.

NEW PARTICLE PRODUCTION BY NEUTRINOS

C. Rubbia
Harvard University, Cambridge Mass. 02138
and
CERN, Geneva, Switzerland

1. INTRODUCTION

I shall discuss the recent observations of the Harvard-Wisconsin-Penn-Fermilab Collaboration[1] with a large electronic detector in the high-energy neutrino beam at Fermilab. I would like to review, in particular, the most recent experimental evidence for the production by and for the subsequent weak decay of new hadrons with a new quantum violated by weak interactions. For these objects we shall use the name Y-particles[2].

The experimental evidence is based on the following observations which I shall discuss:

  i) multilepton final states (dimuon events);
 ii) anomalies in the deep inelastic single-muon production by neutrinos;
iii) violation of the symmetry of charge between ν_μ and $\bar\nu_\mu$ on an I-spin zero target.

Neutrinos offer a unique way of producing particles with new quantum numbers. After the incoming neutrino has changed the nature of one of the elementary constituents of the target (quarks), final-state hadronic interactions will preserve such a new quantum number. Instead, in the case of hadronic or electromagnetic production, a *pair* production is required and the new particles can easily recombine again before leaving the interaction volume. Indeed, according to the simple thermodynamical models, production of relatively heavy objects is expected to be almost totally suppressed. In the case of neutrinos, on the contrary, the cross-section will remain as determined by the initial lepton-quark coupling.

2. MULTILEPTON FINAL STATES

2.1 Which processes are observed?

The following processes with three leptons in the final state are possible:

$$\nu_\mu + N \to \mu^-\mu^+\nu_\mu X \quad \text{observed}$$
$$\mu^-\mu^-\bar{\nu}_\mu X \quad \text{observed}$$
$$\mu^-\mu^-\mu^+ X \quad \text{not observed}$$

$$\bar{\nu}_\mu + N \to \mu^+\mu^-\bar{\nu}_\mu X \quad \text{observed}$$
$$\mu^+\mu^+\nu_\mu X \quad \text{observed, however outside fiducial volume}$$
$$\mu^+\mu^+\mu^- X \quad \text{not observed.}$$

Because of the four-fermion nature of weak interactions and of the inelasticity associated with the process, events have to be interpreted as multiple-step processes.

The experiments recently carried out at Fermilab by our group have collected of the order of 150 events (before fiducial cuts). About two-thirds of them remain within the standard sample. A first, important conclusion is that no three-muon event has been so far observed. This sets an upper limit of $\sim 3 \times 10^{-4}$ to the fractional number of events of three charged muons for neutrino interactions and $\langle E_\nu \rangle \sim 60$ GeV.

2.2 The experimental set-up

The experimental set-up consists of a large liquid calorimeter[3] and of a iron core muon spectrometer[4]. Between the liquid calorimeter and the iron spectrometer we have inserted an iron target, in order to be able to subtract experimentally the possible contribution from decays in flight or ordinary particles. Briefly, two targets of very different densities and roughly the same mass have been used:

i) The liquid scintillator target, 576 g/cm^2, density $\rho = 0.85$ g/cm^3 and $\lambda_{coll} = 125$ cm. A typical event from the liquid is shown in Fig. 1a.

ii) The iron target, 590 g/cm^2, density $\rho = 7.87$ g/cm^3 and $\lambda_{coll} = 25$ cm. A typical event is shown in Fig. 1b.

NEW PARTICLE PRODUCTION BY NEUTRINOS

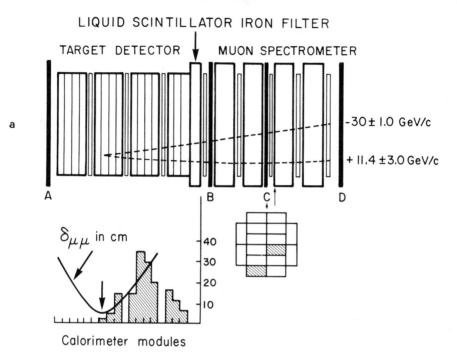

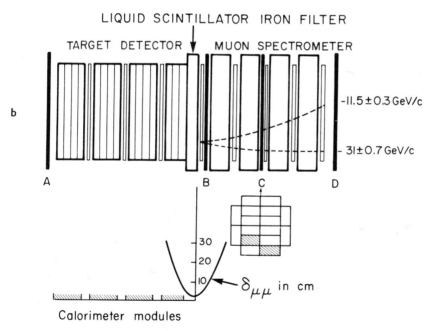

Fig. 1 A typical event in the liquid (1a) or in the iron absorber (1b).

Note that for events of type (i) we measure both the total hadronic (and e.m.) energy deposition and the vector momenta of the muons. In the case of events from iron, only the latter information remains available since no calorimetric analysis is possible.

2.3 Beams

Data are available for three different beam configurations:

a) <u>Horn focusing for ν_μ</u> and short spill time ($\sim$ 20 μsec). Those experimental conditions which have been used in the earlier part of the experiment[2]) are the least favourable because of the large number of relatively low-energy neutrinos and the short duty cycle which introduces an appreciable probability of two neutrino interactions within the resolving time of the detector.

b) <u>Quadrupole triplet focusing</u>[5]). A triplet of quadrupoles focuses present particles of both signs parallel to the decay channel. The beam contains a large number of high-energy ν_μ and $\bar{\nu}_\mu$ with their "natural" mixture and with good spill-time ($\geq$ 1 msec) (see Fig. 2). In these conditions most of the events are due to ν_μ, and the probability of multiple interactions is negligible.

c) <u>Double horn focusing $\bar{\nu}_\mu$</u> with a central plug and short spill-time (20 μsec). An enriched $\bar{\nu}_\mu$ beam with some ν_μ contamination is available (see Fig. 3). Note that flux and consequently the probability of multiple interaction is considerably less than in case of configuration (i).

2.4 Event selection and backgrounds

Events were selected on the basis of standard cuts. The trigger requires an ABC condition and therefore events from the liquid and iron are collected simultaneously. The picture information is used to introduce a fiducial volume cut and that of the dimuon vertex, which is observed by three different stereo angles, be compatible with the calorimeter information. Finally, all electronic information, i.e. timing pulse heights and so on, are required to be consistent with the expected patterns.

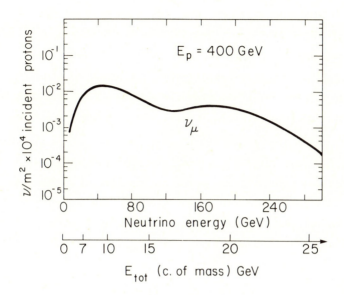

Fig. 2   Neutrino from quadrupole triplet focus.

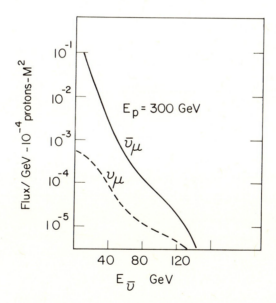

Fig. 3   Spectrum of antineutrinos from horn focusing.

Under these selection criteria, backgrounds are expected to be negligible. Two types of background are of potential danger and have to be considered in detail:

a) An accidental space and time coincidence of two standard muon events. From the number of events in which this coincidence is actually observed we can estimate the number of events in which confusion may arise. This number is completely negligible, except for some early runs when the beam horn focusing ν_μ has been used.

b) Decays of relatively long-lived particles, such as $\pi^\pm$, $K^\pm$, Λ^0 and so on. There are several tests which ensure that also this background contribution is negligible. The rate both from dense and diluted targets is independent of density and proportional to the mass (see Fig. 4). It is apparent from the figure that $\lambda_{coll} \gg \lambda_{dec}$ for the decay process responsible of the dimuon signal. Other independent evidence against this background comes from the fact that the sign distribution and the kinematics are sharply different from the expectations of decays. Finally, also the calculated rates from $K^\pm$, $\pi^\pm$ decays are much smaller than experimental observations.

We conclude that we are in the presence of a genuine new phenomenon of non-trivial origin.

2.5 Main features of the dimuon events

We shall review briefly the main characteristics of the events:

a) Branching ratios. We have selected a restricted sample from the q-pole triplet beam, in order to ensure a reasonable fiducial volume and a good detection efficiency. In order to estimate the event losses in the absence of any known production mechanism, we have used a crude calculation which consists in moving the origin and the azimuthal angles of the dimuon system over the fiducial volume in order to evaluate a detection probability for the events.

Results for the q-pole triplet beam (ν_μ):

$$\frac{R(\mu^-\mu^+\nu_\mu X)}{R(\mu^- X)} = (0.8 \pm 0.2) \times 10^{-2}$$

$$\frac{R(\mu^-\mu^-\bar\nu_\mu X)}{R(\mu^+\mu^-\nu_\mu X)} = (0.1 \pm 0.05)$$

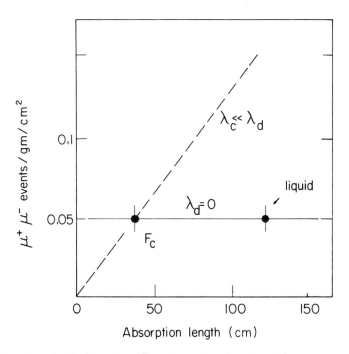

Fig. 4 Infinite density extrapolation for dimuon events.

and for the antineutrino beam:

$$\frac{R(\mu^+\mu^-\bar{\nu}_\mu X)}{R(\mu^+ X)} \simeq 2 \times 10^{-2} \text{ (based on four events)} .$$

b) The production cross-section appears to be smooth for $E_\nu > 50$ GeV. There is no evidence for sharp threshold, although the qualitative statement that the cross-section for $E_\nu < 30$ GeV is substantially smaller (up to a factor of 10) can be made.

c) The hadronic inelasticity is large. On the average, approximately 1/2 of the total viable energy appears in the calorimeter (see Fig. 5). *Note the absence of quasi-elastic or of elastic events.* In other words, the lack of events at $E_h \simeq 0$ is not due to experimental biases, since the trigger does not require hadronic energy deposition.

d) The μ-μ invariant mass $M_{\mu\mu}$ is shown in Fig. 6. It is smooth, with no specific structure, and it extends to about 8 GeV/c². At larger M masses the detection efficiency is sharply cut off.

e) An important observation is the existence of leading particle effects. For q-pole triplet data (mostly ν_μ), we observe generally that $p_{\mu^-} > p_{\mu^+}$ (Fig. 7a). For the $\bar{\nu}_\mu$-horn data (mostly $\bar{\nu}_\mu$) the opposite is observed, i.e. $p_{\mu^+} > p_{\mu^-}$ (Fig. 7b). The small fraction of events in the q-pole run, where $p_{\mu^+} > p_{\mu^-}$, are consistent with the expected $\bar{\nu}_\mu$ constant of the beam.

f) We can define p_T, the projection of the transverse momentum of the muon with respect to the plane defined by the incident neutrino and the leading muon (μ^- for ν_μ and μ^+ for $\bar{\nu}_\mu$). The advantage of choosing such a variable is that it can be constructed from the muon variables only (see Fig. 8). The distribution is flattish and it has an end-point at $p_T \approx 1.6$ GeV/c (Fig. 9). Note that for $\pi^\pm \to \mu^\pm$, $K \to \mu^\pm$, and so on, one expects a distribution of the general form e^{-9p_T}, which is in complete disagreement with our data.

2.6 Possible origins of the dimuon events

Several mechanisms can *a priori* generate events with two muons in the final state. We shall list a number of possible alternatives. All but one can be excluded.

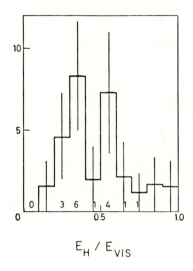

Fig. 5 Visible inelasticity of dimuon events.

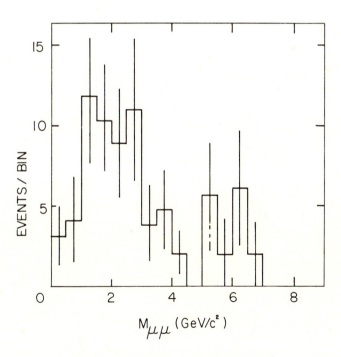

Fig. 6 Dimuon invariant mass.

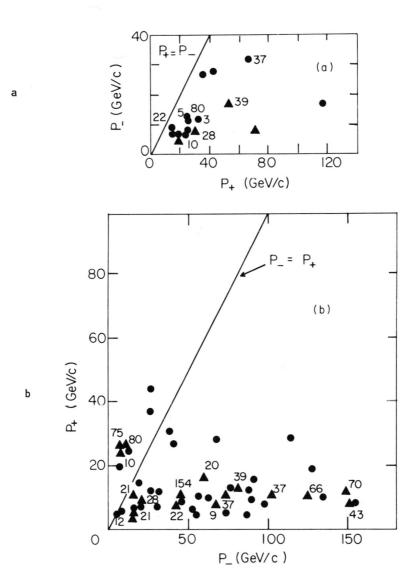

Fig. 7 p_+ versus p_- for neutrinos (7a) and p_- versus p_+ for antineutrinos (7b). Note the leading particle effects.

NEW PARTICLE PRODUCTION BY NEUTRINOS

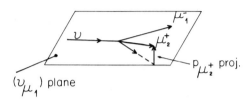

Fig. 8   Transverse momentum variable (definition).

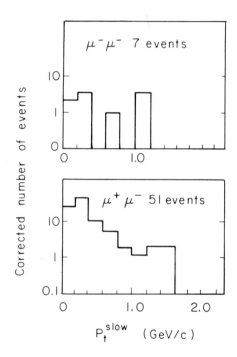

Fig. 9   Distribution of events in the transverse momentum variable.

a) Higher-order weak and electromagnetic process (weak tridents). The graph is shown in Fig. 10a. The process can easily be excluded since: i) the calculated rate is 10^3 times smaller; ii) $M_{\mu\mu}$ mass should peak at very low values (few tens of MeV); and, iii) the process should be mostly elastic, i.e. $E_H \ll E_{vis}$.

b) Production and decay of a neutral W or of other vector mesons:

$$\nu_\mu + N \to V^0 + \nu_\mu + X$$
$$\hookrightarrow \mu^+ + \mu^-$$

where V^0 is a narrow state ($W^0, J, \psi \ldots$). This is excluded by the absence of a peak in the $M_{\mu\mu}$ distribution.

c) Production and decay of a charged W, according to the diagram of Fig. 10b. This is again excluded since: i) this process is expected to be mostly elastic; ii) the p_T should be larger; iii) additional evidence should be expected in normal scattering due to propagator effects; and, iv) $\langle p_{\mu-} \rangle \approx 0.4 \langle p_{\mu+} \rangle$ instead of $p_{\mu-} \gg p_{\mu+}$.

d) Production and decay of a new heavy lepton (L^0) according to diagram of Fig. 10c. Model-dependent calculations exclude the simple production and decay[6]. However, more complicated references in which the L^0 might not be produced directly, i.e.,

$$\nu + N \to L^+ + \text{hadrons}$$
$$\hookrightarrow L^0 + \text{hadrons}$$
$$\hookrightarrow \mu^+ + \mu^- + \nu$$

or

$$\nu + N \to \text{many kinds of leptons}$$

could be eventually be postulated. A more general argument has been developed by Pais and Treiman[7], who have shown that in complete generality on production and decay mechanisms, bounds exist for the quantity

$$R^{(\nu)} = \frac{\langle p_{\mu-} \rangle}{\langle p_{\mu+} \rangle}$$

(for $\bar{\nu}_\mu$, μ^+ and μ^- are exchanged).

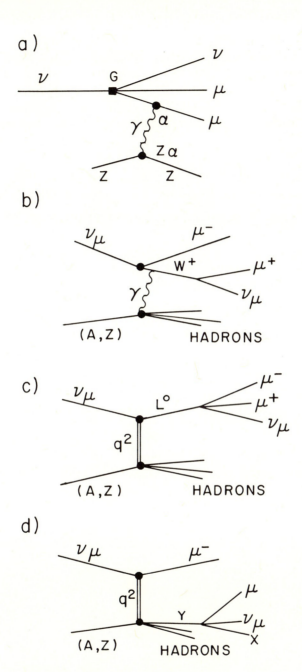

Fig. 10 Feynman diagram of possible processes leading to dimuon production.

They are

$$1/2 < R^{(\nu,\bar{\nu})} < 2 \qquad (V, A)$$

$$\frac{9 - 4\sqrt{2}}{7} < R^{(\nu,\bar{\nu})} < \frac{9 + 4\sqrt{2}}{7} \qquad (S, T, P, A, V) .$$

Experimentally $R^{(\nu)} = 8.5 \pm 1.7$, excluding in a more general way the heavy-lepton hypothesis.

We are then left with the inescapable conclusion than any conceivable mechanism which attributes the source of the effect to the lepton vertex of the interaction is unable to explain the experimental observations. The only possible alternative is then that the origin of the dileptons is related somehow to the hadronic vertex of the interaction (Fig. 10d).

Further evidence discussed below confirms this assignment.

3. ANOMALIES IN THE DEEP INELASTIC SCATTERING

3.1 Definitions

We shall use the standard scaling variables

$$x = \frac{Q^2}{2M\nu}, \qquad y = \nu/E_\nu ,$$

where $\nu = E_h$ is the energy of the hadronic system and M the mass of the initial target. The scattering process is in general a function of the three structure functions, which are only functions of the variable x:

$$\frac{d^2\sigma^{\nu,\bar{\nu}}}{dxdy} = \frac{G^2 M E_\nu}{\pi} \left[\left(1 - y + \frac{y^2}{2}\right) F_2^{\nu,\bar{\nu}}(x) - \frac{y^2}{2} F_L^{\nu,\bar{\nu}}(x) \mp y\left(1 - \frac{y}{2}\right) x F_3^{\nu,\bar{\nu}}(x) \right] ,$$

where G is the Fermi constant.

Positivity requires that $F_2^{\nu,\bar{\nu}}(x) \geq 0$, $F_L^{\nu,\bar{\nu}} \geq 0$, and $F_2^{\nu,\bar{\nu}}(x) - F_L^{\nu,\bar{\nu}}(x) \geq x|F_3^{\nu,\bar{\nu}}(x)|$. We can define

$$B^{\nu,\bar{\nu}}(x) = - \frac{x F_3^{\nu,\bar{\nu}}(x)}{F_2^{\nu,\bar{\nu}}(x)} .$$

The above conditions limit $|B^{\nu,\bar\nu}(x)| \le 1$. Note also that B can be positive or negative since it represents a V, A interference term which changes sign under the exchange $\nu_\mu \leftrightarrow \bar\nu_\mu$. Also we can set

$$|R_L^{\nu,\bar\nu}(x)| = \frac{|2xF_L^{\nu,\bar\nu}(x) - F_2^{\nu,\bar\nu}(x)|}{F_2^{\nu,\nu}(x)}$$

and the cross-section can be re-written as:

$$\frac{d^2\sigma^{\nu,\bar\nu}}{dxdy} = \frac{G^2 ME_\nu}{\pi} F_2^{\nu,\bar\nu}(x) \times$$

$$\times \left\{ 1 - y\left[1 \mp B^{\nu,\bar\nu}(x)\right] + \frac{y^2}{2}\left[1 \mp B^{\nu,\bar\nu}(x)\right] - \frac{y^2}{2}|R_L^{\nu,\bar\nu}(x)|\right\}.$$

3.2 Simple quark-parton model

Assume first a simple quark-parton model with no new particle being produced. We expect

i) charge symmetry invariance, i.e. $F_i^{(\nu)} = F_i^{(\bar\nu)}$ for an isoscalar target;

ii) $R_L^{(\nu,\bar\nu)} \to 0$; note that in general this term produces a *decreasing* term in the y-distribution and it cannot explain an excess of events at large y;

iii) $B^{(\nu)} = B^{(\bar\nu)} = +1.$, i.e. $xF_3(x) = -F_2(x)$.

This leads to the simple distributions

$$\frac{d^2\sigma^\nu}{dxdy} = \frac{G^2 ME_\nu}{\pi} F_2(x),$$

$$\frac{d^2\sigma^{\bar\nu}}{dxdy} = \frac{G^2 ME_\nu}{\pi}(1-y)^2 F_2(x).$$

Results of the experiment are shown in Fig. 11. While the neutrino distributions and the antineutrino distributions below 30 GeV are in agreement with the simple model, the data for $\bar\nu$ and E > 30 GeV are, for $x \le 0.1$, inconsistent with the expectations of the above-mentioned model.

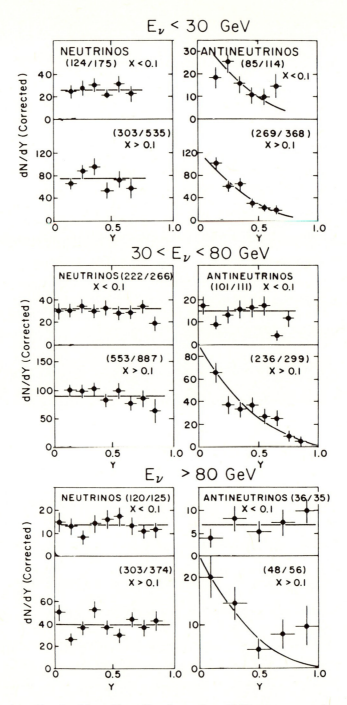

Fig. 11 Scaling distributions for different energies.

3.3 Main features of the y anomaly

The anomaly, which has persisted through several improvements of the experimental technique, has the following qualitative features:

i) It occurs mainly for $E_{\bar{\nu}} > 30$ GeV. To show this, we have plotted the $\langle y \rangle$ for antineutrinos as a function of the energy of the (anti)-neutrinos (Fig. 12). One can see that there is clear evidence of a threshold effect.

ii) The effect is concentrated at low values of x, i.e. $x \leq 0.0$. Small x and high y means large hadron recoiling masses W, since $W^2 = E_{cm}^2 y(1-x)$.

iii) It is impossible to understand simultaneously the B(x) distributions for neutrinos and antineutrinos at small x (see Fig. 13). For $x < 0.1$ $B^\nu \simeq 1$, while $B^{\bar{\nu}} \simeq -1$. Hence charge symmetry is apparently violated. Antiquark mixtures at low x cannot explain the effect since it should still be charge symmetric (see also Fig. 14).

3.4 Can Y particles explain the effects?

Y particles were abundantly produced at high energies and would have a sizeable non-leptonic decay branching ratio, effects would be expected in the deep inelastic channels:

$$\nu + N \to \mu^- + X + Y$$
$$\hookrightarrow \text{hadrons}.$$

Qualitatively we could speculate that it would be easy to explain the anomalies if this were to be the case. In fact, small x should be favoured if it takes a "quark" from the sea to produce the Y, and the charge symmetry violation would be the natural consequence of the presence of a new quantum number.

It is, however, far too early to establish unambiguously a connection between the y anomaly and the Y particles.

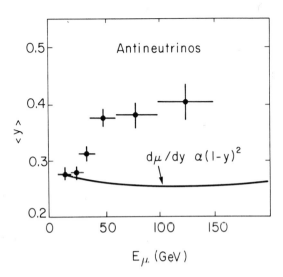

Fig. 12 Average value of y; ⟨y⟩ for $\bar{\nu}$ as function of the incident $\bar{\nu}$ energy.

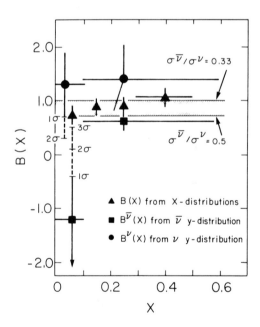

Fig. 13 Charge symmetry violation.

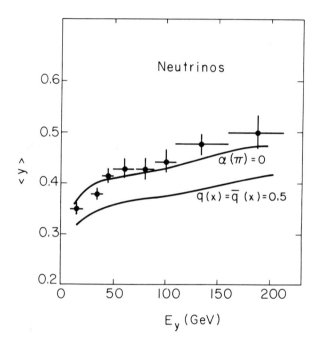

Fig. 14 Average value of y; $\langle y \rangle$ for $\bar{\gamma}$ as function of the incident energy.

4. PROPERTIES OF Y PARTICLES

4.1 Kinematics

If the emission of the second muon is associated with the weak decay of the Y particle, i.e.

$$\nu + N \to \mu^- + X + Y$$
$$\hookrightarrow \mu^+ + \bar{\nu}_\mu + X$$

we can define a number of kinematic variables of the hadronic complex, including the Y particle, recoiling with respect to the μ^-. Because of the unknown energy carried away by the neutrino, only limits can be stated for some of the variables.

With the usual notations, we can write:

$$\nu = x \cdot y = \frac{E_{\mu^-}}{M} \sin^2(\theta_{\mu^-}/2) ,$$

$$E_{vis} = E_{true} - E_{\nu'} = E_{\mu^-} + E_{\mu^+} + E_h ,$$

$$y_{min} = \frac{E_{vis}}{E_{vis} - E_{\mu^-}} \leq y ,$$

$$x_{max} = \nu/y_{min} \geq x ,$$

$$W_{min} = \text{minimum recoil mass against } \mu^- \leq W .$$

4.2 What can we say about Y particles?

The distribution of W_{min} shows a clear threshold at $W_{min} \sim 5 \text{ GeV}/c^2$. This threshold coincides with the threshold for charge symmetry violation[2] observed by our group in the single muon production (Fig. 15). As is well known, charge symmetry violations are expected for production of particles with additional quantum numbers which are not conserved by weak interactions. The distribution in the scaling variable x_{max} shows that dimuons from $\bar{\nu}_\mu$ come mostly from the low-x region. This is exactly the region in which change symmetry violations have been observed. Also the distribution in the scaling variable y_{min} shows that the events,

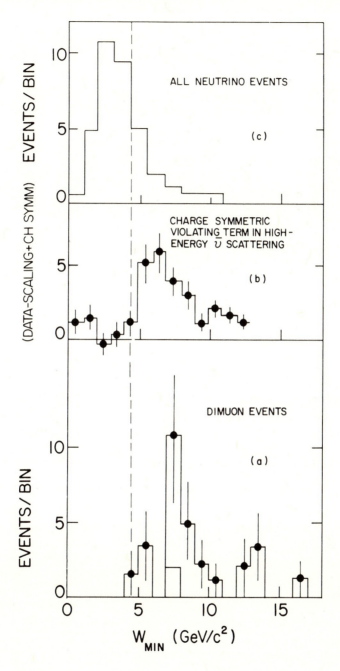

Fig. 15 Comparison between the charge symmetry violation and dimuon events. Note the apparent identical threshold effect.

especially for $\bar{\nu}_\mu$, come mostly from the region at high y. This matches the behaviour of the high-y anomaly in $\bar{\nu}_\mu$ scattering (Fig. 16).

If the *whole* y-anomaly is associated with the dimuons we conclude that:

$$\frac{R(Y \to \text{semi-leptonic})}{R(Y \to \text{non leptonic})} = \frac{0.02}{0.2} = \frac{1}{10} .$$

From the W_{min} and p_T distribution, we can say that the mass or the masses are between 2 and 4 GeV.

The preference of equal-sign muons, $(+-)/(--) \approx 10/1$, indicates the presence of an approximate selection rule in the Y-production and decay, relating to the change of the new quantum number to changes of the charge of the leptons (similar to $\Delta Q = \Delta S$ selection rule for strange decays).

Likewise, the absence of trimuons forbids the decay of Y by neutral currents:

$$Y \not\to \mu^+ + \mu^- + X .$$

4.3 The question of equal-sign dimuons

The experimental observation of equal-sign dimuons with smaller rates is a very important element in the phenomenology of the Y particles. If further experimental investigation will definitely confirm the presence of this effect, a number of possible explanations are possible:

i) The charge selection rule of Y particles may be simply approximate.

ii) Associated production of pairs of Y particles occurs at the level of $\simeq 1/100$ of all ν_μ-induced interactions. This would look very promising for the possible production of these states by hadron or electron beams.

iii) There are as yet more quantum numbers in addition to charm (i.e. friendship, sex, and so on).

It is evident that further investigations are of great interest and they are being pursued at Fermilab and elsewhere.

NEW PARTICLE PRODUCTION BY NEUTRINOS

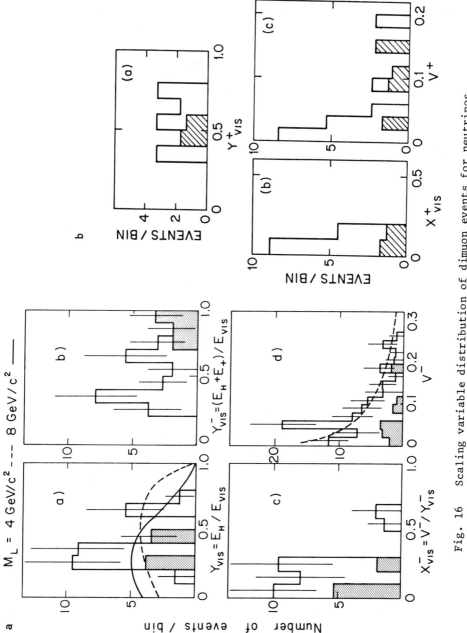

Fig. 16 Scaling variable distribution of dimuon events for neutrinos (16a) and antineutrinos (16b).

5. CONCLUDING REMARKS AND OUTLOOK TO THE FUTURE

It is now well established that Y particles exist. They are most likely *hadronic* states with a new quantum number. They are copiously produced by neutrinos *and* antineutrinos at high energies.

The connection to charm and to the beautiful ψ-complex has yet to be established, although it appears to be likely.

REFERENCES

1) The members of the HPW-Fermilab Collaboration are: A. Benvenuti, D. Cline, W.T. Ford, R. Imlay, T.Y. Ling, A.K. Mann, R. Orr, D.D. Reeder, C. Rubbia, R. Stefanski, L. Sulak and P. Wanderer.

2) B. Aubert et al., Phys. Rev. Letters 33, 984 (1974).

3) A. Benvenuti et al., A liquid scintillator total absorption hadron calorimeter for the study of neutrino interactions, Harvard-Pennsylvania-Wisconsin-FNAL preprint, to be published in Nuclear Instruments and Methods.

4) A. Benvenuti et al., A large-area magnetic spectrometer for the study of high-energy neutrino interactions, Harvard-Pennsylvania-Wisconsin-FNAL preprint, to be published in Nuclear Instruments and Methods.

5) A. Benvenuti et al., Colloquium Int. CNRS 245, 297 (1975).

6) L.N. Chang, E. Oerman and J.N. Ng, Phys. Rev. Letters 35, 6 (1975). C.H. Albright, Fermilab report No. 75-35-THY, 1975.

7) A. Pais and S.B. Treiman, Neutral heavy leptons as a source for dimuon events; a criterion. Submitted to Physics Letters.

DISCUSSION

CHAIRMAN: C. Rubbia

Scientific Secretaries: R.K. Ellis, R. Petronzio

DISCUSSION No. 1

- *GOURDIN:*

Some comments on your analysis. What you call extended CVC includes a V = A type assumption.

- *RUBBIA:*

Yes. Consequently the breakdown of extended CVC does not mean that CVC itself is violated.

- *GOURDIN:*

You ignore the longitudinal part of structure functions. At SLAC the longitudinal part is 10-20% of the total.

- *RUBBIA:*

We impose the Callan-Gross relation from the beginning. It is important to note that the inclusion of an arbitrary amount of $F_1(x)$ can never generate high y events in the antineutrino cross-section.

- *GOURDIN:*

You ignore the $\Delta S = 1$ contribution.

- *RUBBIA:*

Strange particles may be created either by associated production in the final state or by the strangeness violating part of the weak current. A large production of strange particles would produce an experimental situation similar to the one we find (e.g. violation of charge symmetry). However, results from the Fermilab-Michigan State bubble chamber experiment show that the non-associative production of strange particles is less than 4%. Our effects are an order of magnitude larger i.e. about 10-20%. The results could be explained by an anomalously large non-associative production of strange particles at high energy. We favour an explanation in terms of the production of new heavy objects. Because of kinematics, such effects would first be observed at high y.

- *GOURDIN:*

An increase in the Cabibbo angle at high mass would produce a large production of strange particles.

- *RUBBIA:*

That is correct. Of course, it has never been measured at high energy.

- *GOURDIN:*

The absence of antipartons is not proved if charge symmetry is violated.

- *RUBBIA:*

Yes, this was my point.

- *GOURDIN:*

Is it really impossible to fit the Gargamelle and HPWF data for $\langle q^2 \rangle$ with formulae of the type $A + B E$ or $A + B \log (E/M) + C E$?

- *RUBBIA:*

You can always fit the average values. It is the distributions themselves which one cannot fit.

- *YOSHIKAWA:*

If new massive particles are produced why is there no deviation from linearity observed in the total cross-section.

- *RUBBIA:*

The total cross-section measurements are not good enough. They require knowledge of the neutrino flux which is not well understood.

- *YOSHIKAWA:*

How do you explain the violation of charge symmetry by the introduction of massive particles?

- *RUBBIA:*

When a new channel opens and particles with new quantum numbers are produced it is no longer necessary for the cross-section of W^+ on an isospin zero object to be equal to that of W^- on the same object.

- *MILLER:*

CalTech and HPWF seem to disagree in counter experiments. Will the 15 foot bubble chamber filled with neon be able to clear up the situation?

DISCUSSION

- *RUBBIA:*

I do not feel that the disagreement between our own experiment and CalTech is very large. It should be stated that the CalTech apparatus is not designed for the high y region where we observe most of these new phenomena. The neon chamber is now in operation.

- *KIRKBRIDE:*

Would one expect to see the $F_2(x)$ rise for $x < 0.2$ in μ experiments, and if so, why is it not evident in the Chen-Hand data presented at Palermo?

- *RUBBIA:*

Because they are working at lower $\langle Q^2 \rangle$. I do not think that the violations of scaling at low x have anything to do with what we observe.

- *POLITZER:*

Several points of clarification: on the experimental side, to my recollection, the Chen-Hand μ-scattering experiment does include large q^2, i.e. q^2 up to 60 $(GeV)^2$. They observe an enhancement in $\nu W_2(x)$ (over the scaling expectation) for $x \lesssim 0.2$ (or 0.3), and it is greatest for smallest x. However, the magnitude is nowhere nearly as large as that reported in the ν experiments. From the theoretical side, there is an important distinction to be made between electromagnetic and weak scattering. The photon does not change quark type. So long as the proton does not contain an appreciable fraction of exotic, heavy quarks, the existence of charm cannot affect electroproduction. In contrast, the weak current can change quark type and exhibit threshold effects at high energies.

- *TRIANTIS:*

In ν and $\bar{\nu}$ experiments in nuclear targets, like ^{12}C and ^{16}O in your liquid scintillator, a non-negligible contribution to the antiquark spectrum could come in principle from the mesons which provide the nuclear force, since no important contribution is expected from the nucleons themselves. What is the limit for this contribution to be observable?

- *RUBBIA:*

The only nuclear effect which we are concerned with is Fermi motion, which even at infinite energy modifies the values of the scaling variables. The effect of the meson cloud or of collective phenomena should, in principle, be observable at low q^2. For our purposes we assume that such effects can be ignored. You should remember that neutrino experiments are very coarse, and that the cross-section is very small.

A lot of people have asked about the compatibility of our results with the CalTech data. Petronzio has prepared a graph (Fig. 1) of the dN/dy versus y for both the HPWF data and the CalTech data (the latter has been rescaled so that the mean value of the neutrino data is the same). My conclusion, at least for the moment, is that the discrepancy between the HPWF data and the CalTech data is not statistically fully significant.

In fact, within about two standard deviation, the two sets of data can be made consistent. The CalTech points are also consistent with the charge symmetric solution and the equality of the neutrino and antineutrino distributions as y → 0, whereas our points are not.

— ZICHICHI:

Could you make a comment on the experimental reasons for which CalTech might systematically overestimate their events at low y, and the reasons for which your group might underestimate events of this type?

— RUBBIA:

Our beam focusing system does not discriminate between neutrinos and antineutrinos, which therefore hit the target simultaneously. The relative normalizations are, therefore, determined from the data themselves. In order to ensure that we do not systematically defocus a certain type of event, we change the sign of the field in the detecting magnet every few hours. In this situation, both our beam and out detectors are charge symmetric, so if there were any experimental bias, it would appear in both the neutrino and antineutrino data.

Secondly, the data fit very well for $x > 0.1$. At fixed y, the measurement of small x depends primarily on the measurement of θ alone, because $x \propto \theta^2$ at small angles.

— DYDAK:

What is your x resolution for $x = 0.1$ for $E_\mu \sim 50$ GeV?

— RUBBIA:

About 0.3.

— DYDAK:

And for CalTech?

— RUBBIA:

The x resolution for small x is determined by multiple scattering and the chamber resolution. The problem of multiple scattering is worse in iron than in liquid scintillators, so the CalTech resolution will be worse than ours.

DISCUSSION

DISCUSSION No. 2   *(Scientific Secretary : W. Kozanecki)*

- *DYDAK*:

You collected the quoted sample of fifty-seven dimuon events in a quadrupole focused beam. Can you determine whether a particular event originated from a neutrino or an antineutrino interaction?

- *RUBBIA*:

The quadrupole focused beam we used delivered mainly neutrinos, with an antineutrino component of 10%. I think that the bulk of the dimuon events are neutrino-induced. This is supported by the fact that the flux is less abundant at higher energies, whereas the dimuon events show a tendency to occur at high energies.

- *GOTTLIEB*:

You showed data at various energies, testing the validity of the Adler sum rule, and compared it with the respective Gargamelle data. At what energy were the Gargamelle data taken, and do you have any overlap with their energy domain?

- *RUBBIA*:

The average Gargamelle neutrino energy is about 2 GeV. Unfortunately, we do not yet have an overlap in neutrino energy, since the acceptance of our present set-up is not well suited for very low energy events. However, since an overlap with the Gargamelle energy domain is highly desirable, our set-up is being adapted to that purpose. In particular, an iron core magnet, 24 feet in diameter, will be installed to increase the acceptance for low energy, large angle muons.

- *MILLER*:

How did you calibrate your detector in order to be certain that you are measuring y correctly, especially at small y?

- *RUBBIA*:

A liquid scintillator is sensitive to very small energy depositions. The calorimeter was calibrated with hadron beams of known energy, and we continuously monitor the Landau distribution generated by beam associated muons. At very high energies, we have to apply a correction factor of 1.3 to account for partial containment of the hadronic shower, heavy ionizing prongs, nuclear de-excitation, etc. At low energy, the calorimeter is extremely reliable, much more so than iron sampling calorimeters, the resolution of which diverges at low y.

- *MILLER*:

You use a quadrupole train for focusing. Small misalignments could produce a considerable difference in the effective source position for ν and $\bar{\nu}$, and hence a bias in the relative q^2. How can you be sure you are using the correct source position for both neutrinos and antineutrinos?

- *RUBBIA:*

Source positions can be checked in at least two ways: first, by checking the absence of left-right or up-down asymmetry of events in the detector; secondly, by repeating the test of the Adler sum rule for quasi-elastic events. These show a strong form factor dependence on q^2, so that if the ν, $\bar{\nu}$ beams were misaligned, the difference of cross-sections should not stay smoothly at zero when $q^2 \to 0$. In addition, the decay pipe is only one metre wide and is 1 km away; if either π^+ or π^- parents were not aimed properly, we would only lose flux, not move the effective source by very much.

- *PESKIN:*

Which fraction of your total event sample are "anomalous y events" and "dimuon events"?

- *RUBBIA:*

After corrections for acceptance and trigger inefficiency, about 5 to 10% of the total event sample can be regarded as candidates for anomalous y. As for dimuon events, we observe them at a level of about 0.5%, not corrected for efficiency. Assuming a uniform production of these events within our set-up, we estimate the over-all efficiency to be about 50%. So the dimuon production rate is about 1% of the total neutrino cross-sec section above 50 GeV.

- *KOZANECKI:*

How do the lower energy data behave under the Adler sum rule test, at low q^2?

- *RUBBIA:*

The lowest energy is $\langle E_\nu \rangle \sim 15$ GeV. There seems to be an excess of events at low q^2, similarly to what happens in the higher energy data, but the statistical significance is poor.

- *GOURDIN:*

This is a comment rather than a question. The correct way to test the Adler sum rule is to plot the difference $d\sigma^\nu/dq^2 - d\sigma^{\bar{\nu}}/dq^2$ versus incident energy E for fixed q^2. The prediction is:
a) at large energy we get a limit measuring the commutator of the weak charged current with its hermitian conjugate, and
b) that limit is independent of q^2 and has nothing to do with scaling.

- *RUBBIA:*

You are right: instead of considering $\nu \to 0$, we should look for a q^2 independent limit at infinite neutrino energy; but this is impossible to achieve. So our way of interpreting the Adler sum rule could be misleading.

DISCUSSION

- *CALDI:*

Do you throw out any dimuon events as being due to decay rather than direct production?

- *RUBBIA:*

In the $\mu^+\mu^-$ sample, there are two reasons not to reject events. Empirically, the comparisons of rates in two targets of very different density shows it is unnecessary to reject any event. Next, our Monte Carlo calculation suggests a rejection of at most 10% in the low transverse momentum region; this amounts to less than the statistical errors.

The $\mu^-\mu^-$ events are treated on an individual basis, as we have only four of them.

- *WEISSKOPF:*

Basically, all you have shown is that scaling does not work at low q^2. So why do you put so much emphasis on a possible breakdown of the Adler sum rule?

- *RUBBIA:*

Even if you disregard the considerations on the Adler sum rule, the y anomaly persists for $q^2 > 1$ (GeV/c)^2 and up to 5 or 10 (GeV/c)^2. What disturbs me in the test of the so-called Adler sum rule, is that there is a discrepancy between Gargamelle data and ours, whether or not our method really probes correctly this sum rule.

I would like to add that our data agree very well with the Gross-Llewellyn Smith sum rule.

DISCUSSION No. 3

- *DYDAK:*

You always assume in the dimuon events the existence of an unobserved neutrino in the final state. Although this is supported by muon lepton number conservation, there is, so far, no experimental evidence for that.

- *RUBBIA:*

True. We cannot observe an outgoing neutrino by the missing energy method. As far as the muons are concerned, we can definitely rule out the hypothesis of penetrating particles of mass higher than m_μ. This is based on the observation that a substantial fraction of the muons stop inside the apparatus, and the observed range coincides with the expectation for a muon of the observed momentum.

— PETRONZIO:

Just a comment about dimuon events: I think that is would be extremely useful to look at similar threshold effects in neutral current events. In that case, one should observe a decrease in the counting rate of muonless events, since the muon produced at the hadron vertex would simulate charged current events. In the first possible case, there is such a threshold, in the second case there is not. In my opinion, both possibilities are interesting from a theoretical point of view, looking at present models for neutral currents.

If the threshold exists, since neutral currents, as Politzer pointed out, cannot change the type of quark, say the flavour, it would mean they change the colour, and that such a threshold is a colour threshold. The second case, where there is no threshold effect, could be explained by saying that such a threshold is related to a change of type of quark, as in the charm scheme.

— CABIBBO:

In colour models you expect a threshold in neutral current processes, since the neutral current is not a colour singlet. This could be a large effect.

— DIDDENS:

Apart from the anomaly for $x > 0.1$ and the new phenomenon of $\mu^+\mu^-$ production, you showed in the beginning of your lecture evidence for another anomaly, namely an excess of events in $f_2(x)$ for x around 0.2 as compared to SLAC and Gargamelle data.

— RUBBIA:

We have not spent that much effort on the effect you mentioned, but my guess is that SLAC energies are too low for scaling to have been reached, expecially in view of the existence of the ψ or J particles.

RECENT RESULTS FROM THE CALTECH-FERMILAB NEUTRINO EXPERIMENT*

B. C. Barish

California Institute of Technology, Pasadena, California, 91125, USA

It has been just one year since we verified the existence of "neutral currents" ($\nu + N \rightarrow \nu +$ hadrons) in our neutrino experiment at Fermilab. Last year, at the XVII International Conference on High Energy Physics at London[1] and then in a special session on neutral currents here at Erice[2], we showed convincing evidence for the existence of neutrino interactions with no muons in the final state.

The prime focus of the research of our group this past year has been to conduct a neutrino experiment under rather special running conditions, which I will describe in these lectures. The purpose of this experiment was to follow up the exciting discovery of neutral currents with the first of what should be a series of measurements especially designed to understand both the strength of the neutral current interaction and the nature of its coupling.

As an exciting side product of this experiment we have observed neutrino events with two muons in the final state. These events appear to be due to new and as yet unexplained physics! I will describe the observation of these events, their characteristics, and comment on possible physics explanations.

We have also studied our ordinary charged current events ($\nu + N \rightarrow \mu +$ hadrons), especially with antineutrinos, looking for any possible thresholds or anomalies. I will also describe our present understanding and interpretation of this charged current process.

Neutral Current Investigations

Our first experiment on neutral currents was oriented only toward seeing if neutral currents really existed. A great deal of effort was spent to make the experimental search unambiguous--and as much as possible to directly test for any possible source of background.

The basic technique involved identifying neutrino (or antineutrino) induced events and measuring the amount of Fe traversed by the most penetrating particle. In general, if a muon is present in the final state it only loses energy by ionization loss. Therefore, it typically penetrates our Fe target until it either emerges from the end (and goes through our magnet) or is at a large angle and escapes out the sides of the apparatus. The hadrons made in the neutrino collision, on the other hand, interact strongly making a hadronic and electromagnetic cascade in the iron. Typically, the most penetrating particle in a hadronic cascade traverses only about 1 meter of iron. Penetrations significantly longer than this indicate the presence of a muon in the final state.

Figure 1 shows the data from this first experiment where the distributions of the material traversed by the most penetrating particle for both neutrinos and antineutrinos are shown. Note from the scale at the top of the figure that, if a muon is present in the final state, the penetration is inversely related to the muon angle. There is not, however, a one-to-one correspondence because of the finite transverse size of the target. The top scale represents the mean muon angle corresponding to a given penetration before escape out the sides.

The most striking feature of the graphs are the large peaks for $P \sim 10$ collision lengths (~ 1 meter of Fe). These peaks for both ν and $\bar{\nu}$ have been interpreted as a neutral current signal. The peak is typical of the distribution of the most penetrating particle expected if only hadrons exist in the final state.

Our best estimate of the number of charged current events ($\nu + N \rightarrow \mu^-$ + hadrons) under this peak is also shown. This background represents scattering at such large angles that the muon does not penetrate much matter before escaping out the sides of the apparatus. The background curves are drawn using a simple fit to the charged current distributions $P > 13$ C.L. (collision lengths: 1 C.L. = 10 cm of Fe) and extrapolating to wide angles (or short penetrations). The signal far exceeds this estimated charged current background both for neutrinos and antineutrinos.

Since the estimate of charged current "fakes" does involve an extrapolation we have done a final check to see if an anomalous distribution of wide angle muons from charged current interactions could account for the results. This check is possible in our experiment since we use a dichromatic neutrino beam with $< E_\nu > \sim 40$ GeV

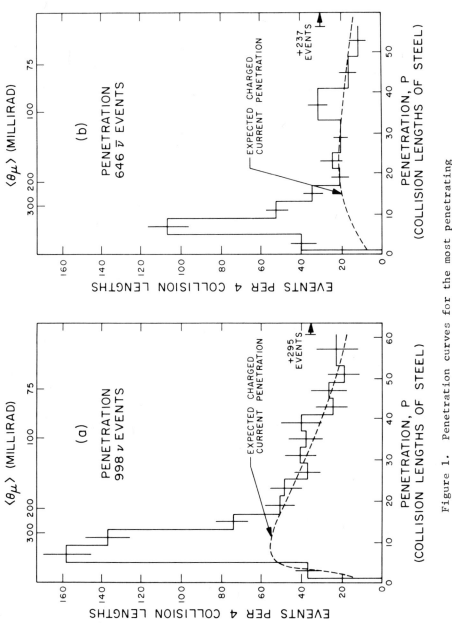

Figure 1. Penetration curves for the most penetrating particle in neutrino and antineutrino collisions.

for neutrinos from π - decay. There are very few very low energy
neutrinos in the beam. Kinematically, as the muon from charged current
interactions goes to larger angles more and more of the neutrino energy
must be carried away by the hadrons. Figure 2a shows the distribution
in hadron energy for all charged current events with an identified
muon (P > 14 C.L.). Note the distribution has a significant number
of events with little hadron energy (low inelasticity). In contrast,
figure 2b shows the hadron energy distribution for charged events
with 18 C.L. $\leq$ P $\leq$ 26 C.L. characteristic of $\langle\theta_\mu\rangle \sim$ 200 mrad. For
this data the hadrons tend to carry away a larger fraction of the
neutrino energy. The solid line is the predicted shape. For charged
current events giving P < 14 C.L. we expect $\langle\theta_\mu\rangle$ to be so large that
$E_{had} \sim E_\nu$ as reflected in the solid curve in figure 2c. The normalization is the expected level of charged current background using the
fits described above.

Note that if the charged current background were underestimated
by a factor of two we would expect the dashed curve in figure 2c.
The data, however, look characteristically more like the shape of 2a.
This means that at least statistically the hadrons are not carrying
away most of the neutrino energy and a larger level of charged current
events with muons at large angles cannot describe the data.

In summary, what was observed in this initial experiment were
too many events with no identifiable muon and evidence for missing
energy in the interaction (presumably carried off by a neutrino).

The "raw" level of the excess in this initial experiment corresponded to

$$R = 0.22$$

$$R_{\bar{\nu}} = 0.33.$$

These are uncorrected ratios and do not represent the ratio of neutral
current to charged current cross sections. However, we estimate that
the ratio of the cross sections are within a factor of two of these
raw ratios.

It should be emphasized that all neutral current experiments
measure only partial cross sections since a minimum observable energy
in hadrons (minimum y) is needed to even observe neutral currents.

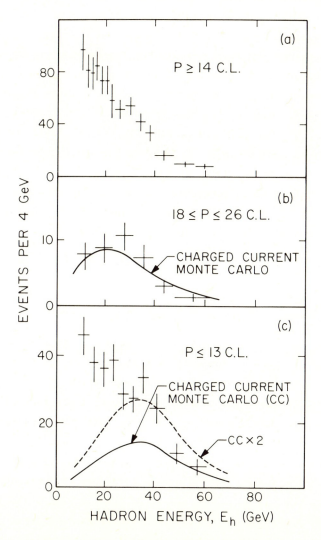

Figure 2: Hadron energy distributions for (a) all charged current events, (b) wide angle or short penetration charged current events and, (c) events with no penetrating particle.

To obtain the ratio of total cross sections σ_{NC}/σ_{CC}, the neutral current hadron energy distribution must be extrapolated to zero hadron energy. Therefore, knowledge of the y-distribution is necessary to extract total cross sections from the partial cross sections that are actually measured.

Early data from Argonne[3] indicated for the reaction $\nu n \to \bar{\nu} \pi^- p$ a large number of events near threshold for the $\pi^- p$ system. This was interpreted as possibly indicating the need for a large S or P coupling. More recent data, however, from the Argonne group[4] indicated a neutron background was probably causing the effect. Nevertheless, rather than just accept the V,A hypothesis the whole question of V,A or S,P,T should be explored experimentally.

It is interesting to note that the angular distribution of the neutrino, and hence the inelasticity distribution, in the neutral current reaction depends on the type of coupling.

Type of Coupling	Distribution ($d\sigma/dy$)
V,A	$a + b(1-y)^2$
S or P	y^2
T	$(2-y)^2$

In order to pursue the physics of neutral currents after the effect was firmly established we conducted a second neutral current experiment where the intent was to obtain information relating to the neutral current y-distribution, and thus to the nature of the neutral current coupling. In order to get a handle on the y-distributions for neutral currents we have started a program of measuring carefully the E_{had} distributions for neutral current events while changing the spectrum of the beam. In this way the change in total neutral current event rate reflects the energy dependence of neutral currents, and the hadron energy distributions reflect the y-dependence.

The idea is to take data with the neutrino beam mis-steered with respect to the apparatus. Since the neutrinos from $K \to \mu\nu$ are produced at wider decay angles than those from $\pi \to \mu\nu$, changing the steering angle varies the relative number of high energy neutrinos (from K's) and lower energy neutrinos (from π's).

I report here on a preliminary analysis of data where the neutrino beam was steered at ~ 1 mr with respect to the apparatus. This yielded a neutrino spectrum giving roughly equal numbers of events from ν_π and ν_k. (At 0 mr there are about 4 ν_π events for every ν_k event.)

We have used this data so far just to look at the extreme cases and see if the coupling is consistent with a <u>dominant</u> S or P, V-A, or V + A type coupling.

In general for neutral currents,

$$\frac{d\sigma_\nu}{dy} = \frac{G_N^2 M E_\nu}{\pi} (\alpha + \beta(1-y)^2 + \gamma y^2)$$

$$\frac{d\sigma_{\bar\nu}}{dy} = \frac{G_N^2 M E_{\bar\nu}}{\pi} (\alpha (1-y)^2 + \beta + \gamma y^2)$$

where α and β change roles for ν and $\bar\nu$. For pure V-A coupling one expects mostly α; pure V + A mostly β; and dominant S or P mostly γ. Other mixtures of V and A or S and T have not yet been investigated.

Each of these couplings produces a different ratio of neutrino to antineutrino total neutral current cross-sections:

$$\text{pure } \alpha \to \sigma_\nu = 3\sigma_{\bar\nu}$$
$$\text{pure } \beta \to 3\sigma_\nu = \sigma_{\bar\nu}$$
$$\text{pure } \gamma \to \sigma_\nu = \sigma_{\bar\nu}$$

Figure 3 shows the neutrino data from this new experiment. The data has been cut at $E_{had} \geq 12$ GeV to insure good efficiency for identifying neutral current events. Figure 3c shows the best fits to the neutrino data for $E_{had} \geq 12$ GeV for pure α, pure β, and pure γ. A total cross section for neutrinos is obtained from the extrapolation to $E_{had} = 0$ in all three cases.

These yield for the ratio of neutral to charged currents (for neutrinos):

$$\alpha : R_\nu = .23$$
$$\beta : R_\nu = .37$$
$$\gamma : R_\nu = .18$$

Note the large difference in the total cross section depending on the y-distribution. The relative normalization for each case has

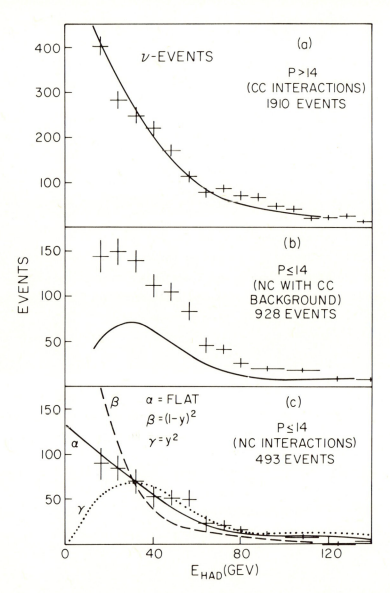

Figure 3: Hadron energy distributions from new experiment. Best fits to NC distributions with CC statistically subtracted are shown in (c). The fits are normalized to $E_{had} > 12$ GeV.

then been used to predict curves for pure α, pure β, and pure γ for antineutrinos. These curves along with the data for antineutrinos are shown in figure 4. Both pure β and pure γ are very bad fits. Pure α is closer but is too low in normalization by about a factor of two.

From these curves, without doing a detailed best fit, we have so far concluded that dominant S or P is ruled out; dominant V + A is also ruled out. Pure V-A appears unlikely because of the normalization, however all corrections to the data should be applied before a firm conclusion can be reached for V-A.

In conclusion, we are now starting to determine the nature of the neutral current coupling and expect to have best fits in a couple of months for this data. Also, in future improved experiments we hope to pursue this interesting question and obtain more precise data and determine the coupling with good accuracy. This will involve more extensive runs with an improved apparatus.

Neutrino Events With Two Muons In The Final State

For some time now we have been observing ν-events with two muons in the final state.[5]

$$\nu + N \rightarrow \mu + \mu + X$$

They occur for about 0.5 - 1% of all neutrino interactions. A possible source for these events, which would not imply new physics, is from the decay of pions or kaon in ordinary neutrino interactions

$$\nu + N \rightarrow \mu + X \ (\pi\text{'s, K's, etc.....})$$
$$\rightarrow \mu + \nu$$

It is difficult to estimate these possible sources of background for several reasons. The multiplicity and distribution of final state hadrons in high energy neutrino interactions is not yet available. That means that some assumptions are necessary in order to estimate these sources of extra muons. Also, the pions and kaons from the initial neutrino collision reinteract making more (but lower energy) secondary pions and kaons, which in turn reinteract until eventually the hadron cascade is absorbed in the Fe-calorimeter target. This

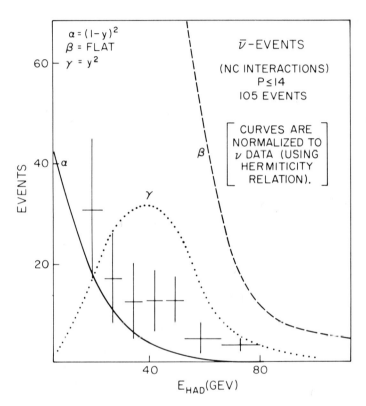

Figure 4: Hadron energy distributions for antineutrino neutral currents. The three curves are the shapes for pure α, β, and γ normalized by the fit to the ν-data and the predicted relative cross sections for ν and $\bar{\nu}$.

means a rather elaborate calculation following the development of hadron cascade and possible decays must be done in order to estimate the total spectrum and number of muons arising from pion or kaon decay.

In order to make the most reliable estimates of these possible backgrounds, I report here on the results of a search for events with an "extra" muon in the neutral current experiment just reported. That particular experiment involved taking particular care to measure the hadron energy distributions resulting from neutrino interactions.

For that experiment we defined a muon as a particle penetrating ≥ 2 meters of steel following the ν-interaction. A two muon event is defined as an event where two particles are found that penetrate ≥ 2 meters. This implies that the minimum energy of a detectable muon is ≥ 3 GeV. For this data sample we found

	ν-Running	$\bar{\nu}$-Running
1 μ	2313	446
2 μ	19	2

In order to estimate the possible backgrounds we used the <u>measured</u> hadron energy distributions from ν-events with 1 μ in the final state (figure 5) and the <u>calculated</u> integral probability of finding a muon from π or K decay with $E_\mu > E$ (figure 6). The calculated decay probability makes the assumption that a ν-interaction with a given E_{had} has the same distribution and multiplicities as a pion induced reaction of the same energy.

Figures 7 and 8 show the resulting expected distributions from π and K decay compared with the observed data. The excess number of neutrino events with a high energy second muon for ν-interactions cannot be explained by π or K decay. However, for antineutrinos no evidence of anomalous production (with limited statistics) has been observed.

We interpret this signal as being indicative of new physics and probably the same phenomena observed in the Harvard-Penn-Wisconsin-Fermilab ν-experiment.[6] At this point we can only get some preliminary idea of the physics properties of these events. Table 1 shows the parameters of nine events where at least one of the muons

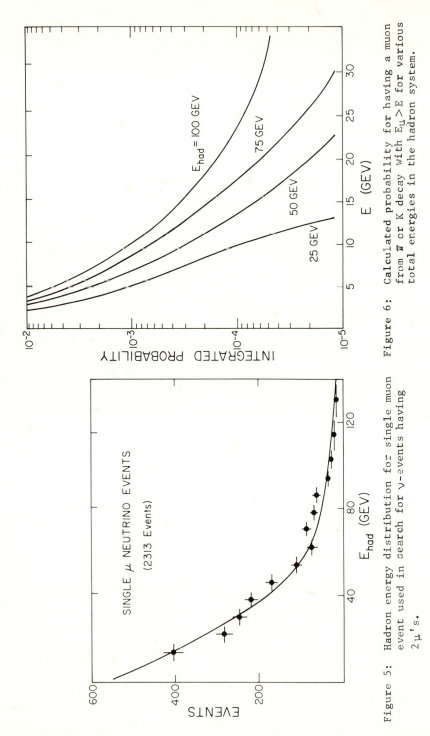

Figure 5: Hadron energy distribution for single muon event used in search for ν-events having 2 μ's.

Figure 6: Calculated probability for having a muon from π or K decay with $E_\mu > E$ for various total energies in the hadron system.

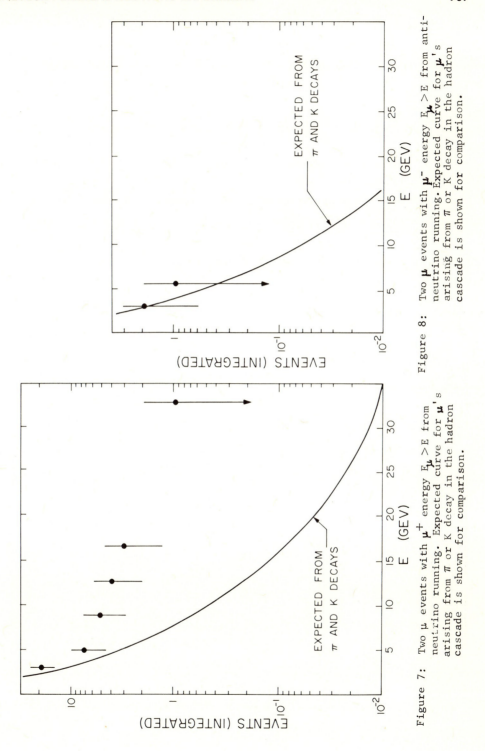

Figure 7: Two μ events with μ^+ energy $E_\mu > E$ from neutrino running. Expected curve for μ's arising from π or K decay in the hadron cascade is shown for comparison.

Figure 8: Two μ events with μ^- energy $E_\mu > E$ from antineutrino running. Expected curve for μ's arising from π or K decay in the hadron cascade is shown for comparison.

was momentum analyzed in our muon spectrometer.

Until the geometric efficiencies, biases, etc. and background levels from π or K decay are better understood, any conclusions are extremely preliminary. However, certain characteristics have already emerged.

(1) The events occur with typical x and y for deep inelastic scattering.
(2) There is no apparent peak in the mass distribution for the 2 μ system.
(3) The events appear to be produced more prolifically with 150 GeV than 50 GeV neutrinos.
(4) The invariant mass distribution appears to favor very high invariant masses recoiling against the $\bar{\mu}$.

Caltech-Fermilab 2 μ Events With One Muon Measured (ν)
All Energy Units in GeV

Event	Type	E_{μ}^{-}	E_2	E_{HAD}	E_{OBS}	ν_{type}	$m_{\mu\mu}$	x	y	W	p	p*
1	I	62.1	+44.8	25.1	132.0	ν_K	5.4	.47	.53	10.4	2.2	4.0
2	I	29.8	+19.0	82.9	131.7	ν_K	0.5	.04	.77	13.6	0.8	1.5
3	I	84.8	+18.8	64.6	168.2	ν_K	3.4	.33	.50	10.9	0.7	1.4
4	I	62.2	+14.1	71.2	148.5	ν_K	2.6	.19	.58	11.6	1.0	1.4
5	II	≥4.4	+10.3	56.6	≥ 71.3 / =150	ν_π / ν_K	>.01	<.94 / .36	/ .45	≤11.1 / 9.0		
6	III	56.1	≥ 6.8	59.5	≥122.4 / =150	ν_K		.17	.63	12.1		
7	III	102.4	≥ 4.3	25.6	>132.3 / =150	ν_K		.04	.32	9.3		
8	III	122.5	= 3.3	13.5	139.3	ν_K		.11	.11	5.3		
9	III	64.1	≥ 3.0	14.7	≥ 67.1 / =150	ν_π / ν_K		.43 / .20	.26 / .57	4.4 / 11.3		

$$\begin{cases} <x>_{obs} \text{ for dimuon events} = 0.21 \\ <x>_{obs} \text{ for all charged current events} = 0.24 \end{cases}$$

p = transverse momentum of second muon relative to hadron (W) direction.

p^* = total momentum of second muon in W rest frame.

What are the possible sources of these neutrino events with an "extra" muon in the final state?

There are basically three classes of possibilities illustrated in figure 9:

(I) Decay of π's of K's - As we saw earlier this is not likely unless the assumptions made to estimate these possible backgrounds are wrong. In particular, if for some reason fast pions (current fragmentation) are made with high probability these would yield more high energy mesons than calculated. Although this is probably an unlikely possibility, there are no empirical data at this time which allows us to rule out this logical possibility.

(II) New Physics at the lepton vertex - An example would be production of a new neutral heavy lepton via exchange of a $Z°$. This lepton, $M°$, if it is ~ 5-8 GeV, left handed, and decays via V-A interaction ($M° \rightarrow \mu^+ + \mu^- + \nu$) would yield rates and distributions similar to those observed.[7]

(III) New particle production in the Hadron System - Charm particles or some other source of particles could decay leptonically and give rise to an extra muon. However, the "classical" charm picture seems unlikely since the events appear with a probability of ~ 1% with typical deep inelastic x and Q^2. This means that they are made of ordinary valence quarks and would be expected to be suppressed by $\sin^2\theta_c$ in production. This means that unless leptonic modes of decay completely dominate the absolute rate appears high.

These three possibilities just represent examples of explanations. As of now we just do not know enough yet experimentally to understand the physics. Some new information should be available soon. There are indications that ~ 10% of the time both μ's have the same sign

(1) Hadron distributions are different for Neutrino collisions

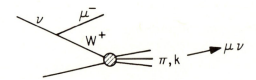

(2) Production of a <u>Neutral</u> Heavy Lepton (Mass 5-8 GeV)

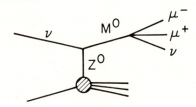

(3) Production of New Particle in Hadrons (eg Charm)

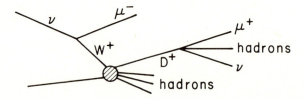

Figure 9: Illustration of some physics possibilities for origin of 2μ events.

from the HPW experiment.[8] If this is really a signal and not π or K decay background it could be an extremely important fact. We expect in the Caltech experiment to have information later this year on production by antineutrinos (if it exists). The present data sample is just not statistically significant.

Although more statistics and the above mentioned results will become available in the near future, it is likely that better experiments will be necessary to follow up and understand this phenomena. Improvements to both our experiment and the HPW experiment will be made over the next year. The improved experiments will have better geometric acceptance and hopefully better handles on understanding the π and K decay problem.

For this year we should feel excited that this new physics has been uncovered! Let's just hope that with hard work over the next year or so the physics implications will become clear.

Charged Current Interactions

The early work on the charged current reactions ($\nu + N \rightarrow \mu +$ hadrons) has been focused on seeing whether any _gross_ effects indicating a new threshold are evident. This appears to be even more important with the recent discovery of a probable threshold evidenced by the observation of ν-events with 2 muons in the final state.

In our work on charged currents we have emphasized tests for indications of new thresholds as the incoming neutrino energy increases. We have approached this problem in two ways. The first has been to obtain normalized cross sections to determine the energy dependence and to compare neutrino and antineutrino scattering, and the second has been to investigate the behavior of the differential cross sections.

Using the dichromatic neutrino beam we are uniquely suited to pursue normalized work. It is possible to accurately monitor and measure the K/π/p ratio in the hadron beam, the energy of the hadrons, the steering of the ν-beam, and the total flux. This means that the largest source of systematic error in determining normalized cross sections can be minimized. A major effort to obtain normalized cross sections will be undertaken this summer and fall. However, a preliminary experiment was conducted last year.

In that experiment the hadron beam was set at 120 GeV. A beam survey was performed to determine the K/π/p ratio and ion chambers were used to directly measure the ν-flux. Nevertheless, in this preliminary run the largest source of systematic error in determining cross sections was still the flux determination.

The 120 GeV hadron beam yielded two bands of neutrinos $E(\nu_\pi) \sim 38$ GeV and $E(\nu_k) \sim 105$ GeV. Both neutrino and antineutrino cross sections were measured, as well as the differential distributions in E_μ, E_{had}, θ_μ, etc. The errors in the knowledge of the deep inelastic distribution also are a significant source of error since corrections for the geometric acceptance of the apparatus are dependent on these distributions.

The results from this experiment,[9] along with previous results from Gargamelle[10] at lower energies, are shown in figure 10.

On the Caltech points the inner error bars represent statistical errors only and the outer error bar includes systematic errors.

In general, the differential cross sections for ν-scattering have the form

$$\frac{d^2\sigma^\nu}{dxdy} = \frac{G^2 M E_\nu}{\pi} [q(x) + (1-y)^2 \bar{q}(x) + (1-y) k(x)] \quad (1)$$

$$\frac{d^2\sigma^{\bar\nu}}{dxdy} = \frac{G^2 M E_{\bar\nu}}{\pi} [\bar{q}(x) + (1-y)^2 q(x) + (1-y) k(x)] \quad (2)$$

assuming scaling (i.e. $q(q^2, \nu) \to q(x)$ etc.) and that the W propagator is heavy $(G/(1 + q^2/M_W^2)^2 \to G^2)$. A deviation from these formulas could imply non-scaling effects such as new particle production in the hadron system or might be due to weak interaction effects.

Integrating these relations we obtain expressions for the total cross sections

$$\sigma_\nu = \frac{G^2 E_\nu}{\pi} [Q + \frac{1}{3}\bar{Q} + \frac{1}{2} K] \quad (3)$$

$$\sigma_{\bar\nu} = \frac{G^2 E_{\bar\nu}}{\pi} [\bar{Q} + \frac{1}{3} Q + \frac{1}{2} K] \quad (4)$$

where $Q = \int_0^1 q(x) \, dx$, etc.

The total cross sections for both ν and $\bar\nu$ grow linearly

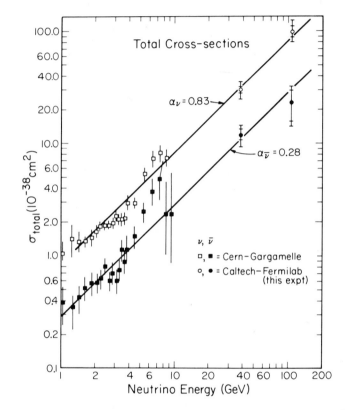

Figure 10: Neutrino and antineutrino total cross sections vs. E_ν for the Gargamelle and Caltech-Fermilab experiments.

with energy and the relative slopes determine the relative magnitudes of Q, $\bar{Q}$, and K. The results in figure 1 are consistent with linearly rising cross sections from ~ 1-100 GeV. Also, from the slopes in figure 10 $\frac{\sigma_{\bar{\nu}}}{\sigma_{\nu}} \sim \frac{1}{3}$ which implies $Q >> \bar{Q}$ and $Q >> K$. K represents the non-spin $\frac{1}{2}$ scattering off and Q represents scattering.

So the conclusions from the early normalized cross section measurements show no major deviations from the simplest picture and in contrast show surprisingly accurate agreement.

More recently, the focus has been on the x and y differential distributions. A so-called "High-y Anomaly" has been reported for antineutrino scattering by the Harvard-Penn-Wisconsin Group.[11] What has been reported is in good agreement with the simplest picture ($q(x) = 0$ and $k(x) = 0$ for $x > 0.1$, that is, $\frac{d\sigma^{\nu}}{dy} \sim$ flat and $\frac{d\sigma^{\bar{\nu}}}{dy} = (1-y)^2$. However, for small x, normalizing to the distributions for $x > 0.1$, they observe that the distribution becomes flat for both ν and $\bar{\nu}$, while the ratio of antineutrino to neutrino cross sections remains $\frac{1}{3}$! This has been interpreted as an excess of events at large y indicating a possible threshold has been crossed. In fact, the effect is only observed for $E_{\bar{\nu}} > 30$ GeV which has been interpreted as indicating that this phenomena might somehow be connected to the production of 2 μ's by neutrinos. (Production of 2 μ's has been observed only for high energy neutrinos).

What should we really expect for the y-distributions at small x? Assuming $k(x) = 0$ the differential cross sections have the form

$$\frac{d\sigma^{\nu}}{dy} \quad q(x) + \bar{q}(x) (1-y)^2$$

$$\frac{d\sigma^{\bar{\nu}}}{dy} \quad \bar{q}(x) + q(x) (1-y)^2$$

Although $\int q(x)dx >> \int \bar{q}(x)dx$, most of the integral for $\bar{q}(x)$ is expected to be at small x in a quark picture and in fact $\bar{q}(o) = q(o)$. So $\bar{q}(x)$ is expected to have a very different shape from $q(x)$ and in fact fall very rapidly with x. (This is also what the Gargamelle results indicate).

Since $\bar{q}(x)$ may be significant for small x, it should not be neglected, a priori. We have included a $\bar{q}(x)$ form in our analysis;

however, since we do not know what shape to use for $\bar{q}(x)$ and the data is too poor to determine it directly we must assume a form and just see whether we are consistent with such a picture. It should be noted that the behavior of the x and y distributions are not independent in these experiments because of the acceptance. The inefficiency at large θ_μ represents the region where x and y are both large.

We have assumed the following form for the x-dependence,

$$F_2(x) = F_2^{ed}(x)$$

$$q(x) + \bar{q}(x) = F_2(x)$$

$$\bar{q}(x) = \frac{F_2(x)}{2} e^{-bx}$$

Then the $\bar{Q}$ component is

$$\alpha = \frac{\int \bar{q}(x)\, dx}{\int F_2(x)\, dx}$$

Figure 11 shows these shapes and figure 12 shows the determination of α, fitting simultaneously data for a muon through the magnet with $x > 0.1$, $x < 0.1$, and E_{had} distribution for events with muon missing the magnet (wide angle).

The best fit is $\alpha = 0.11^{+0.13}_{-0.09}$ for $<E_\nu> \sim 50$ GeV. For $x < 0.1$, this fit gives for the y-dependence

$$\frac{d\sigma^{\bar{\nu}}}{dy} \quad 0.3 + (1-y)^2$$

$$\frac{d\sigma^{\nu}}{dy} \quad 1.0 + 0.3(1-y)^2$$

We conclude that with limited data we are consistent with such a simple quark model having a finite $\bar{q}$ distribution that falls rapidly with x.

It should be noted that this fit to our data agrees with expectations from low energies (Gargamelle) and with the HPW results at large y. For Gargamelle the distributions for $q(x)$ vs. x and $\bar{q}(x)$ vs. x have been determined.[12] For $x < 0.1$ they obtain $\bar{Q}(x)/Q(x) \frac{0.25}{0.85}$ 0.29. Since the ratio of $\bar{Q}/Q$ can be determined

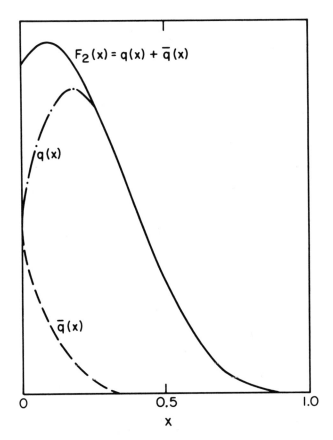

Figure 11: Assumed parameterization for x dependence. $F_2(x) = q(x) + \bar{q}(x) = F_2^{ed}(x)$ and $\bar{q}(x) = \dfrac{F_2(x)}{2} e^{-bx}$. The $\bar{Q}$ component is then $\alpha \equiv \dfrac{\int_0^1 \bar{q}(x)\, dx}{\int F_2(x)\, dx}$. Shown are the shapes of $q(x)$ and $\bar{q}(x)$ for the best fit of $\alpha = 0.11$.

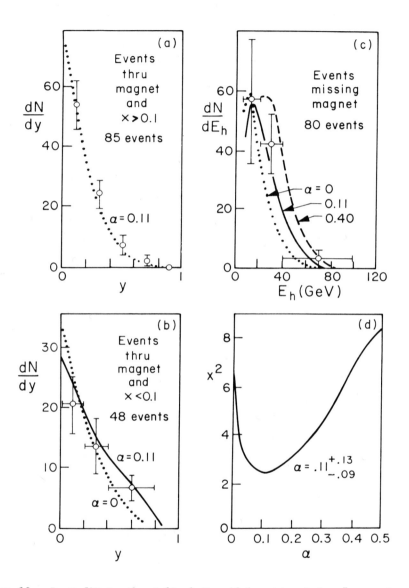

Figure 12: Best fit to the Caltech-Fermilab antineutrino $\bar{\nu}_\pi$ events. The three distributions are fit simultaneously.

by the ratio of the antineutrino to neutrino cross sections at y = 1, this implies

$$d\sigma^{\bar{\nu}}/d\sigma^{\nu} \quad 0.29 \quad \text{G.G.M}$$
$$d\sigma^{\bar{\nu}}/d\sigma^{\nu} \quad 0.3 \quad \text{Caltech} \qquad \text{for y } 1$$
$$d\sigma^{\bar{\nu}}/d\sigma^{\nu} \quad 0.36 \quad \text{HPW}$$

All three experiments are in agreement with a simple quark model with finite $\bar{q}(x)$. That is the determination from GGM and the ratio of antineutrino/neutrino cross sections as $y \to 1$ from the Fermilab experiments.

We conclude that if there is new physics in the antineutrino y-distributions as indicated in the HPW experiment the anomaly is probably at Low-y not at High-y! If the effect is at low-y then it cannot be due to a new threshold in the hadron system, but rather a breakdown of charge symmetry or something else in the weak interaction. At face value the effect is very large and should be easy to check. At small y, where the cross sections are expected to be equal for neutrino and antineutrino, they are found to differ by a factor of 3.

Our data is shown in figure 13. Both the neutrino and antineutrino data for x < 0.1 are normalized to the data for x > 0.1, which is the same procedure used by the HPW group. The Caltech-Fermilab results show no indication of strongly deviating from the picture described above with α = 0.11. However, the data are limited statistically and are not in sharp conflict with the HPW results.

Any conclusions regarding high-y vs. low-y, or whether there is a deficiency at low-y in ν-scattering depends very heavily on the reliability of the normalization procedure. We hope to improve on this and other sources of systematic errors by normalizing the data to externally measured fluxes in an experiment underway this summer, while at the same time acquiring much better statistics.

---

[*] Work supported in part by the U.S. Energy Research and Development Administration. Prepared under Contract AT(11-1)-68 for the San Francisco Operations Office.

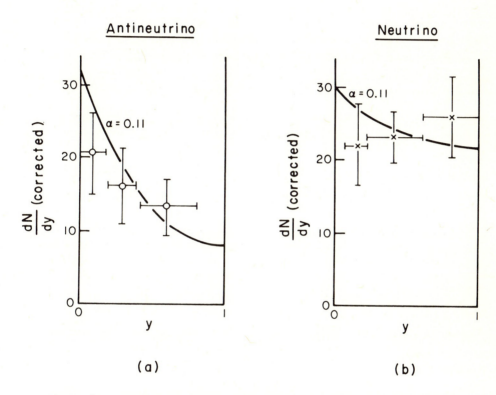

Figure 13: Antineutrino and neutrino y distributions for $x<0.1$. Both sets of data have been normalized to data for $x>0.1$. The curve is the best fit to the overall distributions described in the paper.

REFERENCES

1) B. C. Barish, et.al., Proceedings of XVII International Conference on High Energy Physics, (London, 1974), p. IV-111.

2) B. C. Barish, et.al., Proceedings of International School of Subnuclear Physics, (Erice, Sicily, 1975).

3) P. Schreiner, Proceedings of XVII International Conference on High Energy Physics, (London, 1974), p. IV-123.

4) L. G. Hyman, La Physique du Neutrino a Haute Energie Proceedings, (Paris, 1975), p. 183.

5) B. C. Barish, et.al., La Physique de Neutrino a Haute Energie Proceedings, (Paris, 1975), p. 131.

6) A Benvenuti, et.al., Phys. Rev. Letters $\underline{34}$, 419 (1975).

7) C. Albright, Fermilab-Pub-75/34-THY (1975), unpublished.

8) D. Cline, talk given at Washington APS Meeting (April, 1975).

9) B. C. Barish, et.al., Proceedings of XVII International Conference on High Energy Physics, (London, 1974), p. IV-105.

10) M. Haguenauer, Proceedings of XVII International Conference on High Energy Physics, (London, 1974), p. IV-95.

11) D. Cline, La Physique du Neutrino a Haute Energie Proceedings (Paris, 1975), p. 107.

12) Gargamelle Neutrino Collaboration, in Proceedings of XVII International Conference on High Energy Physics, (London, 1974), p. IV-146.

DISCUSSION

CHAIRMAN: B.C. Barish

Scientific Secretaries: F. Dydak and P. Steinhardt

DISCUSSION

— *KIRKBRIDE:*

What was the distribution of dimuon events along the calorimeter?

— *BARISH:*

The events occurred uniformly along the apparatus, typical for neutrino interactions, with no indication that they are background.

— *HEUSCH:*

Are the dimuon masses quoted by you in conflict with those quoted by the Harvard-Pennsylvania-Wisconsin-Fermilab (HPWF) group? Are these values heavily constrained by the acceptances of the two sets of apparatus?

— *BARISH:*

Our dimuon mass acceptance is rather flat, and drops at about 6 GeV to zero. The HPWF acceptance looks similar, but extends up to about 8 GeV. In both cases, the observed dimuon events are distributed from very low mass to the highest observable mass.

— *HEUSCH:*

You quote only high energy neutrinos inducing dimuon events. Is your relatively small muon acceptance biasing this result?

— *BARISH:*

Although our statistics are poor, it looks like we do not see events at somewhat lower hadron mass W, where our acceptance is comparable. In other words, our acceptance does not seem to favour high energy events.

— *DYDAK:*

Lepton conservation requires a neutrino in the final state. Do you have evidence for missing energy which could be attributed to an unobserved neutrino?

— *BARISH*:

On a single event basis, the energy resolution is not good enough to allow a firm conclusion. The mean observed energy for the events originating definitely from 150 GeV K-decay neutrinos is 145 GeV. For comparison, the corresponding average for single muon events is 152 GeV. If we obtained a sample of about 100 events, we probably would "detect" a missing neutrino by seeing a shift in the mean observed energy of the events.

— *MORPURGO*:

Why do you not have antineutrino induced dimuon events?

— *BARISH*:

The single muon antineutrino sample is about 5 times less than the neutrino sample. If, as we believe, the high energy events are really favoured, the expected dimuon rate should be further diminished, since the high energy antineutrinos are less abundant in our flux.

— *KIRKBRIDE*:

Are there any plans to change the beam energy to include high energy neutrinos from pion decay?

— *BARISH*:

Yes. During this summer's experiments, we will run at a series of energies, from which we hope to obtain more data on the energy dependence of dimuon production.

— *MILLER*:

What is the next experiment on dimuons?

— *BARISH*:

Firstly, this summer we will be collecting more dimuon data, up to ten times the number of events reported here. In particular, we hope to examine closely the antineutrino events. At the same time, the HPWF group will be sorting out their data to try to distinguish neutrino from antineutrino events as best they can. Secondly, within six months, both experiments will be changed in order to increase the muon acceptance. The HPWF group will install a 24 feet diameter magnet, and our group will get several 12 feet diameter magnetized iron target modules with very good acceptance for wide angle and low energy muons.

— *YOSHIKAWA*:

A few years ago, people assumed that scaling would start at $q^2 \sim 4$ GeV^2 for neutrino interactions. Has the situation changed?

DISCUSSION

- *BARISH:*

Our analysis is consistent with scaling setting in at $q^2 \sim 1\ GeV^2$. Also, evidence from Gargamelle indicates that scaling sets in very early.

- *KOZANECKI:*

The W distribution of the HPWF experiment shows a bump at high W, with $E_{\bar{\nu}} > 30\ GeV$ and $x < 0.1$. How is this bump related to the so-called "y anomaly"?

- *BARISH:*

The W plot shows essentially the same effect as the y plot. If you agree to the interpretation that the reported "large y anomaly" is rather "low y anomaly", one has a deficiency of events at low y and $x < 0.1$. From the $q^2-\nu$ plot, drawing the lines of constant W, one learns that the y and the W plots are correlated in that sense that a deficiency of events at low y would show up as a deficiency of events at low W as well, in a W mass plot.

- *RÜCKL:*

Would the use of x' rather than x improve the scaling behaviour?

- *BARISH:*

The difference between x and x' is not important at high energies.

- *CRONIN:*

Could you comment on the possible reasons for the discrepancy between your results and the HPWF data for the y distribution of antineutrino events with $x < 0.1$?

- *BARISH:*

The analysis of low x, low y data is exceedingly difficult. The x resolution is poor, and making a cut $x < 0.1$ has problems. Moreover, it is not clear whether quasi-elastic scatters should be included or not, and how to cut the data on q^2. At this point, the two sets of data have been analysed differently, and a careful comparison is necessary.

- *GOURDIN:*

Did you worry for antineutrino scattering about $|\Delta S| = 1$ transitions?

- *BARISH:*

So far, this $\sim 5\%$ correction has been neglected in the analysis.

- *GOURDIN:*

Consider the ratio $r_c = \sigma^{\bar{\nu}}/\sigma^{\nu}$ and the positivity constraints arising from charge symmetry. Writing $r_c = 1/3 + \delta$, one gets from your data $\delta < 0.08$. This result, together with positivity, in turn imposes upper bounds on the right-handed contribution I_+ via the relation

$$0 \leq I_+/I_- \leq \frac{9\delta}{8 - 3\delta} < 0.093 ,$$

as well as on the longitudinal contribution I_L via the relation

$$0 \leq I_L/I_- \leq \frac{3\delta}{2 - 3\delta} < 0.136$$

on the cross-section. I suspect your parameter α being just I_+/I_-, and I am surprised to see your fitted value

$$\alpha = 0.11 \; {}^{+0.13}_{-0.09}$$

since the upper limit allowed by positivity is 0.093, assuming a vanishing longitudinal contribution.

- *BARISH:*

The best fit result $\alpha = 0.11$ depends somehow on the parametrization for $\bar{q}(x)$ which we took as

$$\frac{1}{2} \times e^{-bx} \times F_2^{SLAC}(x) .$$

In the fit, α was only loosely bound between 0 and 0.5, not by positivity constraints.

- *GOURDIN:*

Again from positivity and taking the same δ as above, one arrives at the relations

$$0.48 \leq \langle y^{\nu} \rangle \leq 0.50, \text{ in agreement with } 0.49 \; {}^{+0.01}_{-0.02} ,$$

and

$$0.25 \leq \langle y^{\bar{\nu}} \rangle \leq 0.304, \text{ in agreement with } 0.31 \; {}^{+0.04}_{-0.05} .$$

- *KIRKBRIDE:*

Could you comment on the fit result $b \sim -0.3$ in your parametrization

$$[1 + b(1 - x)] F_2^{SLAC}(x) \; ?$$

DISCUSSION

- *BARISH:*

The term $b(1-x)$ was introduced to allow a deviation from the SLAC result for $F_2(x)$. The best fit value, $b = -0.3$, seems to indicate a slight deviation, but it should be taken into account that we are not yet sensitive enough on the basis of our data to draw any real conclusions. Especially the events at very low x cause a problem, due to their poor resolution. An effect at very small x, as reported in the Fermilab μ-scattering experiment could not yet be observed for these reasons.

- *BARBIELLINI:*

Suppose your beam is contaminated with some new, short-lived heavy states, e.g. charm states, giving rise to an increased electron neutrino flux. Would this affect your results on neutral currents?

- *BARISH:*

No, since short-lived particles decay before the momentum slit. Any possible background would be subtracted when measuring the background with a closed momentum slit. However, if a new, long-lived particle with a lifetime of about 10^{-8} sec were produced, it would create a new source of neutrinos.

- *ZICHICHI:*

Can you rule out a process like $\nu_\mu + N \rightarrow e +$ hadrons, which would violate the muonic lepton number conservation?

- *BARISH:*

Yes, since we observe in our neutral current event sample the missing energy taken by the outgoing neutrino. For the above process there is no missing energy.

REVIEW OF DIRECT LEPTON PRODUCTION IN NUCLEON-NUCLEON COLLISIONS[†]

James W. Cronin

Enrico Fermi Institute
University of Chicago
Chicago, Illinois 60637

INTRODUCTION

In this lecture we will review the experimental status of the observation of direct lepton production in nucleon-nucleon collisions. The conclusion of these considerations can be stated simply: the yield of direct leptons cannot be understood in terms of identifiable sources. Making realistic estimates of direct leptons from vector mesons, mass 3.1 GeV particles, and Drell-Yan processes, we fail by about a factor of five in predicting the direct lepton yields. In the first section of the lecture, we will discuss the techniques of measurement. We then review the current data, pointing out some discrepancies and inconsistencies at the AGS and PS energies. A review of the known sources of direct leptons is made, followed by speculations on other possible sources and suggestions for further investigation. We will not take the time to give a proper historical view here; a very good review with complete references was given by B. Pope at the recent Palermo Conference.[1]

In the discussion of the experiments, we will not give a balanced review of each experiment, but rather stress notable points. The reader should consult the original papers for a complete understanding of a given experiment.

I. MEASUREMENT TECHNIQUE

Direct leptons are defined as leptons coming from sources other than pion and kaon decay. This definition when referred to muons, refers to all muons whose parent is short-lived compared to pion and K decay. When referred to electrons it refers to sources other than internal pairs, external pairs, and K_{e3} decay.

A. Muons

A direct muon signal must be separated from copious muons arising from pion and K decay. The muon yield from hadron decay can be modulated by varying the distance between the source and a hadron absorber. The desired muons penetrate the absorber while the hadrons are absorbed. A schematic arrangement is shown in Figure 1a. Figure 1b shows the yield curve observed. The curve is distorted because the motion of the hadron absorber varies the apparent target size (because of multiple scattering) and hence the acceptance of the system. The variation of the acceptance must be corrected for, and the size of the correction depends on the specific arrangement, and increases with decreasing momentum. If the acceptance is uncorrected, then an apparent direct yield will result. One group is sufficiently conservative that they extract no information from their experiment which requires a correction. The remaining muon experimenters are bolder.

Assuming that the corrections can be made exactly, one can discuss the sensitivity of an experiment which does not suffer from statistical errors.

The yield of muons from pion decay can be easily calculated at $\theta_{cm} = 90°$. Most experiments have been carried out at $\theta_{cm} \sim 90°$. If the pion yield is parameterized by an invariant cross section

DIRECT LEPTON PRODUCTION IN NUCLEON-NUCLEON COLLISIONS

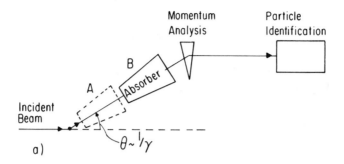

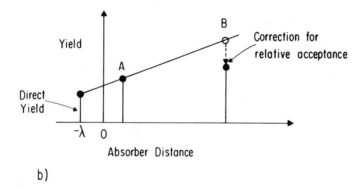

Figure 1. a) Schematic arrangement of an apparatus to detect direct Muons.

b) Idealized yield curve of muons vs. absorber position.

$$E \frac{d^3\sigma}{dp^3} = Ae^{-\alpha p_\perp}$$

then the differential cross section for the yield of muons from pion decay at a distance $d + \lambda$, where λ is the hadron absorption length, is given by

$$\frac{d^2\sigma_\mu}{dpd\Omega} = A \int_{p_\mu}^{1.75\, p_\mu} pe^{-\frac{\alpha p}{\gamma}} \left(\frac{m_\pi(d + \lambda)}{\tau_\pi cp}\right) \frac{dp}{.43p}$$

where $p_\perp \sim p/\gamma$ with γ being the center of mass motion of the colliding nucleon-nucleon system, p is the laboratory momentum of the pion, p_μ is the observed laboratory momentum of the muon, and $\tau_\pi c$ is the mean decay length of the pion in it's rest system.

Evaluating the integral (with some approximation) and dividing by the pion cross section $d^2\sigma_\pi/dpd\Omega$ at the same momentum we find:

$$\left(\frac{N_\mu}{N_\pi}\right) = \left(\frac{2.32}{\gamma}\right) \left(\frac{d+\lambda}{7.8}\right) \frac{\alpha m_\pi}{(\alpha p_\perp)^2} \frac{1}{\left[1 + \frac{1}{\alpha p_\perp}\right]} \left[1 - .472 e^{-.75\alpha p_\perp}\right]$$

If the K meson component is significant, then the number is to be multiplied by $\sim (1 + 2N_K/N_\pi)$. The factor 2 is in fact

$$\left[.43 \left(\frac{m_K}{m_\pi}\right) \left(\frac{\tau_\pi}{\tau_K}\right) B\right], \text{ where } B = \Gamma(K \to \mu\nu)/\Gamma(K \to \text{all})$$

In order to guess an ultimate sensitivity, we say that the decay path cannot be known to be better than ~ 4 cm. This can be due to finite length of targets, uncertainties in the absorption length, etc. The computed sensitivities are given in Figure 2. Note that the sensitivity basically goes as $\frac{1}{\gamma} \left(\frac{1}{p_\perp \alpha}\right)^2$ so that motion of the center of mass is very helpful and the experiments become dramatically more sensitive at high transverse momentum. Recalling that the absorber corrections become very important at low laboratory momentum, one understands the difficulty of these experiments at low $p_\perp$ and low energy.

Occasional criticism of direct muon experiments has centered about the fact that secondaries from the production target can produce muons in the absorber which appear as an apparent direct signal because of the multiplication by interaction of the hadrons in the shield. This effect is totally negligible at $\theta_{cm} = 90°$ because of the steeply

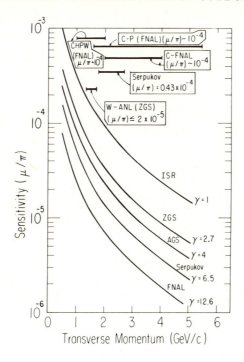

Figure 2. Sensitivity as a function of $p_\perp$ of direct muon experiments at $\theta_{cm} = 90°$ at the various proton accelerators. The $p_\perp$ region explored, the c.m. energy and the approximate (N_μ/N_π) ratio are indicated for each experiment.

falling spectrum coupled with magnetic analysis of the muons emerging from the absorber.

B. Electrons

Figure 3a shows schematically an arrangement to measure direct electrons. Essential features are: 1) low radiation length in the spectrometer, 2) electron identification and hadron rejection, and 3) a veto of pairs if small transverse momenta are to be attained. Figure 3b shows a typical extrapolation to zero external radiator.

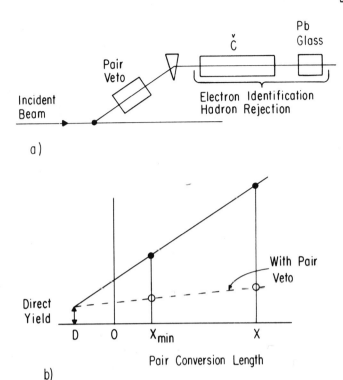

Figure 3. a) Schematic arrangement of an apparatus to detect direct electrons.

b) Extrapolation to zero external radiator as a function of pair

The principal source of apparent direct electrons is from Dalitz decays of π^0. Proceeding in a manner similar to the muon decay, one finds the contribution of electrons from π^0 decay to be

$$\left(\frac{N_e}{N_{\pi^0}}\right) = \frac{2(D + X)}{(\alpha p_\perp)^2} \left(\frac{1}{1 + \frac{1}{\alpha p_\perp}}\right)$$

where D is the Dalitz internal conversion coefficient, X is the thickness of external conversion material measured in <u>pair</u> conversion lengths.

DIRECT LEPTON PRODUCTION IN NUCLEON-NUCLEON COLLISIONS

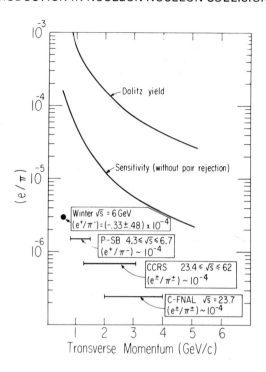

Figure 4. Sensitivity as a function of $p_\perp$ for direct electron experiments at $\theta_{cm} = 90°$ in which pairs are not vetoed. Also indicated is the total yield from internal pairs. The $p_\perp$ region explored, the c.m. energy and the approximate (N_e/N_π) ratio found are indicated for each experiment.

Since pair conversion lengths of materials are accurately known, the ultimate sensitivity in an ideal experiment is controlled by the uncertainty in D. The value of D measured by Samios[2] is $(5.83 \pm 0.23) \times 10^{-3}$. Figure 4 plots this sensitivity as a function of $p_\perp$ with the assumption $\alpha = 4$ and a two standard deviation uncertainty in the internal conversion coefficient. One notes that, in this case, the center of mass motion of the collision does not enter, so that the ISR does not suffer the disadvantage that it does in the muon experiments.

Some experiments have achieved high sensitivity by the vetoing of internal and external pairs. Typically, if one electron of a pair is observed in the spectrometer, the other pair can be detected which removes the event from the single electron category.

Another source of background comes from η decays. The relevant parameters for the η are $D_\eta = 8.1 \times 10^{-3}$, $\Gamma(\eta \to \gamma\gamma)/\Gamma(\eta \to \text{all}) = 0.38$. In an experiment where pairs are not vetoed the ratio (N_e/N_{π^o}) expected at $X = 0$ must be multiplied by $(1 + 0.5\, N_\eta/N_{\pi^o})$. If N_{π^o} is deduced via the decay gamma rays, then the ratio should be divided by $(1 + 0.38\, N_\eta/N_{\pi^o})$.

Finally, the direct electron experiments are subject to background from K^+_{e3} and K^o_{e3}. These backgrounds can be sizable if the magnetic analysis of the electrons is placed at a large distance from the source.

The direct electron experiments can achieve better sensitivity at low transverse momentum and low c.m. energy than the muon experiments. They also have the advantage that the lepton is observed without the distortion introduced by passage through a thick absorber. In this case any sharp structure in the yield is more readily discerned.

II. THE DATA

A. Muons

1) Serpukov Experiment

The first report of the Serpukov experiment[3] on the production of direct muons was presented at the 1972 Batavia Conference.

A careful reading of their paper does not make clear whether the acceptance correction for the moving absorber was made. Also, it was difficult to extract quantitative information from the paper. A more complete paper presented by the same group at the 1974 London Conference,[4] clearly shows the beauty of the experiment. It presents very strong evidence

DIRECT LEPTON PRODUCTION IN NUCLEON-NUCLEON COLLISIONS 937

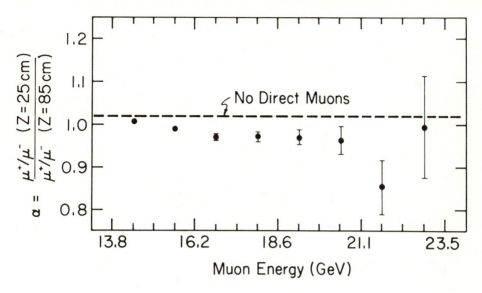

Figure 5. Evidence for direct muons from Serpukov experiment.

for a direct muon signal. The authors are cautious about computing the acceptance corrections for the moving absorber, but note that the charge ratio μ^+/μ^- decreases from 1.75 at a distant absorber position to a much smaller value at the close absorber position. If the muons observed were only from π and K decay, the μ^+/μ^- ratio would remain constant. Figure 5 shows the effect. If the authors assume that the direct muon signal is charge symmetric, they find $N_\mu^\pm/N_{\pi^+} = 2.5 \times 10^{-5}$ or $N_\mu^\pm/N_{\pi^-} = 4.4 \times 10^{-5}$ for the range $1.8 < p_\perp < 2.7$ GeV/c and $\sqrt{s} = 11.5$ GeV.

One should note that their technique for detection of a direct muon signal is exempt from the limiting sensitivity given in Figure 2, since only a contrast between the direct muon charge ratio and the hadron charge ratio need be observed. No acceptance calculations are necessary, and only a small correction is required to account for a tiny dependence on the change in charge ratio with absorber position due to a small difference in absorption between positive and negative particles.

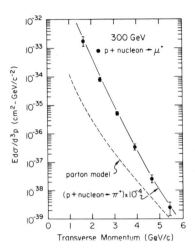

Figure 6. Invariant cross section for direct negative muon production from the Chicago-Princeton experiment.

The Serpukov group has also measured the polarization of the positive muons at $p_\perp \sim 2$ GeV. The result was consistent with the polarization expected from pion decay, but the sensitivity of the measurement was insufficient to establish that a direct muon component of strength $\mu^+/\pi^+ \sim 2.5 \times 10^{-5}$ has an opposite polarization.

2) Chicago-Princeton Experiment[5]

In this experiment, carried out at the Fermilab, direct muons were observed $\sqrt{s} = 23.7$ in a range $1.6 < p_\perp < 5.4$ GeV/c and at $\theta_{cm} \sim 90°$. Acceptance corrections were computed and a direct muon signal was observed for both positive and negative muons. The acceptance corrections were checked by comparing the observed extrapolation slope with the expected slope based on direct observation of the hadrons in the same experiment. The remarkable feature is that the ratio N_μ/N_π remains constant, while the pion yield falls by six decades. Figure 6 shows this result for negative muons. Another feature is that the ratio μ/π is independent of the atomic number of the target. The yield of muons was found to vary as $A^{1.05 \pm 0.1}$ while the pion yields vary as $A^{1.07}$.

3. The Columbia-FNAL Experiment[6]

This experiment measured both direct muons and electrons. The muon measurements were made at $\sqrt{s}$ = 23.7 and $2 < p_\perp < 4$ GeV/c. The corrections for variation of acceptance with absorber position were calculated. The results are consistent with $\sim 10^{-4}$ of pion production. No charge asymmetry was observed.

4. The Chicago-Harvard-Pennsylvania-Wisconsin Experiment[7]

This experiment used the internal target of the Fermilab accelerator and was able to observe direct muons in the range $8 < \sqrt{s} < 23$ GeV and $1 < p_\perp < 2$ GeV/c, because of the continuous energy sweep of the accelerator ramp. The muons were observed at a laboratory angle of ~ 80 mrad so that θ_{cm} varies between 60° and 90° as the energy is ramped. The pion yields were deduced from the slope of the yield as absorber position curves, while the muon yield was evaluated from the intercept. The noteworthy feature of these data are that N_μ/N_π appears to decrease below 1.5 GeV/c transverse momentum, although the effect is of marginal statistical validity. The results are shown in Figure 7. There was no evidence of a charge asymmetry.

5. The Argonne-Wisconsin Experiment[8]

This experiment, carried out at the Argonne Z. G. S. in 1965, and ostensibly a search for the weak intermediate boson, was in fact a search for a direct muon signal. No signal was found and the question ten years later is the extent to which the result quoted is correct. The sensitivity of the experiment can be judged by the error bars on the (null) cross section for the production of a muon at 90° c.m. with $p_\perp = 1.36$ GeV/c on a uranium nucleus. This error is 2.0 x 10^{-32} cm²/

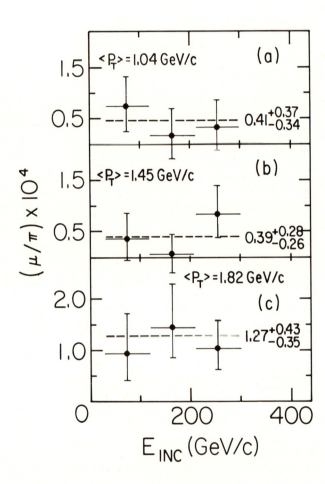

Figure 7. Results of Chicago-Harvard-Pennsylvania-Wisconsin experiment.

DIRECT LEPTON PRODUCTION IN NUCLEON-NUCLEON COLLISIONS 941

GeV-ster. At 12 GeV $\frac{d\sigma}{d\Omega dp}$ for π^- production at $p_\perp = 1.5$ GeV is 1.2×10^{-31}[9] cm^2/GeV-ster. Taking the upper limit to be three times the error, one finds $\frac{\mu^-}{\pi^-} < 3 \left[\frac{2 \times 10^{-32}}{(238) \, 1.2 \times 10^{-29}} \right] = 2 \times 10^{-5}$. This limit is 4 times better than our estimated possible precision for an experiment at the Z. G. S. The paper discusses auxiliary experiments with the absorber to measure more carefully the contribution of pion and muon decay, but gives few details. One is left with some uncertainty as to whether the experiment is truly as sensitive as the figures indicate. The level corresponds to the number of pion decays in ∼ 1 centimeter!

6) Yale-Brookhaven Experiment[10]

This experiment, originally reported in 1969, has recently been reanalyzed.[11] It is unique in that forward muons are observed. The ratio of direct muons to pions was deduced from the ratio of the direct muon signal to the slope of the pion decay muons. The authors estimate that the errors in ratio of positive to negative direct muons is ± 15%. The experiment was carried out with 30 GeV protons incident on a variable density uranium target. If the results are normalized to negative pions, one finds the following results given in the table below:

E_μ	(N_{μ^+}/N_{π^-})	(N_{μ^-}/N_{π^-})
11.6	1.5×10^{-4}	1.3×10^{-4}
20.3	2.9×10^{-5}	2.6×10^{-5}
23.5	1.4×10^{-5}	1.3×10^{-5}
25.1	7.0×10^{-6}	7.0×10^{-6}

There is no evidence for a charge asymmetry. The data at the lowest energy is in good agreement with the results at $\theta_{cm} = 90°$.

B. Electrons

1) The Columbia-FNAL Experiment[6]

This experiment, in addition to muons, measured the direct electron yield by extrapolation to zero radiator thickness and subtracting the known Dalitz contribution. No attempt was made to veto pairs. Measurements were made at both 83 mrad ($\sim 90°$ c.m.) and 50 mrad (60° c.m.). Direct electrons were observed with $N_e/N_\pi \sim 10^{-4}$ at $\sqrt{s}$ = 23.7 GeV and $2 < p_\perp < 3.5$ GeV/c. The ratio N_e/N_{π^0} is measured directly and is insensitive to η decays because N_{π^0} is determined from the γ spectrum.

2) The Cern-Columbia-Rockefeller-Saclay Experiment[12]

This experiment, carried out at the CERN ISR, has recently presented the most complete study of direct electron production at large $\sqrt{s}$. Dalitz pairs and external conversions were vetoed so that they corresponded to only 5.5% and 7.5% of the observed direct electron signal. Hadron background was $\sim 17\%$. Measurements at $\sim 90°$ CM were made for $\sqrt{s}$ between 23.5 and 62.4 GeV for transverse momenta between 1.3 and 3.5 GeV/c. They find that the ratio (N_e/N_π) appears to rise slowly with $\sqrt{s}$, and is again $\sim 10^{-4}$. They also find a slight excess of negative electrons. Calling $\varepsilon = (N_+ - N_-)/(N_+ + N_-)$, they find $\varepsilon = -0.119 \pm 0.039$ for $1.3 < p_\perp < 1.6$ GeV/c and $+ 0.096 \pm 0.080$ for $p_\perp > 1.6$ GeV/c.

The authors of this experiment hesitate to conclude that the asymmetry is real because the hadron background is larger for negative electrons and there is a background ~ -0.02 due to Compton electrons from the γ rays interacting in the vacuum wall of the ISR.

Correlations with other particles emitted in the process have also been studied. The CCRS apparatus is symmetrical about the intersection region. Correlated particles were observed at 180° to the detected electrons. No differences were observed in the features of particles correlated with a direct electron or a π^0. Furthermore, for particles of low momentum, correlated with direct electrons, the particle composition was $\pi/K/p = 67/4/3$ compared to $720/50/53$ for particles correlated with π^0's.

3) The Pennsylvania-Stony Brook Experiment[13]

In this experiment, the yield of direct electrons was measured at 90° c.m. for $\sqrt{s}$ = 4.3, 5.3 and 6.8 GeV/c for $0.8 < p_\perp < 1.5$. The incident positive beam contained 50%, 85%, and $\sim$ 100% protons at the three respective energies. The electrons were produced in a hydrogen target. Only the ratio N_{e^+}/N_{π^-} was measured because of a background of Compton scattering from atomic electrons in the target. The experiment suppresses Dalitz pairs and external pairs such that only 5% of the Dalitz pairs and 3% of the external pairs give a positron in the spectrometer with an undetected electron. These inefficiencies account for 25% of the electrons observed at $.8 < p_\perp < 1.0$ GeV/c which is the lowest momentum bin measured. Hadron contamination and K decay account for 16% of the signal. The essential result of the experiment is that direct electrons are observed with $N_{e^+}/N_{\pi^-} \sim 10^{-4}$ independent of energy and transverse momentum. No strong dependence on incident pions or protons was observed.

4) Neutral Pion Lifetime Experiment

Winter[14] has analyzed a small discrepancy noted in an experiment[15] which measured the π^0 lifetime 12 years ago. This experiment found a direct yield of $e^+/\pi^+ = (8.430\pm0.050) \times 10^{-4}$ where $(7.270\pm0.287) \times 10^{-4}$ was expected on the basis of Dalitz pairs. The error is dominated by the uncertainty in the Dalitz internal conversion coefficient. The excess is $(1.16\pm0.29) \times 10^{-4}$. Winter is able to account for this entire excess by background from K_{e3}^+ and K_{e3}^0 decay, leaving a net yield of

$-(0.2\pm0.29) \times 10^{-4}$. The experimental arrangement had a large drift distance between the target and the magnet. The K decay correction can only be calculated by detailed Monte Carlo. A hand calculation does indicate that the correction is large.

There is no pair suppression in this experiment so that the extraction of a direct yield requires excellent understanding of small details. For example, a small difference in distribution of the energy partition for internal and external pairs can affect the expected result. Also, one notes that the error in the conversion coefficient itself corresponds to a direct yield of $.28 \times 10^{-4}$.

C. <u>Summary of the Data</u>

We have tried to summarize the important features of the data in a series of figures. The direct lepton yields are all expressed as ratios to pions. This is natural since on most experiments it is the quantity directly measured. In cases where there is a π^+ to π^- ratio, significantly different from unity, we have taken the ratio to negative pions.

In Figure 8 we plot $N_\mu\pm/N_{\pi^-}$ as a function of $\sqrt{s}$ for $\theta_{cm} \sim 90°$. In Figure 9 we plot $N_\mu\pm/N_{\pi^-}$ as a function of $p_\perp$ for $\sqrt{s} > 11$ GeV. In Figures 10 and 11 the corresponding quantities are plotted for $N_e\pm/N_{\pi^-}$. In Figure 12 we plot all lepton data as a function of $\sqrt{s}$, while in Figure 13 we plot all lepton data with $\sqrt{s} > 11$ as a function of $p_\perp$. The requirement $\sqrt{s} > 11$ is an attempt to separate any possible $p_\perp$ dependence from an energy dependence.

The data for these figures were taken, in many cases, directly from the figures of various papers. Often direct lepton cross sections and not ratios were presented. In these cases, the data were divided by

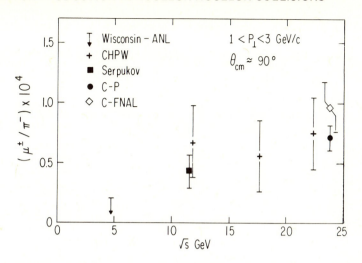

Figure 8. Ratio N_μ/N_{π^-} at $\theta_{cm} \sim 90°$ plotted as a function of $\sqrt{s}$. The data were obtained from the following sources: crosses, ref. 7; solid squares, ref. 4, solid circles, ref. 5, open diamonds, ref. 6. The upper limit was obtained from ref. 8.

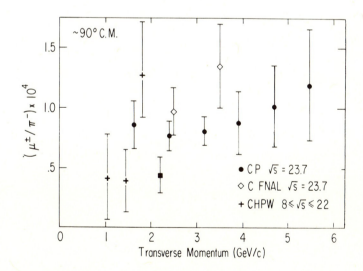

Figure 9. Ratio N_μ/N_{π^-} at $\theta_{cm} \sim 90°$ plotted as a function of $p_\perp$. The references to the original data are the same as Figure 8.

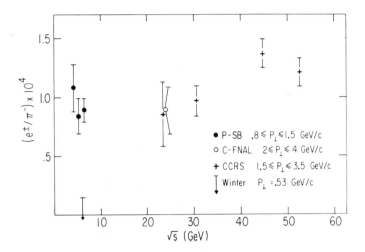

Figure 10. Ratio N_e/N_{π^-} at $\theta_{cm} \sim 90°$ plotted as a function of $\sqrt{s}$. The data were obtained from the following sources: closed circles, ref. 13; open circles, ref. 6; crosses, ref. 12. The upper limit was obtained from ref. 14.

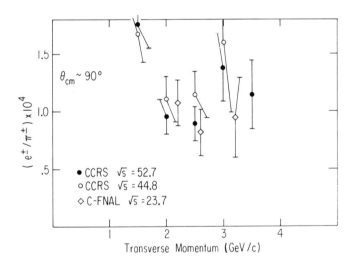

Figure 11. Ratio N_e/N_{π^-} at $\theta_{cm} \sim 90°$ plotted as a function of $p_\perp$. The references to the original data are the same as Figure 10.

DIRECT LEPTON PRODUCTION IN NUCLEON-NUCLEON COLLISIONS

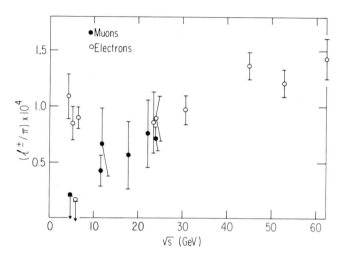

Figure 12. Plot of N_ℓ/N_π- as a function of $\sqrt{s}$. Open circles are electrons, closed circles are muons.

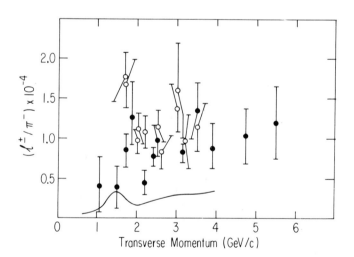

Figure 13. Plot of N_ℓ/N_π- as a function of $p_\perp$. Open circles are electrons, closed circles are muons. The solid curve is the calculated contribution of ρ, ω, and J.

the appropriate negative pion cross section obtained from the British-Scandinavian collaboration.[16] The cross sections measured on Be by the Columbia-FNAL group were divided by the π^- cross sections measured on Be by the Chicago-Princeton group.[17] While this procedure undoubtedly introduces additional errors, it is hoped that conclusions drawn from trends in the data will be correct.

The following features strike one if he looks at the curves:

1) For $\sqrt{s} > 10$ the electron and muon data appear identical, and the charge ratio is consistent with unity.

2) There seems to be an energy dependence in the ratio $N_e^{\pm}/N_{\pi}^-$. Ignoring the Penn-Stony Brook experiment, the data would suggest a threshold for $\sqrt{s} \sim 7$ GeV.

3) There is possibly a sharp increase in the ratio at $p_{\perp} \sim 1.5$ GeV/c.

4) The Penn-Stony Brook experiment contradicts the pattern suggested in points 2 and 3.

It is possible that the sources of direct electrons may be different from direct muons. Further, we should re-emphasize that the two experiments which suggest a threshold are reanalyses of old experiments which push their sensitivity beyond what one might think possible à priori.

III. KNOWN SOURCES OF DIRECT LEPTONS

The most likely sources of direct lepton production are the vector mesons, including recently discovered narrow resonance at 3.1 GeV.[18,19] Many authors[5,6,20] have discussed the role of vector mesons by asking: What cross sections are required for the production of vector mesons in order to "explain" the direct leptons? Here we take the opposite view. We have tried to assemble all the information available on vector meson production in order to predict their contributions to direct lepton yields.

A. Production of ρ and ω

For 12 and 24 GeV incident protons, the best study of inclusive vector meson production in nucleon-nucleon collisions is by Blobel, et al.[21] They find that inclusive ρ production is peaked about the central region (rapidity $y = 0$). They also measure the semi-inclusive cross section $\sigma(p + p \rightarrow \omega +$ charged particles) and find it identical to $\sigma(p + p \rightarrow \rho +$ charged particles). Thus we assume that the ρ and ω cross sections are identical. We take for the branching ratios: $\Gamma(\rho \rightarrow \mu^+\mu^-)/\Gamma(\rho \rightarrow$ all) = 0.42×10^{-4} and $\Gamma(\omega \rightarrow \mu^+\mu^-)/\Gamma(\omega \rightarrow$ all) = 0.83×10^{-4}. These values, assuming electron-muon universality, are obtained from e^+e^- storage ring measurements.[22] Assuming ρ and ω production to be equal, we find that:

$$E \frac{d^3\sigma_{\mu\mu}}{dp^3} = (3.1 \times 10^{-30}) \sqrt{x^2 + (0.065)^2} \, (1-x)^5 \, e^{-3.6 p_\perp^2} \, cm^2$$

gives a good fit to the cross section for production of μ pairs from ρ and ω at $\sqrt{s} = 6.7$ GeV. Here X is the Feynman variable.

Until recently, the data on inclusive ρ production at Fermilab and ISR energies have been scarce. Preliminary results are now available from two sources. First, at Fermilab, inclusive ρ production has been studied with 200 GeV proton incident on the 30" bubble chamber.[23] The features of the production are similar to 24 GeV; the production is concentrated at $y = 0$. The ratio of production of ρ's to pions was found to be 0.08±0.02. At 24 GeV this ratio was 0.085±0.015.

A second experiment by a Columbia-Fermilab-Illinois-Cornell-Hawaii group[24] has provided preliminary information which bears directly on the issue. They have measured muon pair production in the mass range $.650 < M_{\mu\mu} < .900$ GeV by ~ 320 GeV neutrons incident on a beryllium target. They find that the total cross section for production of muon pairs in the above mass range is 3.5×10^{-31} cm^2, and that the cross section can be fit to a form

$$\frac{d\sigma}{dx\, dp_\perp^2} = C\, e^{-4p_\perp} (1-x)^5.$$

The measurements extend over the entire forward hemisphere and reach a $p_\perp$ of ~ 1.7 GeV. The errors are not fully estimated, but are less than a factor 2. Further, the total yield of μ pairs for the remaining mass range kinematically accessible is a small fraction of the ρ and ω mass region.

Using the total cross section and its dependence on $p_\perp$ and X, we parameterize the invariant cross section per nucleon for production of μ pairs in the range $.65 < M_{\mu\mu} < .90$ GeV as,

$$\frac{E\, d\sigma_{\mu\mu}}{dp^3} = (2.7 \times 10^{-30})\, \frac{E}{12.2}\, e^{-4p_\perp} (1-x)^5\, \text{cm}^2$$

One must note that these cross sections are based on measurements that are restricted to rather low transverse momentum; $p_\perp < 1.1$ GeV/c at 24 GeV and $p_\perp < 1.8$ GeV/c at 320 GeV. There is no data on the production of vector mesons at larger $p_\perp$. We have made a guess of the extended $p_\perp$ dependence by noting that ρ production is very similar to K^+ production. Figure 14 shows a plot of invariant cross sections at 90° vs. $\sqrt{s}$ with $X_\perp = p_\perp/p_\perp(\text{max})$ as a parameter. In closed circles are the ρ cross sections of Blobel et al.[21] The dashed curves are from a compilation of K^+ cross sections.[25] The open circles are the cross sections obtained from the 320 GeV parametrization above when divided by 1.25×10^{-4}, the effective muon pair branching ratio. For predictions of single muon yields at higher transverse momentum, we assume that both ρ production and ω production are equal to K^+ production.

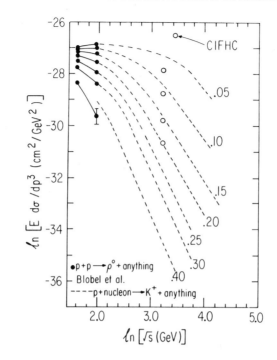

Figure 14. Postulated global plot for ρ production. Solid circles are measured ρ production at $\sqrt{s}$ = 6.7, ref. 21; the dashed line is K^+ production from ref. 25; the open circles are from the fit to ρ production from ref. 24.

B. Production of φ

No significant inclusive production of φ mesons has been observed at 24 GeV[26], or at Fermilab[27], or ISR[12] energies. The Columbia-Fermilab group finds $\sigma_\phi/\sigma_{\pi^-}$ < 0.05 at $p_\perp$ = 2.48 GeV/c and $\sigma_\phi/\sigma_{\pi^-}$ < 0.10 at $p_\perp$ = 3.3 GeV. These limits are for $\sqrt{s}$ = 23.7 GeV. At ISR the CCRS group finds σ_ϕ < .4σ_{π^0} for $p_\perp$ > 1.6 GeV/c at $\sqrt{s}$ = 52.7. Further, there is no evidence in the dimuon production experiment[24] by neutrons for large φ production despite the large branching ratio $\Gamma(\phi \to \mu^+\mu^-)/\Gamma(\phi \to \text{all})$ = 2.5 x 10^{-4}, which makes the experiment twice as sensitive to φ as ρ and ω combined.

In the forthcoming calculations, we will assume ϕ contributes negligibly to the direct muon yield.

C. Production of the J(3.1) Resonance[28]

The discovery of the 3.1 GeV resonance lead many of us to jump to the conclusion that the source of the direct muons was found. Subsequent data leads one to conclude that the J(3.1), while a contributor to the direct leptons, is not the dominant source.

There are three pieces of information about the inclusive production of the 3.1 GeV resonance:

1. At 30 GeV ($\sqrt{s}$ = 7.5) Aubert et al.[18] find
$\sigma(p + p \to J + \text{anything}) = 10^{-34}$ cm^2 on the assumption the cross section has the form

$$B \frac{d\sigma}{dy dp_\perp^2} = C\, e^{-6p_\perp}$$

where B is the branching ratio of the J into muons or electrons.

2. At ISR 9 events of the form $J \to e^+ e^-$ were observed by Büsser, et al.[29] with $\sqrt{s} \geq 30.5$ GeV. These led to a cross section:

$$B \frac{d\sigma}{dy}(y = 0) = (0.75 \pm 0.25) \times 10^{-32} \text{ cm}^2$$

The result was based on the assumption

$$B \frac{d\sigma}{dy dp_\perp^2}(y = 0) = C\, e^{-3p_\perp}$$

3. Knapp, et al.[30] produced the J(3.1) by $\sim$ 260 GeV incident neutrons. They find

$$B\sigma = 1.7 \times 10^{-33} \text{ cm}^2 \text{ for } |X_L| > .32$$

where $X_L = p_\parallel / p_\parallel(\text{max})$ and $p_\parallel$ is the laboratory momentum.

DIRECT LEPTON PRODUCTION IN NUCLEON-NUCLEON COLLISIONS

Further, they find the dependence of the cross section on X_L and $p_\perp$ is given by

$$B \frac{d\sigma}{dX_L \, dp_\perp^2} \sim e^{-10X_L} \, e^{-2p_\perp^2}$$

Recently, preliminary results[24] of a similar measurement at ~ 320 GeV neutron energy gives

$$B\sigma = 1.6 \times 10^{-33} \text{ cm}^2 \text{ for } |X_L| > .32$$

and

$$B \frac{d\sigma}{dX_L \, dp_\perp^2} \sim e^{-12X_L} \, e^{-1.5p_\perp^2}$$

We have combined the two measurements into a parameterization form:

$$B \frac{d\sigma}{dy \, dp_\perp^2} = 1.5 \times 10^{-30} \, X_L \, e^{-11X_L} \, e^{-3p_\perp} \text{ cm}^2$$

The conversion from a $p_\perp^2$ to a $p_\perp$ dependence in the exponent preserves $\langle p_\perp \rangle = .66$ GeV found in the experiment and agrees with the $\langle p_\perp \rangle$ found at the ISR. At $y = 0$ and $X_L = .128$ this form gives:

$$B \frac{d\sigma}{dy} (y = 0) = 1.0 \times 10^{-32} \text{ cm}^2$$

in good agreement with the ISR value, 0.75×10^{-32} cm^2.

At 30 GeV ($\sqrt{s} = 7.5$) we have parameterized the cross section preserving the shape in longitudinal momentum of the 300 GeV cross section. We have arbitrarily changed the $p_\perp$ dependence $e^{-6p_\perp}$ assumed to $e^{-3p_\perp}$. We do this by assuming that the experiment measures the cross section only in the neighborhood of $y = 0$ and $p_\perp = 0$. The resulting parameterization is

$$B \frac{d\sigma}{dy \, dp_\perp^2} = 1.1 \times 10^{-33} \, e^{-3p_\perp} \, (1 - x)^4 \text{ cm}^2$$

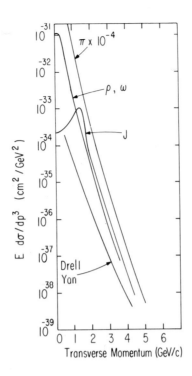

Figure 15. Invariant cross sections at $\theta_{cm} = 90°$ for direct lepton production at $\sqrt{s} \sim 24$ GeV. Shown are the contributions of the J, ρ-ω, and the Drell-Yan process. For comparison the invariant π^- cross section at $\sqrt{s} = 23.7$ is also plotted.

Then

$$B \frac{d\sigma}{dy}(y=0) = 2.4 \times 10^{-34} \text{ cm}^2,$$ a factor 40 less than at $\sqrt{s} \sim 30$.

D. Calculation of Direct Electron Yields

We have calculated the invariant cross sections for inclusive lepton production using the above parameterizations at $\sqrt{s} \approx 24$ GeV. (Fig. 15). Included on the curve, in addition to the J and ρ,ω contributions, is an estimate of the Drell-Yan process.[30] It should be stressed that the input data used in these calculations do not extend beyond $p_\perp \approx 1.8$ GeV,

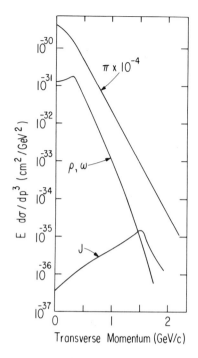

Figure 16. Invariant cross sections at $\theta_{cm} = 90°$ for direct lepton production at $\sqrt{s} \sim 7$ GeV. Shown are the contributions of ρ and ω, and J production.

so that for very large $p_\perp$ the conclusions may not be completely reliable. If one assumes that ρ and ω production resembles K^+ production, then the yield is somewhat less for $p_\perp > 2$ GeV than shown in the graph. The yield from these sources, expressed as a ratio to π^- is also plotted on Figure 13.

Figure 16 shows the results of the calculation for $\sqrt{s} \sim 7$ GeV.

Figure 17 shows the prediction for yields in the forward direction at $\sqrt{s} \sim 7$ GeV. These were computed by integrating the cross sections over $p_\perp$ and plotting the ratio of the calculated invariant lepton cross section from ρ and ω decay to the integrated negative pion cross

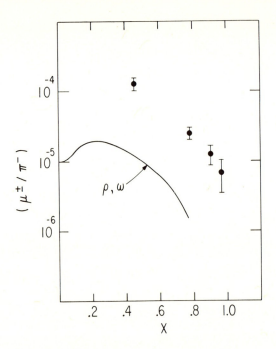

Figure 17. Plot of $(N_\mu^\pm/N_\pi)$ produced by 28 GeV incident protons in the forward direction. Points with errors are from ref. 11. Solid curve is expected yield from ρ and ω.

sections given by Blobel et al.[9] Shown on the same graph are the measured points of the Yale-Brookhaven experiment.[10] The predicted results for the ratio (μ/π) are sensitive to the amount of transverse momentum integrated over. One can see from Figure 16 that there is a depletion of leptons at the lowest transverse momentum. We assume that the large amount of multiple scattering effectively integrates over the transverse momentum distribution.

In all of these calculations the yields of direct leptons from ρ and ω decay are about an order of magnitude less than the observations. The J resonance can account for about 25% of the direct leptons but only in the neighborhood of 1.5 GeV/c. One might note that for $p_\perp$ > 3 GeV the J contribution is again about 25%, but this is an artifact of the calculation since one does not know the $p_\perp$ dependence of the J beyond $\sim$ 1.5 GeV.

Below $p_\perp$ = 2 GeV the calculations are based on actual measurements and it is in this region one can reliably conclude that the known yield of ρ, ω and J are insufficient to account for the direct leptons observed.

IV. CONCLUSIONS

All the <u>qualitative</u> features of the data support the hypothesis that the direct leptons are the debris of strongly interacting particles. A detailed examination of the known sources does not give <u>quantitative</u> support to this idea.

We stress that much of the data which served as input came from direct measurement of dileptons. Hence, there is no guess work about the proportion of ρ, ω, and ϕ, etc. We anticipate in the future improved dilepton experiments which have much broader acceptance. These can trigger on a single lepton and search with good efficiency for a second lepton if it is present.[32]

The evidence is strongly suggestive that these single leptons are not produced in pairs. This conclusion cannot be definitive without more careful measurements on the entire dimuon mass spectrum.

A tempting hypothesis is the notion that heavy charged stable particles are produced and have weak decay modes which sometimes contain a lepton and a neutrino. There is some evidence for the existence of such objects in the neutrino experiments of the Harvard-Penn-Wisconsin group.[33] They observe dimuon events which are interpreted as a charged current reaction (accounting for one muon) in which the hadronic shower contains a direct muon about 1% of the time. If this interpretation is correct, then an example of a direct single muon has already been found.

In hadronic interactions these sources, be they hadrons or heavy leptons, must be produced in pairs. What is disturbing is an apparent lack of threshold, since direct positron emission is observed as low as $\sqrt{s}$ = 4.3 GeV. This provides only 2.4 GeV available for particle production. It seems of crucial importance to verify these measurements at low energy. It will be important to verify that direct muon and electron yields remain the same at low energies.

If the direct leptons are not produced in pairs, there is no absolute reason that they should show a charge symmetry and more careful searches may discover such an asymmetry.

But if our inferences are correct, one should expect direct neutrino production with rather similar characteristics. Such experiments have been proposed at Fermilab.[3,4] One can look for neutrino production when the proton beam interacts in a dense target. Or one can search for missing energy when a high energy proton strikes a massive calorimeter equipped to measure the energy of everything but escaping neutrinos.

The resolution of this direct lepton problem shows promise of revealing some new form of matter: a new quantum number for hadrons, or heavy leptons. But, the failure to see a threshold gives one the nagging feeling that some more mundane processes have been forgotten. After all, we are drawing our conclusions only by exhaustion of possibilities.

DIRECT LEPTON PRODUCTION IN NUCLEON-NUCLEON COLLISIONS

ACKNOWLEDGEMENTS

I wish to thank Dr. Irwin Gaines and Professor W. Lee for valuable discussions and use of their data in advance of publication. I also wish to thank my colleagues at the University of Chicago for help and criticism.

ADDED REMARKS

A visit to CERN and discussion at Erice promt me to add a few remarks. At CERN I learned of additional results from the experiment of the Serpukov[4] group. They have now measured direct μ production at $\theta_{c.m.} = 90°$ at 50 and 35 GeV as well as at 70 GeV. These results were learned from B. A. Dolgoshein in private discussions at the Palermo Conference. Their new results are quoted in the table below:

Energy (GeV)	P (GeV/c)	$Ed^3\sigma_\mu/dp^3$ (cm^2/GeV^2)	(Ratio (μ/π^-))
70	2.25	$(1.2 \pm 0.2) \times 10^{-35}$	$(3.6 \pm 0.6) \times 10^{-5}$
50	2.25	$(4.0 \pm 1.0) \times 10^{-36}$	$(2.8 \pm 0.7) \times 10^{-5}$
35	1.75	$< 2.5 \times 10^{-35}$	$< 1.9 \times 10^{-5}$

The direct muon yields were measured both by the ratio technique discussed in Section II and by directly correcting the yield vs absorber position curves for acceptance. Both techniques agree and hence these results are consistent with equal yields of positive and negative direct muons.

These data further emphasize the discrepancy between the muon data, which seems to show a threshold for $\sqrt{s} \approx 7$ GeV and the direct electron results which seem to show no threshold down to $\sqrt{s}$ = 4.3 GeV. One must also note that at 28 GeV in the forward direction, muons have been observed.

This added information emphasizes strongly the need to measure muon and electron production at the A.G.S., the CERN P.S., and the Argonne Z.G.S. to resolve the experimental discrepancies.

If it turns out that indeed a threshold is present, then a measurement at small transverse momentum and large energy will be important. If the lepton source is from two-body decays of a mass m, then the ratio ℓ/π should drop sharply for $P_\perp < m/2$. Hence, if the leptons are coming from a 2 GeV object, the ratio ℓ/π should drop sharply for $P_\perp < 1.0$ GeV/c.

REFERENCES

[†]Supported by the National Science Foundation

1. B. G. Pope, <u>Lepton Production by Hadrons</u>, review talk presented at the International Conference on High Energy Physics, Palermo, June 1975.
2. N. P. Samios, Phys. Rev. <u>121</u>, 275 (1961).
3. G. B. Bondarenko, <u>et al</u>.as cited by V. Lebedev in Proceedings of the Sixteenth International Conference on High Energy Physics, The University of Chicago and National Accelerator Laboratory, 1972, edited by J. D. Jackson and A. Roberts (National Accelerator Laboratory, Batavia, Ill., 1973), Vol. 2, p. 329.
4. V. V. Abramov, <u>et al</u>., reported in the Proceedings of the Seventeenth

International Conference on High Energy Physics, London, 1974 (Rutherford Lab, Chilton, Didcot, 1974), p. V-53.

5. J. P. Boymond, et al., Phys. Rev. Lett. 33, 112 (1974).
6. J. A. Appel, et al., Phys. Rev. Lett. 33, 722 (1974).
7. D. Bintinger, et al., Phys. Rev. Lett., 35 72 (1975).
8. R. C. Lamb, et al., Phys. Rev. Lett. 15, 800 (1965).
9. V. Blobel, et al., Nucl. Phys. B69, 454 (1974).
10. P. J. Wanderer, et al., Phys. Rev. Lett. 23, 729 (1969); C. M. Ankenbrandt et al. Phys. Rev. D 3, 2582 (1971).
11. L. B. Leipuner, et al., Phys. Rev. Lett. 34, 103 (1975).
12. F. W. Büsser et al., Phys. Lett., 53B, 212 (1974); F. W. Büsser. et al., Paper B-06 presented to International Conference on High Energy Physics, Palermo, June 1975.
13. E. W. Beier, et al., Paper B-05 submitted to International Conference on High Energy Physics, Palermo, June 1975.
14. K. Winter, A Search for Prompt Production of Electrons in Proton-Pt. Collisions at 19 GeV/c, Phys. Lett., to be published.
15. G. von Dardel, et al., Phys. Lett. 4, 51 (1963).
16. B. Alper, et al., Nucl. Phys. B87, 19 (1975).
17. J. W. Cronin, et al., Physical Review D, to be published
18. J. E. Augustin, et al., Phys. Rev. Lett. 33, 1406 (1974).
19. J. J. Aubert, et al., Phys. Rev. Lett. 33, 1408 (1974).
20. G. R. Farrar and R. D. Field, California Institute of Technology preprint, CALT 68-501; F. Halzen and K. Kajantie, University of Wisconsin, Madison preprint Coo-881-460.
21. V. Blobel, et al., Phys. Lett. 48B, 73 (1974).
22. The ρ branching ratio comes from Benaksas, et al., Phys. Lett. 39B,

239 (1972), The ω branching ratio comes from Benaksas, et al., Phys. Lett. 42B, 507 (1972), Phys. Lett. 42B, 511 (1972).

23. P. Singer, Argonne National Laboratory, private communication.
24. I. Gaines; private communication.
25. J. W. Cronin, Processes at High Transverse Momentum, talk given at SLAC Summer Institute for Physics, August 1974.
26. e.g. V. Blobel, et al., Nucl. Phys. B69, 237 (1944). At 24 GeV exclusive ϕ production is < 2% of exclusive ρ production in pp collisions.
27. J. A. Appel, et al., submitted to Phys. Rev. Letters.
28. We refer to the 3.1 GeV resonance as J and not Ψ only for the reason that the object, when discovered via pp collisions, was called J. We do not wish to offend either our Eastern or Western colleagues who have done so much to advance physics through their beautiful experiments.
29. F. W. Büsser, et al., Phys. Lett. 56B, 482 (1975).
30. B. Knapp, et al., Phys. Rev. Lett. 34, 1044 (1975).
31. G. R. Farrar, Nucl. Phys. B77, 429 (1974).
32. The only reported experiment which has a hint of this desirable feature is the CCRS experiment (ref. 29). If the direct leptons were entirely from J production, they expected to detect a second lepton in the opposite arm, for 5.3% of the detected single leptons. In fact, they found such coincidences only .53% of the time, demonstrating that J production accounts for only 10% of the single leptons.
33. A Benvenuti, et al., Phys. Rev. Lett. 34, 1419 (1975).
34. Fermilab Exp. 28A; Fermilab Exp. 379.

DISCUSSION

CHAIRMAN: J.W. Cronin

Scientific Secretary: T. Ferbel

DISCUSSION No. 1

- WILSON:

The Russian experiment assumes that direct μ^+ and μ^- are equal. This might not be true for low energies. If not, can the Russian experiment be consistent with the low-energy positron experiments?

- CRONIN:

In principle they could be consistent. To be sure, the Pennsylvania-Stony Brook experiment will need to analyse their negative electron data. This has a large Compton electron background, and may be difficult to analyse.

- WILSON:

Do the electron-positron pairs produced in the radiators have the same energy and angle distribution as Dalitz pairs and, if not, does this affect the analysis of the experiments with pair rejection?

- CRONIN:

In the Pennsylvania-Stony Brook experiment, careful attention was paid to the differences between the details of Dalitz pairs and external pairs, which leads to slightly different inefficiencies for suppression of the two processes.

- DRELL:

In the calculation of the Drell-Yan process, for the observation of single muons, what p_T distribution was assumed for the partons?

- CRONIN:

No transverse momentum was assumed for the partons.

- DRELL:

Has direct lepton production been observed away from 90° in the centre of mass?

— *CRONIN:*

In addition to the observation of direct muons in the forward direction by the Yale-Brookhaven group in an old experiment, the Columbia-FNAL group has observed direct electrons at $\theta_{cm} = 60°$, and find $(e/\pi) \approx 10^{-4}$.

— *POLITZER:*

The constancy of the μ/π ratio and its value suggest that perhaps the only difference between the processes is that one goes via a virtual photon and the other via a gluon, e.g. $(\alpha/\alpha_s)^2 \sim 10^{-4}$, where $\alpha_s \sim \frac{1}{2}$ is the quark gluon coupling. While I cannot really make sense of this, maybe there is something to it for large energy and large p_T. Is there anything obviously wrong with this description?

— *CRONIN:*

I noted, on seeing my own data, that 10^{-4} is about α^2. The critical test, of course, is whether the μ's are produced in pairs. The data suggests (though by no means conclusively) that pairs do not account for the muons.

— *DRELL:*

In most quantum electrodynamic calculations, it is α/π and not α that appears.

— *CALDI:*

In trying to explain yields of direct leptons, you only included decays from ρ, ω, ϕ, and J. What about possible states above the J, as possibly seen at SPEAR; is there some way to include these to make up the difference?

— *CRONIN:*

A given state will contribute to the direct yield only for $p_T > m/2$, where m is the mass of the decaying state. Since much of the data are below $p_T = 2$ GeV, states decaying to muon pairs above 4 GeV will not help. However, a state of ~ 2 GeV decaying to muons and neutrinos instead of muon pairs could be responsible for the effect.

— *CREUTZ:*

The constancy of the μ/π ratio suggests a low mass parent for the leptons. Are the μ-pair experiments sensitive to invariant masses as small as, say, a few pion masses, and in e^+e^- pairs, are masses as high as m_π rejected?

— *CRONIN:*

This is a difficult region for the μ-pair experiments, but the data, such as they are, do not provide evidence for a significant low mass component. In the case of electron pairs, the CCRS experiment rejects masses as high as $2m_\pi$ by a factor of 3.

DISCUSSION

- *DRELL:*

Are experiments on lepton pair production underway for different incident particles?

- *CRONIN:*

Yes, at present a muon pair experiment with incident pions and protons is underway at the Fermilab by another Chicago-Princeton group.

- *ZICHICHI:*

By how much will the antiquark content of a π^- help the $\mu^+\mu^-$ production?

- *DRELL:*

By orders of magnitude.

- *CRONIN:*

Most calculations I have seen predict the muon pair production from pions to be about ten times larger than proton-proton collisions.

METHODS FOR A THEORETICAL UNDERSTANDING OF NEUTRAL CURRENTS

M. Gourdin
Laboratoire de Physique Théorique et Hautes Energies
Université Pierre et Marie Curie, PARIS, FRANCE

The aim of these lectures is to present general methods for a future theoretical understanding of weak neutral currents. Experimental evidences are there but the data are generally analyzed and interpreted in a model dependent way. This procedure is partially justified by the lack of accuracy of the experiments but our goal must be a rigorous determination of the symmetry properties of the neutral current without any reference to a particular gauge theory. Such a phenomenological approach has to be supplemented with a theory of neutral currents which will be part of a theory of weak interactions. This question will not be discussed here.

The two problems we shall be interested in are :

i) the Lorentz structure of the neutral current,

ii) the internal symmetry structure of the hadronic neutral current.

They are studied in PART I and PART II with a special reference to inclusive reactions induced by neutrinos and antineutrinos on a nucleon target

$$\nu(\bar{\nu}) + p \Rightarrow \nu(\bar{\nu}) + X$$

the hadronic system X is observed but not analyzed.

Two elastic reactions, on a nucleon target and on electron target are considered in PART III and the results obtained in PARTS I and II are applied to these simple situations.

Neutral currents have been observed in reactions induced by neutrinos and antineutrinos. They also will affect other processes like those associated to charged leptons where the weak neutral and the one photon exchange amplitudes can interfere. But we must keep in mind that these effects of weak nature are expected to be small. If the neutral current is not a pure Lorentz vector a good signature of their presence will be a violation of parity generally exhibited in polarization measurements. Up to now neutral currents in neutrinoless reactions have not been observed.

Neutral currents must also be present in other parts of physics

   i) Nuclear Physics

  ii) Atomic Physics

 iii) Astrophysics.

Although of fundamental importance they will not be considered in these lectures because of the lack of time.

THEORETICAL UNDERSTANDING OF NEUTRAL CURRENTS

/ PART I - LORENTZ STRUCTURE OF THE NEUTRAL CURRENT /

I. Differential cross section.

II. Scaling assumptions.

III. Crossing properties.

IV. V/A Analysis of Inclusive reactions experiments.

We have extremely good evidences for a Lorentz vector structure of the electromagnetic interactions and in particular quantum electrodynamics works nicely.

All present low energy processes in weak interactions involving charged currents can be explained by a V-A theory and in particular the conserved vector current hypothesis is very successful . At high energy and large momentum transfer the situation is not so clear. It seems that neutrino and antineutrino induced reactions are still describable by an effective Fermi Hamiltonian involving a V-A charged current. But other possibilities are not excluded and new types of experiments must be performed before making a unique statement on this question[6].

Weak interactions proceeding via neutral current exchange have recently be observed and we have to study the corresponding mechanism. It is tempting to proceed by analogy and to assume for the neutral current the existence of a vector and an axial vector components. But this way of thinking can be due to a lack of imagination and the only rigorous way to solve the problem of the Lorentz structure of the weak current is to determine, from experiment only, a unique solution that could be a linear superposition of V and A or something else.

We then have strong motivations for studying the consequences of a general S,P,T,V,A structure of weak currents in reactions induced by neutrinos and antineutrinos. Our approach is not limited to neutral currents and the considerations of sections I and II are applicable to charged currents and more generally to spin 0 and 1 exchange. Section III can also be used for the part of charged current where charge symmetry is applicable.

The problems related to polarization are not investigated here. They are very important from a theoretical point of view but accurate experiments with a polarized target are now and for the immediate future unrealistic.

The considerations of this part are concentrated on inclusive reactions. The kinematics is shown on Fig.I and it is described in Appendix I.

I. DIFFERENTIAL CROSS-SECTION

1°) We consider the general case where the weak current is a superposition of scalar S, pseudoscalar P, vector V, axial vector A and tensor T Lorentz covariants. The simplest way to analyze the problem is to look at the t channel[+] (See Appendix II).

The annihilation amplitude is a superposition of spin zero and spin one exchange contributions.

It follows that the differential cross section will be a second order polynomial in $\cos \Theta_t$ where Θ_t is the t channel centre of mass scattering angle. The kinematical factor is obtained by a trivial calculation

$$\frac{d^2\sigma}{dq^2 dW^2} = \frac{G^2 q^2}{8\pi M^2 E^2} \times [\ ]$$

and the bracket is written as[4]

$$[\ ] = a_0(q^2, W^2) + a_1(q^2, W^2) \cos \Theta_t + a_2(q^2, W^2) \cos^2 \Theta_t \quad (1)$$

[+] This method has been introduced in the study of many photon exchange in charged lepton scattering.

M. GOURDIN and A. MARTIN Preprint CERN TH 4804 (1962).

M. GOURDIN Diffusion des Electrons de Haute Energie Chap. VIII Masson Ed. Paris (1966).

It has also been applied to the neutrino case with (V,A) charged currents

G. CHARPAK and M. GOURDIN Lectures given at the Cargese Summer School (1963)

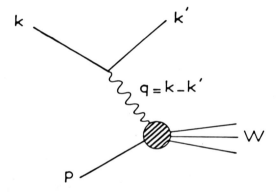

FIG. 1. Kinematics of Inclusive Reactions.

It is easy to compute $\cos \Theta_t$ in terms of S channel laboratory variables

$$\cos \Theta_t = \frac{E+E'}{\sqrt{\nu^2 + q^2}}$$

2°) The origin of the three structure functions a_0, a_1 and a_2 is the following

$$a_0 = \sum' |\text{Spin } 0|^2 + \sum' |\text{Spin } 1|^2$$

$$a_1 = \text{Interference (Spin 0, Spin 1)}, (1^+, 1^-)$$

$$a_2 = \sum' |\text{Spin } 1|^2$$

or, in an equivalent language:

$$a_0 \Rightarrow SS, PP, TT ; VV, AA$$

$$a_1 \Rightarrow ST, PT ; VA$$

$$a_2 \Rightarrow TT ; VV, AA$$

In particular for pure Scalar and/or Pseudoscalar currents we have only one structure function $a_0(q^2, W^2)$ and the differential cross section is independent of $\cos \Theta_t$ as expected for a pure spin zero exchange.

3°) In the physical region of the S channel

$$q^2 \geq 0 \qquad W^2 \geq M^2$$

the quantity $\cos \Theta_t$ is positive and larger that unity.

From the requirement of positivity of the differential cross section we can derive constraints on the structure funtions $a_0(q^2, W^2)$ $a_1(q^2, W^2)$ and $a_2(q^2, W^2)$.

THEORETICAL UNDERSTANDING OF NEUTRAL CURRENTS

Let us consider the three dimensional a_0, a_1, a_2 space. The equation

$$a_0 + a_1 \cos\Theta_t + a_2 \cos^2\Theta_t = 0$$

represents a family of planes whose envelope is the cone

$$a_1^2 = 4 a_1 a_2$$

This cone has its top at the origin and its revolution axis is defined by

$$a_1 = 0 \qquad a_2 = a_0$$

The limiting planes correspond to the extreme values of the parameter $\cos\Theta_t$.

for backward scattering
$$a_0 + a_1 + a_2 = 0$$
$$\cos\Theta_t = 1 \quad ;$$

for forward scattering
$$a_2 = 0$$
$$\cos\Theta_t \Rightarrow +\infty .$$

The positivity domain D in the a_0, a_1, a_2 space is then characterized by the following constraints:

$$a_0(q^2, W^2) + a_1(q^2, W^2) + a_2(q^2, W^2) \geq 0$$

$$a_2(q^2, W^2) \geq 0$$

and when $\quad a_2(q^2, W^2) \leq a_0(q^2, W^2)$

$$a_1(q^2, W^2) \geq -2\sqrt{a_0(q^2, W^2) a_2(q^2, W^2)}$$

4°) In the particular (V,A) case we have only spin-one exchange in the annihilation channel and the three functions a_0, a_1, a_2 are simply related to the total absorption cross sections $\sigma_\lambda(q^2, W^2)$ where λ is the helicity of the weak current

The result is[+] :

$$a_0(q^2, W^2) = \frac{M\sqrt{\nu^2+q^2}}{2\pi}\left[\sigma_T(q^2, W^2) - \sigma_L(q^2, W^2)\right]$$

$$a_1(q^2, W^2) = \frac{M\sqrt{\nu^2+q^2}}{2\pi}\varepsilon\left[\sigma_{-1}(q^2, W^2) - \sigma_{+1}(q^2, W^2)\right] \quad (2)$$

$$a_2(q^2, W^2) = \frac{M\sqrt{\nu^2+q^2}}{2\pi}\left[\sigma_T(q^2, W^2) + \sigma_L(q^2, W^2)\right]$$

where $\varepsilon = +1$ for neutrino induced reactions and $\varepsilon = -1$ for antineutrino induced reactions.

The positivity of total cross sections

$$\sigma_\lambda(q^2, W^2) \geq 0 \qquad \lambda = \pm 1, L$$

defines a domain $\mathcal{D}^{VA}$ in the a_0, a_1, a_2 space limited by three planes

$$|a_1(q^2, W^2)| \leq a_0(q^2, W^2) + a_2(q^2, W^2)$$

$$a_0(q^2, W^2) \leq a_2(q^2, W^2)$$

[+] In terms of the Californian structure functions W_1, W_2, W_3 we have

$$a_0 = \frac{M}{2}\left[2W_1 - (1 + \frac{\nu^2}{q^2})W_2\right]$$

$$a_1 = -\frac{\varepsilon}{2}\sqrt{\nu^2+q^2}\, W_3$$

$$a_2 = \frac{M}{2}(1 + \frac{\nu^2}{q^2})W_2$$

THEORETICAL UNDERSTANDING OF NEUTRAL CURRENTS

and which is obviously a subspace of $\mathbb{D}$.

5°) Conclusion

With the only information of the differential cross section for inclusive reactions it will never be possible to exclude the presence of (S, P, T) currents. More explicitly two situations can occur

i) the point a_0, a_1, a_2 belong to the domain $\mathbb{D}^{VA}$ for all the values of q^2, W^2 in the physical range. Then a pure (V,A) solution is possible and it is explicitly given by equations (2). But a pure (S, P, T) solution can also be found or various mixtures of (S, P, T) and (V, A).

ii) the point a_0, a_1, a_2 belong to the complementary set of $\mathbb{D}^{VA}$ in the domain $\mathbb{D}$, at least for some values of q^2, W^2 in the physical range. Then a pure (V, A) solution is excluded and (S, P, T) currents must exist.

Let us remark that the positivity domain $\mathbb{D}^{SP}$ for pure (S, P) coupling defined by

$$a_1 = 0 \qquad a_2 = 0 \qquad a_0 \geqslant 0$$

does not belong to $\mathbb{D}^{VA}$

$$\mathbb{D}^{SP} \cap \mathbb{D}^{VA} = \emptyset$$

It follows that pure (S, P) and pure (V,A) are two incompatible situations.

6°) Polarization

A natural question to ask now is : what is the difference between (S, P, T) and (V,A) currents ? Such a difference lies in the transfer of helicity at both the leptonic and the hadronic vertices[33, 41].

The leptonic vertex is studied in Appendix II and the main result is the following : with (V,A) couplings the final and initial neutrinos (or antineutrinos) have the same helicity ; with (S, P, T) couplings they have opposite helicities. Therefore, if the latest situation is the correct ones, we must have in nature neutrinos and antineutrinos of both helicities

$$\nu_L, \bar{\nu}_R \quad ; \quad \nu_R, \bar{\nu}_L$$

An analogous mechanism will occur at the hadronic vertex and therefore polarization measurements are needed in order to distinguish between (S, P, T) and (V,A) couplings. This type of experiment is extremely hard to perform and the only result we shall be able to get in the near future is the compatibility of the data with (V,A) couplings without excluding the presence of (S, P, T) ones.

II. SCALING ASSUMPTIONS

1°) We assume Bjorken scaling properties for the structure functions a_0, a_1, a_2 such that, in the high energy limit-and when the local Fermi interactions is still valid-, the total cross section rises linearly with the incident laboratory energy E. We have the following property

$$\lim_{\substack{q^2, W^2 \Rightarrow +\infty \\ x = \frac{q^2}{2M\nu} \text{ fixed}}} a_\alpha(q^2, W^2) = a_\alpha(x) \qquad \alpha = 0, 1, 2$$

It is now very easy to compute the asymptotic forms for the various types of cross sections and the results are

$$\frac{d^2\sigma}{d\xi dx} \Rightarrow \frac{G^2 ME}{\pi} \propto \left[(1-\xi)^2 a_0(x) + (1-\xi^2) a_1(x) + (1+\xi)^2 a_2(x)\right]$$

$$\frac{d\sigma}{d\xi} \Rightarrow \frac{G^2 ME}{\pi} \left[(1-\xi)^2 I_0 + (1-\xi^2) I_1 + (1+\xi)^2 I_2\right] \qquad (3)$$

$$\sigma_{TOT}(E) \Rightarrow \frac{G^2 ME}{\pi} \frac{\bar{I}_0 + 2\bar{I}_1 + 7\bar{I}_2}{3} \qquad (4)$$

where the first moment integrals $\bar{I}_\alpha$ of the scaling functions are defined by

$$\bar{I}_\alpha = \int_0^1 x \, a_\alpha(x) \, dx$$

2°) Positivity Domain For The $\bar{I}_\alpha's$

The differential cross section (3) is a second order positive definite form in ξ for values of ξ in the physical range $0 \leq \xi \leq 1$. Therefore we define in the $\bar{I}_0, \bar{I}_1, \bar{I}_2$ space a domain $\mathcal{D}$ where the physical requirement of positivity is fullfilled. The derivation of $\mathcal{D}$ is analogous to that given for D in the previous section and the result is formally identical

$$\bar{I}_0 + \bar{I}_1 + \bar{I}_2 \geq 0$$

$$\bar{I}_2 \geq 0$$

and when

$$\bar{I}_0 \geq \bar{I}_2$$

$$\bar{I}_1 \geq -2\sqrt{\bar{I}_0 \bar{I}_2}$$

We have represented on Fig. 2 a section of the domain $\mathcal{D}$ by a plane $\bar{I}_0 + \bar{I}_2 = \text{constant}$ orthogonal to the revolution axes of the cone $\bar{I}_1^2 = 4 \bar{I}_0 \bar{I}_2$

3°) Averaged Hadronic Energy

The total energy E_H carried out by the hadrons is related to the parameter ξ by energy conservation. In the high energy limit when $M \ll E$ we simply have

$$y = \frac{E_{HAD}}{E} = 1 - \xi \qquad (5)$$

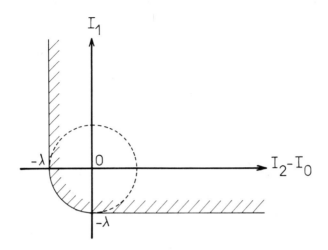

(a) $I_0 + I_2 = \lambda \geqslant 0$

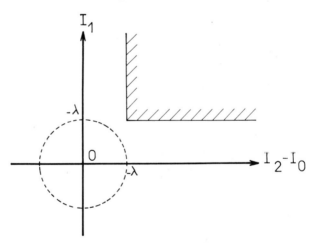

(b) $I_0 + I_2 = \lambda \leqslant 0$

FIG. 2. Positivity domain in the I_0, I_1, I_2 space. Plane sections $I_2 + I_0 = \lambda$ (a) $\lambda \geqslant 0$ (b) $\lambda \leqslant 0$

THEORETICAL UNDERSTANDING OF NEUTRAL CURRENTS

For simplicity we shall continue to work with the quantity ρ and the relation (5) has to be used in order to translate the results in the y language.

The averaged value of ρ is defined as the first moment of the ρ distribution

$$\bar{\rho}\, \sigma_{TOT} = \iint \rho\, d^2\sigma$$

Using the expression (3) for the differential cross section we obtain

$$\bar{\rho} = \frac{1}{4}\frac{I_0 + 3 I_1 + 17 I_2}{I_0 + 2 I_1 + 7 I_2} \tag{6}$$

The point I_0, I_1, I_2 being in the positivity domain $\mathcal{D}$ non trivial bounds will follow for $\bar{\rho}$.

We remark that equation (6) represents a family of planes having in common the straight line L defined by

$$I_0 - 13 I_2 = 0 \qquad I_1 + 10 I_2 = 0$$

It is now easy to check that the extreme values of $\bar{\rho}$ are obtained when the plane is tangent to the cone $I_1^2 = 4 I_0 I_2$. The result of a trivial algebraic calculation is[47]

$$\tfrac{1}{2}\left(1 - \tfrac{1}{\sqrt{3}}\right) \leq \bar{\rho} \leq \tfrac{1}{2}\left(1 + \tfrac{1}{\sqrt{3}}\right) \tag{7}$$

$$\simeq .21 \qquad\qquad \simeq .79$$

The same bounds are obviously valid for $\bar{y}$.

4°) Pure (S, P) Case

In the pure S, P case it has been already noticed that only one structure function is present. Therefore the ρ distribution is of the pure $(1-\rho)^2$ type or equivalently pure y^2 type

$$\frac{d\sigma}{d\rho} \Rightarrow \frac{G^2 M E}{\pi} I_0 (1-\rho)^2$$

The averaged values of ϱ and y are easily obtained from equation (6)

$$\overline{\varrho} = \frac{1}{4} \qquad \overline{y} = \frac{3}{4}$$

Such a situation is probably excluded by the preliminary results of the CALTECH experiment at FERMILAB.

5°) <u>Pure (V,A) Case</u>

In the Bjorken scaling limit the scaling functions associated to a given helicity of the weak current are defined by

$$\text{BJ Limit} \quad \frac{M\sqrt{\nu^2+q^2}}{\pi} \mathcal{G}_\lambda(q^2, W^2) = \overline{F}_\lambda(x) \quad \lambda = \pm, L$$

From equations (2) we get

$$a_0(x) = \frac{1}{2} \left[\overline{F}_T(x) - \overline{F}_L(x) \right]$$

$$a_1(x) = \frac{1}{2} \varepsilon \left[\overline{F}_-(x) - \overline{F}_+(x) \right]$$

$$a_2(x) = \frac{1}{2} \left[\overline{F}_T(x) + \overline{F}_L(x) \right]$$

and the high energy limit of the double differential cross sections is written as

$$\frac{d^2\sigma^\nu}{d\varrho\, dx} \Rightarrow \frac{G^2 ME}{\pi} x \left[\varrho^2 \overline{F}_+^\nu(x) + 2\varrho \overline{F}_L^\nu(x) + \overline{F}_-^\nu(x) \right] \qquad (8)$$

$$\frac{d^2\sigma^{\overline{\nu}}}{d\varrho\, dx} \Rightarrow \frac{G^2 ME}{\pi} x \left[\varrho^2 \overline{F}_-^{\overline{\nu}}(x) + 2\varrho \overline{F}_L^{\overline{\nu}}(x) + \overline{F}_+^{\overline{\nu}}(x) \right]$$

Introducing now the first moment of the scaling functions

$$\overline{I}_\lambda = \int_0^1 x\, \overline{F}_\lambda(x)\, dx \qquad \lambda = \pm, L$$

THEORETICAL UNDERSTANDING OF NEUTRAL CURRENTS

we write the fixed g cross section (3) in the form

$$\frac{d\sigma^\nu}{dg} \Rightarrow \frac{G^2 ME}{\pi} \left[g^2 I_+^\nu + 2g I_L^\nu + I_-^\nu \right] \tag{9}$$

$$\frac{d\sigma^{\bar\nu}}{dg} \Rightarrow \frac{G^2 ME}{\pi} \left[g^2 I_-^{\bar\nu} + 2g I_L^{\bar\nu} + I_+^{\bar\nu} \right]$$

For the total cross sections (4) measured in units $G^2 ME/\pi$ the slope parameters A are given by

$$A^\nu = \tfrac{1}{3} I_+^\nu + I_L^\nu + I_-^\nu \tag{10}$$

$$A^{\bar\nu} = \tfrac{1}{3} I_-^{\bar\nu} + I_L^{\bar\nu} + I_+^{\bar\nu}$$

The positivity of the moments I_λ's allows us to define, in the $\bar I_0, \bar I_1, \bar I_2$ space a positivity domain $\mathcal{D}^{VA}$. From the relations

$$\bar I_0 = \tfrac{1}{2}(I_T - I_L)$$

$$\bar I_1 = \tfrac{1}{2}\varepsilon(I_- - I_+) \tag{11}$$

$$\bar I_2 = \tfrac{1}{2}(I_T + I_L)$$

we obtain

$$|\bar I_1| \le \bar I_0 + \bar I_2$$

$$\bar I_0 \le \bar I_2$$

Therefore the bounds on $\bar g$ and $\bar y$ become more restrictive and the result is now

$$\frac{1}{2} \leq \bar{g} \leq \frac{3}{4}$$

$$\frac{1}{4} \leq \bar{y} \leq \frac{1}{2}$$

III. CROSSING PROPERTIES

1°) In order to obtain a relation between neutrino and antineutrino cross sections, we make assumptions concerning the Hermitian character of the neutral current. In the absence of second class neutral currents[31,45]. The same structure functions describe neutrino and antineutrino processes with the following correspondence[45].

$$a_0^{\bar{\nu}}(q^2, W^2) = a_0^{\nu}(q^2, W^2)$$
$$a_1^{\bar{\nu}}(q^2, W^2) = -a_1^{\nu}(q^2, W^2) \qquad (12)$$
$$a_2^{\bar{\nu}}(q^2, W^2) = a_2^{\nu}(q^2, W^2)$$

Only the interference function a_1 changes sign going from neutrino to antineutrino scattering. The obvious immediate consequences are

a) the difference $d^2\sigma^{\nu} - d^2\sigma^{\bar{\nu}}$ is pure $\cos\Theta_t$

b) the sum $d^2\sigma^{\nu} + d^2\sigma^{\bar{\nu}}$ does not contain the $\cos\Theta_t$ term

and for the g distributions at high energy

a) the difference $\dfrac{d\sigma^{\nu}}{dg} - \dfrac{d\sigma^{\bar{\nu}}}{dg}$ is purely $1 - g^2$

b) the sum $\dfrac{d\sigma^{\nu}}{d\xi} + \dfrac{d\sigma^{\bar{\nu}}}{d\xi}$ does not contain the $1-\xi^2$ term.

We then have simple tests of the relations (12).

2°) Positivity Domains

In the a_0, a_1, a_2 space the positivity domain is now the intersection of the neutrino and antineutrino domains e.g. the symmetric part $\mathbb{D}_s$ in the exchange $a_1 \to -a_1$ of the previous domain $\mathbb{D}$. We easily obtain the conditions for $\mathbb{D}_s$

$$|a_1(q^2,W^2)| \leq a_0(q^2,W^2) + a_2(q^2,W^2)$$

and when $a_2(q^2,W^2) \leq a_0(q^2,W^2)$ the additional constraint

$$|a_1(q^2,W^2)| \leq 2\sqrt{a_0(q^2,W^2)\, a_2(q^2,W^2)}$$

For the first moment integrals I_0, I_1, I_2 of the scaling limits of the functions a_0, a_1 and a_2 we get similar restrictions

$$|I_1| \leq I_0 + I_2$$

and when $I_2 \leq I_0$

$$|I_1| \leq 2\sqrt{I_0 I_2}$$

The positivity domain $\mathcal{D}_s$ is limited by two half planes and half a cone. A section of $\mathcal{D}_s$ by a plane $I_0 + I_2 = \text{constant}$ is represented on Fig. 3.

3°) Bounds For r_N

The first immediate consequence of the crossing symmetry properties is the existence of bounds for the ratio r_N of antineutrino total cross sections

$$r_N = \frac{I_0 - 2I_1 + 7I_2}{I_0 + 2I_1 + 7I_2} \tag{13}$$

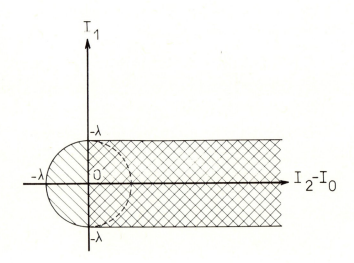

$$I_0 + I_2 = \lambda \geqslant 0$$

FIG. 3. Positivity domain in the I_0, I_1, I_2 space with crossing. Plane section $I_0 + I_2 = \lambda$ with $\lambda \geqslant 0$

Equation (13) represents, in the I_0, I_1, I_2 space a family of planes having in common the straight line

$$I_0 + 7 I_2 = 0 \qquad I_1 = 0$$

The constraints on r_N are obtained when the plane is tangent to the cone and the result is[47,48]

$$\frac{\sqrt{7}-2}{\sqrt{7}+2} \leq r_N \leq \frac{\sqrt{7}+2}{\sqrt{7}-2}$$

$$\simeq .14 \qquad\qquad \simeq 7.2$$

Obviously, in the pure (S, P) case, the interference function a_1 is absent and we get $r_N = 1$

4°) Relation Between $\bar{\rho}_\nu, \bar{\rho}_{\bar{\nu}}$ and r_N

The averaged energies for neutrino and antineutrino processes depend on the same three quantities I_0, I_1 and I_2

$$\bar{\rho}_\nu = \frac{1}{4} \frac{I_0 + 3 I_1 + 17 I_2}{I_0 + 2 I_1 + 7 I_2} \qquad (14)$$

$$\bar{\rho}_{\bar{\nu}} = \frac{1}{4} \frac{I_0 - 3 I_1 + 17 I_2}{I_0 - 2 I_1 + 7 I_2} \qquad (15)$$

The equations (13, (14) and (15) can be considered as a system of three linear homogenous equations in I_0, I_1 and I_2. Non trivial solutions exist if and only if the total determinant is zero.

A convenient way to proceed is to solve the system of equations (14), (15) in I_0, I_1, I_2. To that purpose we compute the three minors

$$\Delta_0 = -51 + 110(\bar{\varsigma}_\nu + \bar{\varsigma}_{\bar{\nu}}) - 224\,\bar{\varsigma}_\nu\bar{\varsigma}_{\bar{\nu}}$$

$$\Delta_1 = 20(\bar{\varsigma}_{\bar{\nu}} - \bar{\varsigma}_\nu) \qquad (16)$$

$$\Delta_2 = 3 - 10(\bar{\varsigma}_\nu + \bar{\varsigma}_{\bar{\nu}}) + 32\,\bar{\varsigma}_\nu\bar{\varsigma}_{\bar{\nu}}$$

and the solutions for I_0, I_1, I_2 are determined up to an arbitrary multiplicative factor

$$\frac{I_0}{\Delta_0} = \frac{I_1}{\Delta_1} = \frac{I_2}{\Delta_2} \qquad (17)$$

From equation (13) the vanishing of the total determinant is simply written as

$$(\Delta_0 - 2\Delta_1 + 7\Delta_2) - r_N(\Delta_0 + 2\Delta_1 + 7\Delta_2) = 0$$

or using equations (16)

$$(3 - 8\bar{\varsigma}_\nu) - r_N(3 - 8\bar{\varsigma}_{\bar{\nu}}) = 0 \qquad (18)$$

The ratio r_N being a positive number, the relation (18) implies correlated range for $\bar{\varsigma}_\nu$ and $\bar{\varsigma}_{\bar{\nu}}$ as follows

$$\bar{\varsigma}_\nu \text{ and } \bar{\varsigma}_{\bar{\nu}} \in \left[\tfrac{1}{2}(1 - \tfrac{1}{\sqrt{3}}),\, \tfrac{3}{8}\right]$$

or

$$\bar{\varsigma}_\nu \text{ and } \bar{\varsigma}_{\bar{\nu}} \in \left[\tfrac{3}{8},\, \tfrac{1}{2}(1 + \tfrac{1}{\sqrt{3}})\right]$$

and the value $3/8$ is reached simultaneously for $\bar{\varsigma}_\nu$ and $\bar{\varsigma}_{\bar{\nu}}$.

5°) Positivity Domain in the $\bar{\varsigma}_\nu, \bar{\varsigma}_{\bar{\nu}}$ Plane

More precise correlations between averaged values of ς for neutrino and antineutrino reactions can be obtained as a consequence of positivity. Starting from the domain $\mathcal{D}_5$ in the I_0, I_1, I_2 space we get the positivity region in the $\bar{\varsigma}_\nu, \bar{\varsigma}_{\bar{\nu}}$ plane using equations (17). We then introduce three curves

C_α : to the cone $\overline{I}_1^2 = 4\overline{I}_0\overline{I}_2$ corresponds the curve $\Delta_1^2 = 4\Delta_0\Delta_2$ which is a quartic having a double point C at
$$\overline{\rho}_\nu = \overline{\rho}_{\bar\nu} = \frac{3}{8}$$

C_β : to the plane $\overline{I}_0 + \overline{I}_1 + \overline{I}_2 = 0$ corresponds a rectangular hyperbola $\Delta_0 + \Delta_1 + \Delta_2 = 0$

C_γ : to the plane $\overline{I}_0 - \overline{I}_1 + \overline{I}_2 = 0$ corresponds a second rectangular hyperbola $\Delta_0 - \Delta_1 + \Delta_2 = 0$

The junction points of these curves are easily computed

C_α and C_β $\overline{\rho}_\nu = \frac{3}{4}$ $\overline{\rho}_{\bar\nu} = \frac{1}{2}$

C_α and C_γ $\overline{\rho}_\nu = \frac{1}{2}$ $\overline{\rho}_{\bar\nu} = \frac{3}{4}$

C_β and C_γ $\overline{\rho}_\nu = \overline{\rho}_{\bar\nu} = \frac{2}{3}$

The positivity domain is shown on Fig. 4.

The relation (18) is now very easy to interpret graphically : on a straight line passing through the double point C of the quartic C_α, the value of r_N is fixed and that value is precisely the slope of that straight line. We are then able to derive bounds for $\overline{\rho}_\nu$ and $\overline{\rho}_{\bar\nu}$ corresponding to a fixed value of r_N.

The lower bounds are always given by the quartic C_α and the result is

$$\overline{\rho}_\nu \geq \frac{3(15 + r_N) - 5\sqrt{22\, r_N - 3(1 + r_N^2)}}{112} \qquad (19)$$

$$\overline{\rho}_{\bar\nu} \geq \frac{3(1 + 15\, r_N) - 5\sqrt{22\, r_N - 3(1 + r_N^2)}}{112\, r_N}$$

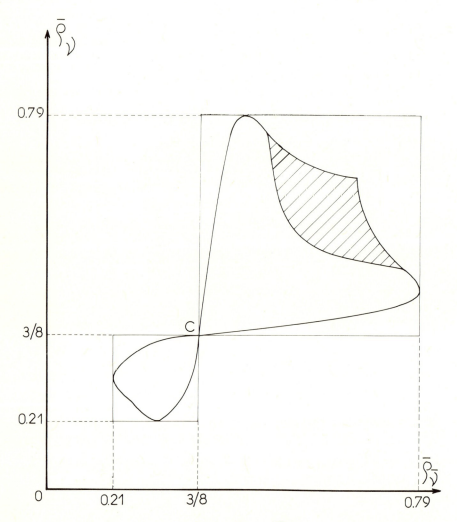

FIG. 4. Positivity domain in the $\bar{\varsigma}_\nu$, $\bar{\varsigma}_{\bar\nu}$ plane. The dashed area corresponds to (V,A) currents.

The upper bound has different analytic expressions depending on the range of r_N

<u>a</u> For $\dfrac{\sqrt{7}-2}{\sqrt{7}+2} \leq r_N \leq \dfrac{1}{3}$ and $3 \leq r_N \leq \dfrac{\sqrt{7}+2}{\sqrt{7}-2}$ it is given by the quartic C_α

$$\bar{\rho}_\nu \leq \frac{3(15+r_N)+5\sqrt{22 r_N - 3(1+r_N^2)}}{112}$$

$$\bar{\rho}_{\bar{\nu}} \leq \frac{3(1+15 r_N)+5\sqrt{22 r_N - 3(1+r_N^2)}}{112 \, r_N} \qquad (20)$$

<u>b</u> For $\dfrac{1}{3} \leq r_N \leq 1$ it is given by the rectangular hyperbola C_γ

$$\bar{\rho}_\nu \leq \frac{3 r_N + 5}{12}$$

$$\bar{\rho}_{\bar{\nu}} \leq \frac{15 r_N + 1}{24 \, r_N} \qquad (21)$$

<u>c</u> For $1 \leq r_N \leq 3$ it is given by the rectangular hyperbola C_β

$$\bar{\rho}_\nu \leq \frac{15+r_N}{24}$$

$$\bar{\rho}_{\bar{\nu}} \leq \frac{3+5 r_N}{12 \, r_N} \qquad (22)$$

6°) The (V,A) Case

If the final neutrino (antineutrino) is identical to the initial one, there is no place for second class currents[49] and the relations (12) hold. In the language of total cross sections $\sigma_\lambda(q^2, W^2)$ we simply have

$$\sigma_\lambda^{\bar{\nu}}(q^2, W^2) = \sigma_\lambda^{\nu}(q^2, W^2) \qquad \lambda = \pm, L$$

It follows that in the right-hand side of equations (8), (9) and

(10), the upper index ν or $\bar{\nu}$ can be disregarded.

The domain $\mathcal{D}^{VA}$ being a subspace of $\mathcal{D}_\beta$, we get more restrictive constraints on the ratio r_N of total cross sections

$$\frac{1}{3} \leq r_N \leq 3$$

Using equations (10) we easily see that the upper bound 3 for r_N is reached when only the I_+ parameter is present in the distributions. The lower bound $\frac{1}{3}$ for r_N is obtained with β distributions containing only the I_- parameter. These two extreme cases seem to be excluded by the preliminary results of the CALTECH experiment at FERMILAB.

The region allowed by positivity in the $\bar{\beta}_\nu$, $\bar{\beta}_{\bar\nu}$ plane is obviously smaller than in the general case. The quartic C_α : $\Delta_1^2 = 4\Delta_0\Delta_2$ is replaced by the rectangular hyperbola C_β : $\Delta_2 - \Delta_0 = 0$ associated to the vanishing of the longitudinal contributions. The positivity domain of the (V,A) case is shown on Fig. 4.

As an immediate consequence, the lower bounds for $\bar{\beta}_\nu$ and $\bar{\beta}_{\bar\nu}$ at fixed r_N are determined by the hyperbola C_β and equations (19) are replaced by

$$\bar{\beta}_\nu \geq \frac{3}{32}(5+r_N)$$

$$\bar{\beta}_{\bar\nu} \geq \frac{3}{32 r_N}(1+5r_N)$$

(23)

The upperbounds (21) and (22) are unchanged.

The normalized energy distributions

$$f(\beta) = \frac{1}{\sigma_{TOT}} \frac{d\sigma}{d\beta}$$

THEORETICAL UNDERSTANDING OF NEUTRAL CURRENTS

are also restricted by the knowledge of the ratio r_N of total cross section.

The first limit is associated to the vanishing of the longitudinal contribution and it is given by

$$\frac{3}{8}\left[(3-r_N) + (3r_N-1)\varsigma^2\right] \quad \text{for } f^\nu(\varsigma)$$

$$\frac{3}{8 r_N}\left[(3r_N-1) + (3-r_N)\varsigma^2\right] \quad \text{for } f^{\bar{\nu}}(\varsigma) \qquad (24)$$

For $\frac{1}{3} \leq r_N \leq 1$ the other relevant limit corresponds to the vanishing of the right handed part and one obtains

$$\frac{1}{2}\left[3(1-r_N) + (3r_N-1)2\varsigma\right] \quad \text{for } f^\nu(\varsigma)$$

$$\frac{1}{2 r_N}\left[(3r_N-1)2\varsigma + 3(1-r_N)\varsigma^2\right] \quad \text{for } f^{\bar{\nu}}(\varsigma)$$

For $1 \leq r_N \leq 3$ the second limit is now associated to the vanishing of the right handed part and we get

$$\frac{1}{2}\left[(3-r_N)2\varsigma + 3(r_N-1)\varsigma^2\right] \quad \text{for } f^\nu(\varsigma)$$

$$\frac{1}{2 r_N}\left[3(r_N-1) + (3-r_N)2\varsigma\right] \quad \text{for } f^{\bar{\nu}}(\varsigma)$$

Experiments on electroproduction performed at SLAC and on neutrino and antineutrino reactions performed at CERN and FERMILAB strongly suggest that in the deep inelastic region the longitudinal component of the electromagnetic and charged weak current play a minor role when compared with the transverse one.

If such a situation is also present with the weak neutral current it is interesting to look at the simple case where the longitudinal component of the current is neglected. The main consequences of such an assumption are :

a) absence of linear ξ term in the neutrino and antineutrino energy distribution

b) the averaged energies are totally determined by γ_N as given by the lower bound (23) of γ_N. The allowed points in the $\bar{\xi}_\nu, \bar{\xi}_{\bar\nu}$ plane are located on the hyperbola $\Delta_2 - \Delta_0 = 0$ or more explicitly

$$\left[16\bar{\xi}_\nu - \tfrac{15}{2}\right]\left[16\bar{\xi}_{\bar\nu} - \tfrac{15}{2}\right] = \tfrac{9}{4} \qquad (25)$$

c) the normalized energy distributions are predicted by the value of γ_N following the equation (24). It is amusing to remark that, independently of γ_N, they have a fixed point[50]

$$f^\nu\!\left(\tfrac{1}{\sqrt{3}}\right) = f^{\bar\nu}\!\left(\tfrac{1}{\sqrt{3}}\right) = 1$$

IV. V/A ANALYSIS OF INCLUSIVE REACTIONS

1°) At high energy the total cross sections for inclusive reaction are described by the slope parameters A introduced in equation (10). Let us first define two quantities averaged over proton and neutron

$$\Sigma_N = \tfrac{1}{4}\left[A_{NC}^{\nu p} + A_{NC}^{\nu n} + A_{NC}^{\bar\nu p} + A_{NC}^{\bar\nu n}\right] \qquad (25$$

$$\Delta_N = \tfrac{1}{4}\left[A_{NC}^{\nu p} + A_{NC}^{\nu n} - A_{NC}^{\bar\nu p} - A_{NC}^{\bar\nu n}\right]$$

The symbol NC means neutral current and CC will be used later for charged current.

THEORETICAL UNDERSTANDING OF NEUTRAL CURRENTS

We <u>assume</u> for the neutral current a (V,A) structure. The difference Δ_N of neutrino and antineutrino total cross sections measures the parameter $2 I_\Delta = I_- - I_+$ representing an interference between vector and axial vector amplitudes. The sum Σ'_N of neutrino and antineutrino total cross sections measures the combination $I_0 + 7 I_2 = 4 I_T + 3 I_L$ which is a sum of diagonal vector and axial vector contributions.

We then write

$$\Sigma'_N = V^N + A^N \qquad \Delta_N = I^N$$

and we introduce an angle ϕ_N, $0 \leq \phi_N \leq \frac{\pi}{2}$ in order to evaluate the relative importance of the vector and axial vector parts in Σ_N

$$V^N = \Sigma_N \cos^2 \phi_N \qquad A^N = \Sigma_N \sin^2 \phi_N$$

The key of the problem is the Schwarz inequality

$$[I^N]^2 \leq V^N A^N$$

which is equivalently written as

$$4 \Delta_N^2 \leq \Sigma_N'^2 \sin^2 2\phi_N$$

The reality of ϕ_N implies a constraint on the ratio Δ_N / Σ_N which is nothing but the well-known inequality

$$\tfrac{1}{3} \leq r_N \leq 3$$

Assuming we know Δ_N and Σ_N we parametrize the data with an angle δ_N, $0 \leq \delta_N \leq \pi/4$

$$\sin 2\delta_N = 2 \frac{|\Delta_N|}{\Sigma_N} = 2 \frac{|1 - r_N|}{1 + r_N}$$

and the constraint due to positivity is simply[51]

$$\delta_N \leq \phi_N \leq \frac{\pi}{2} - \delta_N$$

2°) From the Gargamelle bubble chamber experiment performed at CERN, the value of γ_N is estimated to be

$$\gamma_N = .71 \pm .15$$

and we have two standard deviation effects for pure vector or pure axial vector neutral current.

The angle δ_N is computed to be

$$\delta_N = 10° \pm 4°$$

and the positivity constraint is not extremely severe.

The Harvard-Pennsylvania-Wisconsin-Fermilab collaboration has published the value

$$\gamma_N = 1 \pm 0.2$$

which is consistent with a pure vector or pure axial vector neutral current.

Obviously new accurate measurements are needed before drawing any definite conclusion on that question.

/ PART II - INTERNAL SYMMETRY STRUCTURE OF THE HADRONIC NEUTRAL CURRENT /

  I. Isospin Structure.
 II. Isovector Assumption.
III. Isoscalar Assumptions.
 IV. Algebraic Structure.
  V. Conservation Properties of the Neutral Current.

In this part a (V,A) structure is <u>assumed</u> for the hadronic neutral current and we now investigate the quantum numbers of the neutral current associated to quantities conserved in strong interactions. More precisely we look at the transformation properties of the current under operations of the Internal Symmetry Group.

The first problem is related to isotopic spin. On a phenomenological basis[+] we only retain two pieces for the neutral current, one isovector and one isoscalar. With respect to $SU(3)$ symmetry, while the isovector components belong to an octet, we have various possibilities for the isoscalar ones
  i) pure octet like the electromagnetic current ;
 ii) pure singlet like the baryonic current ;
iii) superposition of octet and singlet weights.

Higher symmetries are also considered and in all cases -but one- a connexion is made between the electromagnetic, weak charged and weak neutral currents.

[+] Experimentally the $|\Delta S| = 1$ neutral currents are strongly suppressed. From upper limits or measured rates on $K^+ \Rightarrow \pi^+ \nu \bar{\nu}$; $K^+ \Rightarrow \pi^+ e^+ e^-$; $K_L \Rightarrow \mu^+ \mu^-$ we can estimate the following order of magnitude for the strangeness violating transitions involving neutral currents

$$G_{NC}(|\Delta S|=1) \leq 10^{-3} G_{CC}(\Delta S = 0)$$

We shall present a method as model independent as possible and the quark parton model will not be used except in particular cases [26,27,30,38,44,51].

I. ISOSPIN STRUCTURE

1°) We take for granted the experimental fact that strangeness changing neutral currents are not present. Therefore we disregard the possibility to have $|\Delta \vec{I}| = \frac{1}{2}$ or $\frac{3}{2}$ components for the neutral current and we make the following assumption :

/The hadronic neutral current has only isoscalar and isovector components/

2°) For a nucleon target, in addition to relations (25) averaged over proton and neutron we introduce two new combinations

$$D_N = \frac{1}{4} \left[A_{NC}^{\nu p} - A_{NC}^{\nu n} + A_{NC}^{\bar{\nu} p} - A_{NC}^{\bar{\nu} n} \right]$$

$$S_N = \frac{1}{4} \left[A_{NC}^{\nu p} - A_{NC}^{\nu n} - A_{NC}^{\bar{\nu} p} + A_{NC}^{\bar{\nu} n} \right]$$

(26)

Experimental data are not yet available on D_N and S_N. Nevertheless, we shall use them in our theoretical analysis and make predictions on D_N for particular situations.

3°) As a consequence of the previous isospin assumption, and for a (V,A) structure of the weak neutral current, the four parameters Σ_N, Δ_N, D_N and S_N can be written in the following way

$$\Sigma_N = V_3^N + A_3^N + V_0^N + A_0^N$$

$$\Delta_N = I_3^N + I_0^N$$

$$D_N = J_V^N + J_A^N$$

$$S_N = K_{30}^{VA} + K_{03}^{VA}$$

The V'_S and A'_S terms are positive, representing modulus squared of matrix elements. The I'_S correspond to a vector-axial vector interference- The J'_S described an isoscalar-isovector interference. The K'_S are associated to mixed type interferences.

4°) In order to get information on the internal symmetry structure of the hadronic neutral current we shall use the following method : we make assumptions relating the NC components with the CC and the electromagnetic ones.

For the weak charged current we use the $V-A$ Cabibbo current whose strangeness-conserving piece is pure isovector. For the pure vector electromagnetic current we have an isoscalar component and an isovector component related to charged current by the conserved vector current hypothesis (CVC).

Inclusive reactions for inelastic lepton scattering in the deep inelastic region can be analyzed in a model-independent way. The method is the same as that used in Section IV of PART I and we only sketch the various steps ; the notations are analogous to those previously introduced[44,51].

$$\Sigma'_c = V^c_3 + A^c_3 \qquad \Delta_c = I^c_3$$

$$\Sigma'_Q = V^Q_3 + V^Q_0 \qquad \mathbb{D}_Q = J^Q_V$$

+)

With the CVC hypothesis

$$V^c_3 = 2 V^Q_3$$

we define an angle ϕ_c by

$$V^c_3 = \Sigma_c \cos^2 \phi_c \qquad A^c_3 = \Sigma_c \sin^2 \phi_c$$

+) The electromagnetic "total cross-sections" are defined after removing the photon propagator in a complete analogy with the weak ones.

and we apply the Schwarz inequality in order to get the restrictions

$$\delta_1 \leq \phi_c \leq \delta_2 \tag{27}$$

where the angles δ_1 and δ_2 parametrize the experimental data

$$\cos^2 \delta_1 = \frac{\Sigma_Q + \sqrt{\Sigma_Q^2 - D_Q^2}}{\Sigma_c'}$$

$$\cos^2 \delta_2 = \frac{\Sigma_c' - \sqrt{\Sigma_c'^2 - 4\Delta_c^2}}{2\Sigma_c}$$

Using the SLAC data for electroproduction and the Gargamelle results for charged current reactions we obtain the following numerical values for δ_1 and δ_2

$$\delta_1 = 40° \pm 5°$$
$$\delta_2 = 56° {}^{+3°}_{-4°} \tag{28}$$

The experimental data are compatible with the CVC constraint $\delta_1 \leq \delta_2$ and we remark that the value $\phi_c = 45°$ used in the simple quark parton model lies in the positivity range (27).

II. ISOVECTOR ASSUMPTION

1°) The most natural assumption we can make for the isovector part of the hadronic neutral current is the following

> The isovector neutral current components belong to the same isospin triplets as the isovector strangeness-conserving charged currents.

We therefore introduce two parameters α and β

$$V_\mu^{3,N} = \alpha V_\mu^3 \qquad A_\mu^{3,N} = \beta A_\mu^3$$

where the set of indices $1+i2, 3, 1-i2$ describes an isospin triplet.

2°) A typical example for measuring the isovector part of the hadronic neutral current is given by the reactions with a deuterium target

$$\nu + d \Rightarrow \nu + \pi^\circ + d$$
$$\bar\nu + d \Rightarrow \bar\nu + \pi^\circ + d \qquad (29)$$

<u>a</u> Non-vanishing cross sections will prove the existence of isovector components for the neutral current.

<u>b</u> A non-vanishing difference $d\sigma^\nu - d\sigma^{\bar\nu}$ will measure the vector-axial vector isovector interference.

<u>c</u> A comparison of the reactions (29) with the corresponding charged current ones.

$$\nu_\mu + d \Rightarrow \mu^- + \pi^+ + d$$
$$\bar\nu_\mu + d \Rightarrow \mu^+ + \pi^- + d \qquad (30)$$

will directly test the isovector assumption.

Unfortunately the experimental data are not yet available.

3°) <u>Inclusive Reactions</u>[38,51]

The consequences for inclusive reactions of the isovector assumption are the following

$$V_3^N = \tfrac{1}{2}\alpha^2 V_3^c \qquad A_3^N = \tfrac{1}{2}\beta^2 A_3^c \qquad I_3^N = \tfrac{1}{2}\alpha\beta\, I_3^c$$

We now study the domain allowed by experiment in the α, β plane using positivity constraints. Let us first isolate the isoscalar part

$$\Sigma_o \equiv V_o^N + A_o^N = \Sigma_{iN} - \frac{1}{2}[\alpha^2 V_3^c + \beta^2 A_3^c]$$

$$\Delta_o \equiv I_o^N = \Delta_N - \frac{1}{2}\alpha\beta \Delta_c$$

We parametrize as previously the charged current matrix elements

$$V_3^c = \Sigma_c \cos^2\phi_c \qquad A_3^c = \Sigma_c \sin^2\phi_c$$

with

$$\delta_1 \leq \phi_c \leq \delta_2$$

An obvious requirement of positivity is $\Sigma_o \geq 0$ but a stronger constraint can be obtained using the Schwarz inequality

$$[I_o^N]^2 \leq V_o^N A_o^N$$

which has the consequence

$$2|\Delta_o| \leq \Sigma_o$$

We are then led to introduce the following curves in the α, β plane

<u>a</u> An ellipse $E(\phi_c)$ representing $\Sigma_o = 0$

<u>b</u> A rectangular hyperbola H representing $\Delta_o = 0$

<u>c</u> Two ellipses $E^\pm(\phi_c)$ corresponding respectively to the conditions $\Sigma_o \pm 2\Delta_o = 0$

At a given value of ϕ_c, the allowed domain $D_1(\phi_c)$ is simply the intersection of the interiors of the two ellipses $E^\pm(\phi_c)$

$$D_1(\phi_c) = E^+(\phi_c) \cap E^-(\phi_c)$$

The domain D_1 for $\phi_c = 45°$ has been represented on Fig. 5.

The maximal values of left handed g_L and right handed g_R couplings defined by

$$g_L = \frac{\alpha + \beta}{2} \qquad g_R = \frac{\alpha - \beta}{2}$$

are computed using the CERN - Gargamelle data.

The result is

$$|g_L| \leq 0.63 \pm 0.05$$
$$|g_R| \leq 0.44 \pm 0.05$$

Finally the complete domain represented on Fig. 6 is the union of the $D_1(\phi_c)$'s for ϕ_c varying in its range $[\delta_1, \delta_2]$.

Finally let us analyze the experimental data with a simple two-parameter model, called <u>MODEL I</u> where the hadronic neutral current is a pure isovector. We obtain two sets of solutions for α, β located at the corners of the domain $D_1(\phi_c)$ because of the vanishing of the isoscalar contributions $\Sigma_0 = 0$, $\Delta_0 = 0$.

For $\phi_c = 45°$ we obtain

$$g_L^2 = \frac{\Sigma_N}{\Sigma_c} + \frac{\Delta_N}{\Delta_c} \qquad g_R^2 = \frac{\Sigma_N}{\Sigma_c} - \frac{\Delta_N}{\Delta_c}$$

and using the Gargamelle data the numerical values for g_L and g_R are

$$|g_L| = 0.62 \pm 0.05 \qquad |g_R| = 0.43 \pm 0.05$$

We have the two obvious predictions

$$D_N = 0 \qquad S_N = 0$$

A model of this type has been proposed by Adler and Tuan[39] with

---

N.B. : For clarification, the experimental errors have not been represented on Figs. 5 and 6.

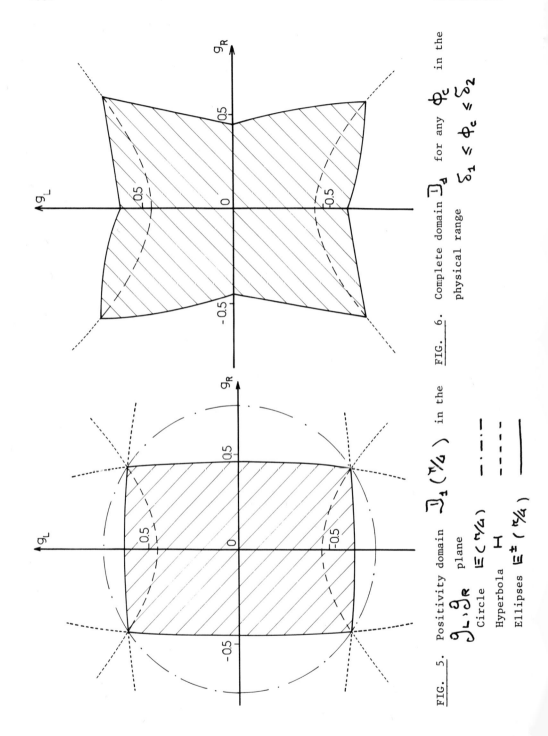

FIG. 5. Positivity domain $\mathcal{D}_1(\pi/4)$ in the g_L, g_R plane
Circle $E(\pi/4)$ —·—·—
Hyperbola H - - - -
Ellipses $E^{\pm}(\pi/4)$ ———

FIG. 6. Complete domain $\mathcal{D}_1$ for any ϕ_c in the physical range $\delta_1 \leq \phi_c \leq \delta_2$

THEORETICAL UNDERSTANDING OF NEUTRAL CURRENTS

the additional constraint of a pure V-A interaction : $g_R = 0$
Such a model is out by few s.d. if one believes the Gargamelle data.

III. ISOSCALAR ASSUMPTIONS

1°) A clean set of reactions for measuring the isoscalar part of the hadronic current is the elastic scattering on a deuteron target[26]

$$\nu + d \Rightarrow \nu + d$$
$$\bar{\nu} + d \Rightarrow \bar{\nu} + d$$

<u>a</u> Non-vanishing cross sections will prove the existence of isoscalar components for the neutral current.

<u>b</u> A non-vanishing difference $d\sigma^\nu - d\sigma^{\bar{\nu}}$ will measure the vector-axial vector isoscalar interference.

Again in this case the experimental data are not available.

2°) The general form of the isoscalar hadronic neutral current $J_\mu^{0,N}$ is a linear combination for both vector and axial vector pieces of the known isoscalar currents, baryonic B, strangeness S and possibly of new ones like for instance charm current C, etc... At this stage the symmetry group for strong interactions from which currents are constructed plays a crucial role.

With only our limited experimental information on neutral currents it is not possible to get useful results without making simple assumptions as for instance[26,38,51].

<u>a</u> Relate $V_\mu^{0,N}$ to the isoscalar part of the electromagnetic current

$$V_\mu^{0,N} = \gamma \, V_\mu^{0,Q}$$

<u>b</u> Relate also $A_\mu^{0,N}$ to electromagnetic quantum numbers

$$A_\mu^{0,N} = \delta \, A_\mu^{0,Q}$$

<u>c</u> Add a second component

$$V_\mu^{0,N} = \gamma V_\mu^{0,Q} + 6 V_\mu^{B}$$
$$A_\mu^{0,N} = \delta A_\mu^{0,Q} + \zeta A_\mu^{B}$$

The hadronic matrix elements of $A_\mu^{0,N}$ are new quantities one can measure in neutral current experiments. In the present case where we try to determine what can be the parameters $\gamma, \delta, 6, \zeta$ etc..., they will be computed using symmetry relations or a quark parton model.

3°) <u>MODEL II</u>

We use only known currents

$$J_\mu^N = \beta (V_\mu^3 + A_\mu^3) + (\alpha - \beta) V_\mu^{0,Q}$$

In this model there is no axial vector isoscalar neutral current and the interference term Δ_0 vanishes. In the α, β plane we get two sets of solutions located on the rectangular hyperbola H. For $\phi_c = 45°$ we obtain

$$g_L^2 = \frac{1}{4}\left[\frac{\Sigma_N}{\Sigma_Q} + (8 - \frac{\Sigma_c}{\Sigma_Q})\frac{\Delta_N}{\Delta_c}\right] \qquad g_R^2 = \frac{1}{4}\left[\frac{\Sigma_N}{\Sigma_Q} - \frac{\Sigma_c}{\Sigma_Q}\frac{\Delta_N}{\Delta_c}\right]$$

and using the Gargamelle data the numerical values for g_L and g_R are

$$|g_L| = 0.60 \pm 0.05 \qquad |g_R| = 0.39 \pm 0.05$$

We are able to predict the ratio D_N/D_Q

$$\frac{D_N}{D_Q} = 2 g_R (g_L + g_R)$$

and using again the Gargamelle data we obtain

$$\frac{D_N}{D_Q} = 0.78 \pm 0.14 \qquad \text{if } g_L g_R > 0$$

$$\frac{D_N}{D_Q} = -0.16 \pm 0.04 \qquad \text{if } g_L g_R < 0$$

THEORETICAL UNDERSTANDING OF NEUTRAL CURRENTS

A particular case of this class of models is the simple extension to hadrons of the Salam-Weinberg gauge theory[9,14,15].

The coupling constants g_L and g_R are given in terms of the mixing angle θ_W

$$g_L = \cos^2 \theta_W \qquad g_R = -\sin^2 \theta_W$$

and we immediately obtain

$$\sin^2 \theta_W = \frac{1}{2} - \frac{\Delta_N}{\Delta_C}$$

Moreover the angle ϕ_c can be determined imposing a consistency condition

$$\cos^2 \phi_c = \frac{1}{2} \frac{1 - 2\frac{\Sigma_N}{\Sigma_C}}{1 - 2\frac{\Delta_N}{\Delta_C}} + \left(1 - 2\frac{\Delta_N}{\Delta_C}\right) \frac{\Sigma_Q}{\Sigma_C}$$

Numerical values for θ_W and ϕ_c are now computed with the Gargamelle data and the results are

$$\sin^2 \theta_W = 0.40 \pm 0.04$$

$$\phi_c = 45° \pm 3°$$

The prediction for the ratio J_N/J_Q turns out to be

$$\frac{J_N}{J_Q} = -0.16 \pm 0.05$$

4°) <u>MODEL III</u>

The quantum numbers of the weak neutral current are the same as those of the electromagnetic current

$$J_\mu^N = \alpha V_\mu^Q + \beta A_\mu^Q$$

We have two sets of solutions in α, β which in the case $\phi_c = 45°$ have the following expression

$$g_L^2 = \frac{1}{4}\frac{\Sigma_N}{\Sigma_Q} + \frac{9}{10}\frac{\Delta_N}{\Delta_Q} \qquad g_R^2 = \frac{1}{4}\frac{\Sigma_N}{\Sigma_Q} - \frac{9}{10}\frac{\Delta_N}{\Delta_Q}$$

The numerical values of g_L and g_R with the Gargamelle data are

$$|g_L| = 0.58 \pm 0.04 \qquad |g_R| = 0.38 \pm 0.05$$

The prediction for $\dfrac{\mathcal{J}_N}{\mathcal{J}_Q}$ is independent on ϕ_c

$$\frac{\mathcal{J}_N}{\mathcal{J}_Q} = \frac{\Sigma_N}{\Sigma_Q}$$

e.g. with the Gargamelle data

$$\frac{\mathcal{J}_N}{\mathcal{J}_Q} = 0.96 \pm 0.19$$

To that class of "electromagnetic" models belong

<u>a</u> the Mathur-Okubo-Kim model[36] where on phenomenological bases the neutral hadronic current is assumed to be an U spin scalar.

<u>b</u> the Bég-Zee gauge theory[16] where the non-coloured part of the neutral current is proportional to the electromagnetic current

$$(g_L = g_R)$$

<u>c</u> the generalization of Bég-Zee theory made by Viallet[52] where, in addition, an axial vector part is present.

5°) <u>MODEL IV</u>

The neutral current is assumed to be purely isoscalar and of the form

$$J_\mu^N = \gamma \bar{V}_\mu^{0,Q} + \delta A_\mu^{0,Q}$$

Again we obtain two sets of solutions for γ and δ ($\alpha=\beta=0$). Using the $SU(3)$ quark parton model we obtain

$$(\gamma \pm \delta)^2 = \frac{\Sigma_N}{\Sigma_Q - \frac{1}{4}\Sigma_e} \pm 36 \frac{\Delta_N}{\Delta_e}$$

and using the Gargamelle and SLAC data

$$\left|\frac{\gamma+\delta}{2}\right| = 1.55 \pm 0.15 \qquad \left|\frac{\gamma-\delta}{2}\right| = 0.75 ^{+0.22}_{-0.33}$$

As for MODEL I we have the predictions

$$D_N = 0 \qquad S_N = 0$$

6°) MODEL V

An interesting suggestion made by Sakurai[22] is to relate the hadronic neutral current to the baryonic number B

$$J^N_\mu = \sigma V^B_\mu + \tau A^B_\mu$$

We use the $SU(3)$ quark parton model in order to compute the hadronic matrix elements of V^B_μ and A^B_μ and we obtain the following expressions for the parameters σ and τ

$$(\sigma \pm \tau)^2 = \frac{\Sigma_N}{\Sigma_Q - \frac{1}{6}\Sigma_e} \pm 9 \frac{\Delta_N}{\Delta_e}$$

The numerical values of σ and τ are computed with the Gargamelle and SLAC data to be

$$\left|\frac{\sigma+\tau}{2}\right| = 0.88 \pm 0.07 \qquad \left|\frac{\sigma-\tau}{2}\right| = 0.56 ^{+0.08}_{-0.10}$$

Again the neutral current is purely isoscalar and we have the two predictions

$$J_N = 0 \qquad S_N = 0$$

Let us recall that the original Sakurai[22,34,43] model was purely Lorentz vector ($\tau = 0$).

7°) MODEL VI

The last particular model we now consider is the correct extension to hadrons of the Salam-Weinberg theory using the Glashow-Iliopoulos-Maiani mechanism[4] for eliminating unwanted strangeness violating neutral currents [+]. This model is based on the internal symmetry group $SU(4)$ for hadrons and implies the existence of a new quantum number called charm

$$J_\mu^N = (V_\mu^3 + A_\mu^3) - \tfrac{1}{2}(V_\mu^B + A_\mu^B) - 2\sin^2\theta_W V_\mu^Q \tag{31}$$

The weak charged current and the electromagnetic current have new pieces related to charm [++] and the relations between the various currents

---

[+] The light-cone approach and the quark parton formalism for the model can be found in the following references[11,12,13,18,19,20,21]

[++] The electromagnetic current is simply the reflection of the Gell-Mann Nishijima generalized relation

$$V_\mu^Q = V_\mu^3 + \tfrac{1}{2}(V_\mu^B + V_\mu^S + V_\mu^C)$$

For the charged weak current, in addition to the Cabibbo part which is charm conserving we have a charm changing component. The simplest way to write this current without referring to the $SU(4)$ generators is to use the quark language and we have

$$J_\mu^C = \tfrac{i}{2}\left[\bar{p}\,\gamma_\mu(1+\gamma_5)n_c + \bar{p}'\,\gamma_\mu(1+\gamma_5)\lambda_c\right]$$

with

$$n_c = n\cos\theta_c + \lambda\sin\theta_c$$
$$\lambda_c = -n\sin\theta_c + \lambda\cos\theta_c$$

where θ_c is the Cabibbo angle

are somewhat more complicated than before.

Nevertheless some general results emerge

<u>a</u> same mixing angle as previously

<u>b</u> same prediction for $D_N/D_\oplus$

<u>c</u> a positivity constraint in the $\dfrac{\Sigma_N}{\Sigma_c}, \dfrac{\Delta_N}{\Delta_c}$ plane is written as[+)]

$$\frac{\Sigma_N}{\Sigma_c} \geq 4 \frac{\Sigma_\oplus}{\Sigma_c}\left(\frac{\Delta_N}{\Delta_c}\right)^2 + \left(1 - 4\frac{\Sigma_\oplus}{\Sigma_c}\right)\frac{\Delta_N}{\Delta_c} + \frac{5}{11}\left(4\frac{\Sigma_\oplus}{\Sigma_c} - \frac{1}{2}\right)$$

This inequality is associated to a parabola whose minimum gives a lower bound for Σ_N/Σ_c

$$\frac{\Sigma_N}{\Sigma_c} \geq \frac{3}{11}\left[3\frac{\Sigma_\oplus}{\Sigma_c} - 1\right] - \frac{1}{16}\frac{\Sigma_c}{\Sigma_\oplus}$$

It is obviously questionable to use the Gargamelle data to evaluate these quantities because of the absence of evidence for the production of charmed particles in neutrino reactions involving charged currents. If we disregard this difficulty we obtain the parabola shown on Fig. 7 and the lower bound for Σ_N/Σ_c takes the value

$$\frac{\Sigma_N}{\Sigma_c} \geq 0.234 \qquad (\text{in 1 s.d.})$$

consistent with the Gargamelle result for neutral currents of

$$\frac{\Sigma_N}{\Sigma_c} = 0.29 \pm 0.03$$

[+)] Now, Σ_c and Δ_c are the full cross sections including strange particle and charmed particle production and not only the strangeness conserving part as before.

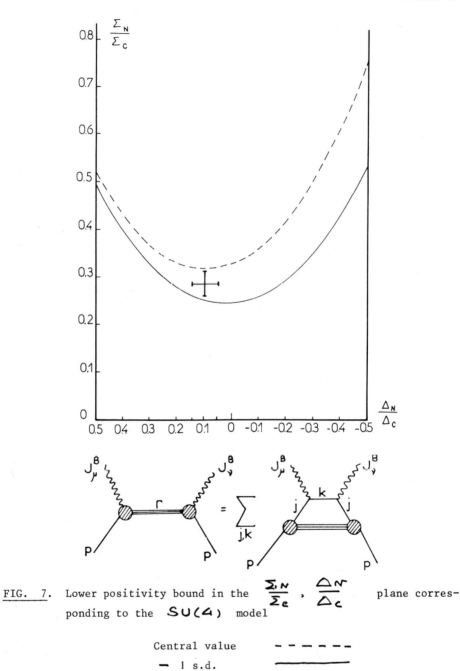

FIG. 7. Lower positivity bound in the $\frac{\Sigma_N}{\Sigma_c}$, $\frac{\Delta_N}{\Delta_c}$ plane corresponding to the $SU(4)$ model

Central value   – – – – –
— 1 s.d.   ─────────
Gargamelle point.   ⊢┼┤

FIG. 8. Quark Parton model amplitude.

IV. ALGEBRAIC STRUCTURE

1°) The neutral current is a particular weight or a sum of weights of irreducible representations of an internal symmetry group for strong interactions. Such a group is generally a Lie group and the corresponding Lie algebra has definite commutation properties. We then construct an algebra of current components from which we can derive sum rules.

As an example we can consider the Gross-Llewellyn-Smith sum rule based on the anticommutator of the current with its Hermitian conjugate $[J_\mu, J_\nu^*]_+$. It is an asymptotic sum rule valid in the $q^2 \Rightarrow \infty$ limit and it depends on both

- the Lie group algebra
- the representation chosen for the basic constituent fields.

2°) A quick way to derive the Gross-Llewellyn-Smith sum rule is to use a quark parton model. Let us briefly remember the basic ingredients of a quark parton model.

<u>a</u> quarks are labelled with a positive integer index j and antiquarks with $\bar{j}$

<u>b</u> the pointlike couplings between the quarks j, k and the current of helicity λ are called a_{jk}^λ

<u>c</u> the distribution function $D_j(x)$ describes the probability to find, in the hadron, a quark of type j carrying a fraction x of the hadron momentum.

It is then easy to compute the scaling functions $F(x)$ for positive and negative helicity using the graphical representation of the parton model drawn on Fig. 8. The result is

$$2 F_\pm(x) = \sum_{j,k} \left[(a_{jk}^L)^2 D_{\bar{k}}(x) + (a_{jk}^R)^2 D_j(x) \right]$$

$$2\bar{F}_-(x) = \sum_{j,k} \left[(a_{jk}^L)^2 D_j(x) + (a_{jk}^R)^2 D_{\bar{k}}(x) \right]$$

and for the (V,A) interference scaling function $\bar{F}_3 \equiv \bar{F}_- - \bar{F}_+$ we obtain

$$2\bar{F}_3(x) = \sum_{j,k} \left[(a_{jk}^L)^2 - (a_{jk}^R)^2 \right] \left[D_j(x) - D_{\bar{k}}(x) \right]$$

Couplings of the quarks with a current B and its Hermitian conjugate $\bar{B}$ are simply related by

$$(a_{jk}^{\bar{B}})^2 = (a_{jk}^B)^2$$

The Gross-Llewellyn-Smith sum rule is obtained using the normalization condition

$$\int_0^1 D_j(x) \, dx = \langle N_j \rangle$$

the quantity $\langle N_j \rangle$ being interpreted as the averaged value of the number of quarks of type j in the hadron.

The result is

$$2 \int_0^1 \left[\bar{F}_3^B(x) + \bar{F}_3^{\bar{B}}(x) \right] dx =$$

$$= \sum_{j,k} \left[(a_{jk}^L)^2 - (a_{jk}^R)^2 \right] \left[\langle N_j - N_{\bar{j}} \rangle + \langle N_k - N_{\bar{k}} \rangle \right]$$

For an Hermitian current

$$\bar{F}_3^{\bar{B}}(x) = \bar{F}_3^B(x)$$

it reduces to

$$2 \int_0^1 \bar{F}_3(x) \, dx = \sum_{j,k} \left[(a_{jk}^L)^2 - (a_{jk}^R)^2 \right] \langle N_j - N_{\bar{j}} \rangle$$

The differences $\langle N_j - N_{\bar{j}} \rangle$ are expected to be finite integrals because they correspond to linear combinations of conserved quantum numbers entering in the Cartan algebra like the baryonic charge B, the

THEORETICAL UNDERSTANDING OF NEUTRAL CURRENTS

electric charge Q, the strangeness S and possibly the charm C, etc...

3°) Let us apply these results to the particular case of the neutral current introduced in MODEL VI.

Equation (31) is equivalently written as

$$\bar{J}_\mu^N = 2 \bar{J}_\mu^3 - 2 \sin^2\theta_W V_\mu^Q$$

with the following quark representation

$$\bar{J}_\mu^3 = \frac{i}{4}\left[\bar{p}\gamma_\mu(1+\gamma_5)p + \bar{p}'\gamma_\mu(1+\gamma_5)p' - \bar{n}\gamma_\mu(1+\gamma_5)n - \bar{\lambda}\gamma_\mu(1+\gamma_5)\lambda\right]$$

$$V_\mu^Q = \frac{i}{3}\left[2\bar{p}\gamma_\mu p + 2\bar{p}'\gamma_\mu p' - \bar{n}\gamma_\mu n - \bar{\lambda}\gamma_\mu\lambda\right]$$

In the $SU(4)$ quark parton model the differences $\langle N_q - N_{\bar{q}}\rangle$ are given by

$$\langle N_p - N_{\bar{p}}\rangle = B + Q + C \qquad \langle N_n - N_{\bar{n}}\rangle = 2B - Q + S$$

$$\langle N_{p'} - N_{\bar{p}'}\rangle = C \qquad \langle N_\lambda - N_{\bar{\lambda}}\rangle = -S$$

and the Gross-Llewellyn-Smith sum rule has the general expression[13]

$$\int_0^1 \bar{F}_3(x)dx = \frac{3}{2}B - \frac{2}{3}\sin^2\theta_W\left[4B + Q + S - 2C\right]$$

In the particular case of a nucleon target

$$B = 1 \qquad S = C = 0$$

the right-hand side of the sum rule has the following values

$$\frac{3}{2} - \frac{10}{3} \sin^2\theta_W \qquad \text{for a proton target}$$

$$\frac{3}{2} - \frac{8}{3} \sin^2\theta_W \qquad \text{for a neutron target}$$

$$3\left[\frac{1}{2} - \sin^2\theta_W\right] \qquad \text{for the average}$$

Comparing with the Gross-Llewellyn-Smith sum rule of the $SU(3)$ case

$$\int_0^1 F_3(x)\,dx = \frac{1}{2}(3B+S) - \frac{2}{3}\sin^2\theta_W[4B+Q+S]$$

we see that the predictions are the same in the nucleon case and more generally for any $S = C = 0$ target.

V. CONSERVATION PROPERTIES OF THE NEUTRAL CURRENT

1°) We know that the matrix elements between physical states of the divergence of the electromagnetic current are zero

$$\langle \,|\, \partial^\mu J_\mu^Q \,|\, \rangle = 0$$

This property is directly related to the conservation of the electric charge and it is in fact true separately for the isovector and the isoscalar parts.

2°) The strangeness-conserving part of the weak charged current is an isovector. For the vector part we have a conserved current and this result is precisely the CVC hypothesis. The axial vector part is obviously not conserved but the $PCAC$ hypothesis gives a way to compute the matrix elements of $\partial^\mu A_\mu^c$ by relating them to the matrix elements of the charged π meson field.

A nice application of this last result is the Adler test which relates forward neutrino and antineutrino cross sections involving charged current with $\pi^\pm$ total cross sections

$$\left.\frac{d\sigma^{\nu,\bar\nu}}{dq^2 dW^2}\right|_{q^2=0} \propto \sigma_{TOT}(\pi^\pm)$$

THEORETICAL UNDERSTANDING OF NEUTRAL CURRENTS

3°) Let us now describe the Adler test for inclusive or semi-inclusive reactions involving neutral currents[26,32]

$$\nu + p \Rightarrow \nu + X.$$

The relevant matrix element is

$$\langle k' | J^\mu | k \rangle \langle X | J_\mu^N | p \rangle$$

In the forward direction, $q^2 = 0$ and the four-vectors k, k', q are lightlike and proportional. Therefore

$$ME \propto q^\mu \langle X | J_\mu^N | p \rangle$$
$$ME \propto \langle X | \partial^\mu J_\mu^N | p \rangle$$

Following Adler[32], forward q^2 means

$$q^2 \lesssim 2 m_\pi^2 \simeq 0.04 \, GeV^2$$

4°) The two problems for neutral currents are the following

<u>a</u> Is the vector part V_μ^N conserved ?

<u>b</u> If yes, what is the divergence of the axial vector part ?

PROBLEM a :

We study the variation with small q^2 of parity violating effects. We have two possibilities for $q^2 \neq 0$

α The parity violating effects are absent when $q^2 \neq 0$ and no statement concerning $\partial^\mu V_\mu^N$ can be made ; it is for instance the case if the hadronic neutral current is pure V or pure A.

β The parity violating effects are present when $q^2 \neq 0$; both V and A pieces of the current are present and two cases may occur

i) the P violating effects are also present at $q^2 = 0$ and V_μ^N -and also A_μ^N - is not conserved,

ii) all the P violating effects vanish when $q^2 \Rightarrow 0$ and V_μ^N is probably a conserved current.

PROBLEM b :

If V_μ^N is conserved we apply the Adler test in order to measure the divergence of the axial vector part $\partial^\mu A_\mu^N$. It is then interesting to compare the result with the corresponding ones obtained in charged current reactions.

<u>α</u> If it is different -up to known multiplicative factors- a model of type II is wrong . Either the isovector assumption is not correct or, more probably, an isoscalar axial vector current is needed.

<u>β</u> If it is the same for all semi-inclusive processes then two currents A_μ^N and A_μ^C have good chances to be proportional like their divergences.

THEORETICAL UNDERSTANDING OF NEUTRAL CURRENTS

PART III – ELASTIC REACTIONS

I. Nucleon Target.

II. Electron Target.

We study in this part the simplest exclusive neutrino reactions : the elastic scattering

$$\nu(\bar{\nu}) + A \Rightarrow \nu(\bar{\nu}) + A$$

Two realistic cases for the target A are

a) a nucleon (or a nucleus) ;

b) an atomic electron.

With a nucleon target and when experiments will be available we shall complete our information on the hadronic neutral current.

With an electron target we directly study the leptonic neutral current. The analysis of these reactions is relatively clean and we then can improve considerably our knowledge of the Lorentz structure of the neutral current.

The kinematics of elastic reactions is shown on Fig. 9 and it is described in Appendix III.

The relation between neutrino and antineutrino scattering takes a simple form in the elastic case even if the initial and final neutrinos are not identical. We consider the two processes

$$\nu_1 + A \Rightarrow \nu_2 + A$$
$$\bar{\nu}_1 + A \Rightarrow \bar{\nu}_2 + A$$

the corresponding t channel amplitudes are

$$\nu_1 + \bar{\nu}_2 \Rightarrow \bar{A} + A$$
$$\bar{A} + A \Rightarrow \bar{\nu}_2 + \nu_1$$

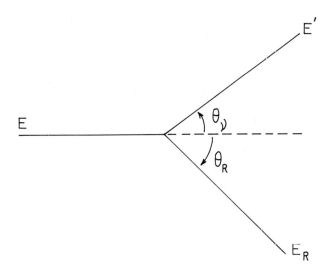

FIG. 9. Kinematics of Elastic Reactions.

THEORETICAL UNDERSTANDING OF NEUTRAL CURRENTS

Therefore if the transformation properties of these amplitudes under time reversal is unique the crossing relations (12) hold.

I. NUCLEON TARGET

1°) In the elastic case $W^2 = M^2$ and the structure functions a_0, a_1, a_2 depend only on the variable q^2. Using again the $\cos\Theta_t$ expansion method we easily get the general form of the differential cross section allowing the presence of S, P, T, V, A, weak currents

$$\frac{d\sigma}{dq^2} = \frac{G^2 q^2}{8\pi E^2}\left[a_0(q^2) + a_1(q^2)\cos\Theta_t + a_2(q^2)\cos^2\Theta_t\right] \tag{32}$$

A more familiar form for equation (32) is obtained using the laboratory lepton scattering angle θ_ν.

$$\cos^2\Theta_t = 1 + \frac{1}{1+\zeta}\operatorname{ctg}^2\frac{\theta_\nu}{2}$$

where the dimensionless variable ζ is defined by

$$\zeta = \frac{q^2}{4M^2}$$

We then get

$$\frac{d\sigma}{dq^2} = \frac{G^2}{2\pi}\frac{\cos^2\frac{\theta_\nu}{2}}{1+\zeta}\frac{E'}{E}[\ ]$$

with

$$[\ ] = a_2(q^2) + (1+\zeta)[a_0(q^2) + a_2(q^2)]\operatorname{tg}^2\frac{\theta_\nu}{2}$$
$$+ \frac{E+E'}{2M}\sqrt{\frac{1+\zeta}{\zeta}}\,a_1(q^2)\operatorname{tg}^2\frac{\theta_\nu}{2}$$

The forward cross section is obtained in the limit $q^2 \to 0$ with E fixed. If the structure functions are not singular at $q^2 = 0$ +) we get the result

$$\lim_{q^2 \to 0} \frac{d\sigma}{dq^2} = \frac{G^2}{2\pi} a_2(0) \tag{33}$$

The forward cross section is independent of the incident energy and it is governed by the structure function a_2.

The same feature does not occur for the backward cross section when q^2 tends to its maximum value q^2_M given by

$$q^2_M = \frac{2EM}{1 + \frac{M}{2E}}$$

In this case we obtain a cross section

$$\lim_{q^2 \to q^2_M} \frac{d\sigma}{dq^2} = \frac{G^2}{2\pi} \frac{M}{2E} \frac{1}{1 + \frac{M}{2E}} \left[a_0(q^2_M) + a_1(q^2_M) + a_2(q^2_M) \right] \tag{34}$$

which vanishes at high energy.

2°) The integrated cross section is defined by

$$\sigma = \int_0^{q^2_M} \frac{d\sigma}{dq^2} dq^2$$

+) In fact the sufficient conditions for the result (33) to be valid are

$$\lim_{q^2 \to 0} q^2 a_0(q^2) = 0 \qquad \lim_{q^2 \to 0} \sqrt{q^2}\, a_1(q^2) = 0$$

$$\lim_{q^2 \to 0} a_2(q^2) \neq 0$$

THEORETICAL UNDERSTANDING OF NEUTRAL CURRENTS

Let us make a very conservative assumption concerning the large behaviour of the elastic structure functions. If there exist $\varepsilon_\alpha > 0$ such that

$$\lim_{q^2 \to \infty} [q^2]^{\varepsilon_\alpha} a_\alpha(q^2) = 0 \qquad \alpha = 0, 1, 2$$

then

$$\lim_{E \to \infty} \sigma(E) = \frac{G^2 M^2}{2\pi} \int_0^\infty \frac{4 a_2(q^2)}{1+\eta} d\eta \qquad (35)$$

The integrated cross section tends to a constant at high energy[+] and this constant is determined by the structure function a_2 only.

3°) We now relate neutrino and antineutrino elastic scattering by assuming the crossing properties (12)

$$a_0^{\bar\nu} = a_0^\nu \qquad a_1^{\bar\nu} = -a_1^\nu \qquad a_2^{\bar\nu} = a_2^\nu$$

The general features discussed in PART I obviously occur and, in addition, the structure function a_2 being invariant under crossing

— The forward cross section is the same for neutrino and antineutrino scattering ;

— The integrated cross section takes the same asymptotic value for neutrino and antineutrino scattering.

4°) <u>The (V,A) Case</u>

In the (V,A) case the elastic structure function can be expressed in terms of the usual elastic form factors : electric G_E, magnetic G_M

[+] In this language, high energy is restricted to the region where the Fermi local theory is still applicable

$$M^2 \ll 2ME \ll M_Z^2$$

where M_Z is the neutral vector boson mass.

and axial vector G_A

$$a_0(q^2) = (1+2) G_A^2(q^2) + 2 G_M^2(q^2) - G_E^2(q^2)$$
$$a_1(q^2) = 4\sqrt{2(1+2)}\, G_A(q^2) G_M(q^2) \qquad (36)$$
$$a_2(q^2) = (1+2) G_A^2(q^2) + 2 G_M^2(q^2) + G_E^2(q^2)$$

The forward cross section is proportional to $a_2(0)$

$$a_2(0) = G_A^2(0) + G_E^2(q^2)$$

and it is interesting to compare neutral current and charged current forward cross sections. In the latter case

$$a_2^c(0) = g_A^2 + 1$$

The ratio of forward cross sections defined for a nucleon target $N = p, n$ by

$$R_{FORW}^N = \left. \frac{d\sigma(\nu N \Rightarrow \nu N)}{d\sigma(\nu_\mu n \Rightarrow \mu^- p)} \right|_{q^2 = 0}$$

takes the simple form[7]

$$R_{FORW}^N = \frac{|G_A^N(0)|^2 + |G_E^N(0)|^2}{g_A^2 + 1} \qquad (37)$$

We give in Table 1 the predictions for R_{FORW}^N corresponding to the various models considered in PART II - Details are given in Appendix IV.

THEORETICAL UNDERSTANDING OF NEUTRAL CURRENTS

Table 1

	$R_{FORWARD}$		R_{TOTAL}	
	PROTON	NEUTRON	PROTON	NEUTRON
MODEL I	0.11 ± 0.02 0.17 ± 0.02		0.18 ± 0.02 0.10 ± 0.01	
MODEL II	0.32 ± 0.05 0.18 ± 0.03	0.011 ± 0.007 0.25 ± 0.03	0.32 ± 0.04 0.11 ± 0.02	0.08 ± 0.01 0.16 ± 0.02
MODEL III	0.37 ± 0.03 0.22 ± 0.03	$0.0036^{+0.0029}_{-0.0021}$ $0.089^{+0.013}_{-0.012}$	0.35 ± 0.05 0.14 ± 0.02	0.068 ± 0.009 0.058 ± 0.008
MODEL IV	$0.53 ^{+0.12}_{-0.16}$ 0.095 ± 0.033		$0.34 ^{+0.06}_{-0.07}$ $0.038 ^{+0.031}_{-0.024}$	
MODEL V	> 0.69 > 0.017		> 0.38 > 0.010	

For the integrated cross section we make the approximate assumption of a universal q^2 behaviour of the elastic form factors and we use a dipole formula

$$G(q^2) = \frac{G(0)}{[1+\tau_2]^2}$$

We define two integrals

$$I = 4 \int_0^\infty \frac{d\varrho}{[1+r\varrho]^4} = \frac{4}{3r}$$

$$J = 4 \int_0^\infty \frac{d\varrho}{(1+\varrho)[1+r\varrho]^4}$$

$$J = 4\left[\frac{1}{3}\frac{1}{r-1} - \frac{1}{2}\frac{1}{(r-1)^2} + \frac{1}{(r-1)^3} - \frac{1}{(r-1)^4} \log r \right]$$

and the high energy limit of the integrated cross section takes the expression

$$\lim_{E \Rightarrow \infty} \sigma(E) = \frac{G^2 M^2}{2\pi} \left[G_A^2(0) I + G_M^2(0)(I-J) + G_E^2(0) J \right] \quad (38)$$

Again we compare neutral current and charged current cross sections and we define for a nucleon target the asymptotic ratio of integrated cross sections

$$R_{TOTAL}^N = \lim_{E \Rightarrow \infty} \frac{\sigma(\nu N \Rightarrow \nu N)}{\sigma(\nu_\mu n \Rightarrow \mu^- p)}$$

Using equation (38) we obtain[54]

$$R_{TOTAL}^N = \frac{|G_A^N(0)|^2 I + |G_M^N(0)|(I-J) + |G_E^N(0)|^2 J}{g_A^2 I + (\mu_p - \mu_n)^2 (I-J) + J} \quad (39)$$

We give in Table 1 the predictions for R_{TOTAL}^N associated to the different models introduced in PART II. The computation of the integrals I and J is done with $r = 5.03$ which is the commonly accepted value for electromagnetic nucleon form factors.

We the get[54]

$$I = 0.265 \qquad J = 0.244$$

In particular the high energy limit of the charged current elastic cross section obtained with these assumptions agrees very well with experiment

$$\lim_{E \Rightarrow \infty} \frac{\sigma(\nu_\mu n \Rightarrow \mu^- p)}{\sigma(\bar\nu_\mu p \Rightarrow \mu^+ n)} = 0.78 \cdot 10^{-38} \, cm^2$$

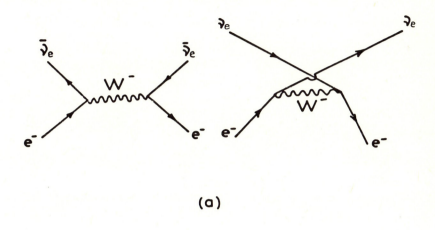

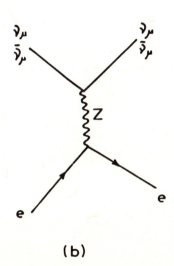

FIG. 10. Lowest order diagram for elastic neutrino and antineutrino scattering on an electron target
  a) charged current exchange
  b) neutral current exchange.

II. ELECTRON TARGET

1°) Elastic $\nu_e e^-$ and $\bar{\nu}_e e^-$ scattering can occur via changed current exchange in the s or u channel as shown on Fig. 10. The corresponding cross sections are computed in the V - A theory as usual and the results are

$$\frac{d\sigma^{\nu_e}}{d\rho} = 2\frac{G^2 m_e E}{\pi} \qquad \sigma^{\nu_e} = 2\frac{G^2 m_e E}{\pi}$$

$$\frac{d\sigma^{\bar{\nu}_e}}{d\rho} = 2\frac{G^2 m_e E}{\pi}\rho^2 \qquad \sigma^{\bar{\nu}_e} = \frac{2}{3}\frac{G^2 m_e E}{\pi} \qquad (40)$$

The experimental result for elastic $\bar{\nu}_e e^-$ scattering obtained with the Savannah River reactor is compatible with this prediction. But the errors are large enough to allow, in addition, a neutral current contribution accociated to a t channel exchange.

Let us remark that the two other reactions of the same type

$$\nu_\mu e^- \Rightarrow \nu_e \mu^-$$
$$\bar{\nu}_e e^- \Rightarrow \bar{\nu}_\mu \mu^-$$

will proceed only via charged current exchange. Their observation would be very interesting.

2°) When the neutrino and antineutrino belong to the μ family the elastic scattering processes

$$\nu_\mu e^- \Rightarrow \nu_\mu e^-$$
$$\bar{\nu}_\mu e^- \Rightarrow \bar{\nu}_\mu e^-$$

are due to neutral current exchange in the t channel as shown on Fig. 10. Assuming local Fermi interaction with S, P, T, V, A, pointlike couplings we easily write the general form of the differential cross section in the high energy limit $m_e \ll E$

$$\frac{d\sigma}{d\rho} = \frac{G^2 m_e E}{4\pi}\left[C_0(1-\rho)^2 + C_1(1-\rho^2) + C_2(1+\rho)^2\right] \qquad (41)$$

THEORETICAL UNDERSTANDING OF NEUTRAL CURRENTS

Equation (41) exhibits two important features

<u>α</u> linear rising with the incident energy E ;

<u>β</u> second order positive definite form in ρ .

An immediate consequence of β is the existence of the bounds (7) for the averaged energy $\bar{\rho}$ [48)].

Integrating over ρ we obtain a total cross section which rises linearly with the incident energy E

$$\sigma_{TOT}(E) = \frac{G^2 m_e E}{\pi} \frac{C_0 + 2C_1 + 7C_2}{12}$$

3°) An explicit calculation of the coefficients C_0, C_1, C_2 in terms of the coupling constants g_S, g_P, g_T, g_V, g_A for non derivative interactions gives the result [5,35)].

$$C_0 = g_S^2 + g_P^2 + g_V^2 + g_A^2$$
$$C_1 = 4[g_S g_T + g_P g_T + g_V g_A] \qquad (43)$$
$$C_2 = 8 g_T^2 + g_V^2 + g_A^2$$

It is interesting to remark the characteristics of two simple cases

<u>α</u> Pure (S,P) $d\bar{\sigma}^\nu = d\bar{\sigma}^{\bar\nu}$ = pure $(1-\rho)^2$

$$\bar{\rho}_\nu = \bar{\rho}_{\bar\nu} = \frac{1}{4} \quad ; \quad r_N = 1$$

$\underline{\beta}$ Pure T $\qquad d\bar{\sigma}^{\nu} = d\bar{\sigma}^{\bar{\nu}} =$ pure $(1+\rho)^2$

$$\bar{\rho}_\nu = \bar{\rho}_{\bar\nu} = \frac{17}{28} \quad ; \quad r_N = 1$$

It is obvious that we can find an (S,P,T) combination equivalent to a (V,A) one. A necessary condition is the absence of linear ρ term in the differential cross section, e.g. the relation

$$g_S^2 + g_P^2 = 8 g_T^2$$

Therefore with only the angular distribution for $\nu_\mu e^-$ and $\bar\nu_\mu e^-$ elastic scattering it will never be possible to prove the (V,A) structure of the leptonic neutral current.

The situation is somewhat more favourable with $\nu_e e^-$ and $\bar\nu_e e^-$ elastic scattering. In this case assuming the initial and final neutrinos -or antineutrinos- to be the same, the charged current and neutral current amplitudes add coherently in the (V,A) case whereas they add incoherently in the other situations like for instance the (S,P,T) ones. Such a feature may help for solving the problem of the Lorentz structure of the leptonic neutral current.

Let us finally remember for completeness two constraints of the (V,A) case already obtained in PART I

$$\frac{1}{2} \leq \bar\rho_\nu \text{ and } \bar\rho_{\bar\nu} \leq \frac{3}{4}$$

$$\frac{1}{3} \leq r_N \leq 3$$

4°) **Low Energy Corrections in the (V,A) Case**

In the (V,A) case the differential cross sections with identical initial and final neutrinos or antineutrinos are easily computed using equations (36) with the pointlike normalization form factors

$$G_E(0) = G_M(0) = g_\nu \qquad G_A(0) = g_A$$

THEORETICAL UNDERSTANDING OF NEUTRAL CURRENTS

The result is simply[8,10]

$$\frac{d\sigma^\nu}{d\xi} = \frac{G^2 m_e E}{2\pi}\left[(g_V+g_A)^2 + \xi^2(g_V-g_A)^2 - \frac{m_e}{E}(1-\xi)(g_V^2-g_A^2)\right] \quad (44)$$

$$\frac{d\sigma^{\bar\nu}}{d\xi} = \frac{G^2 m_e E}{2\pi}\left[(g_V-g_A)^2 + \xi^2(g_V+g_A)^2 - \frac{m_e}{E}(1-\xi)(g_V^2-g_A^2)\right]$$

If the initial and final neutrinos or antineutrinos are different we get the same structure for the differential cross section[46,53]

$$\frac{d\sigma}{d\xi} = \frac{G^2 m_e E}{2\pi}\left[A + \xi^2 B - \frac{m}{E}(1-\xi)C\right] \quad (45)$$

with the crossing properties

$$A^{\bar\nu} = B^\nu \qquad B^{\bar\nu} = A^\nu \qquad C^{\bar\nu} = C^\nu$$

The relation $C^2 = AB$ valid for equation (44) becomes an inequality $C^2 < AB$. Therefore the study of the $\frac{m_e}{E}$ terms in the recoil energy spectrum will be an interesting test of the identity of initial and final neutrinos or antineutrinos.

These considerations obviously apply separately for $\nu_\mu e^-$, $\bar\nu_\mu e^-$ elastic scattering and $\nu_e e^-$, $\bar\nu_e e^-$ elastic scattering.

5°) $\mu-e$ Universality[17,21]

Let us restrict to the (V,A) case with identical neutrinos or antineutrinos in the initial and final states. Consider first $\nu_\mu e^-$ and $\bar\nu_\mu e^-$ scattering. The coefficients A and B of equation (45) are simply given in terms of left-handed and right-handed neutral current coupling constants by equation (44)

$$A^\nu = B^{\bar\nu} = (g_V+g_A)^2 = 4g_L^2$$
$$B^\nu = A^{\bar\nu} = (g_V-g_A)^2 = 4g_R^2$$

In the case of $\nu_e e^-$ and $\bar\nu_e e^-$ scattering we add the contribution due to charged neutral current exchange which, after Fierz transformation, affects only the left-handed amplitude. Assuming μ-e universality we use the same g_L and g_R as previously and we get

$$A^\nu = \bar{B}^\nu = 4(g_L+1)^2$$
$$B^\nu = A^{\bar\nu} = 4 g_R^2$$

The four elastic scattering reactions on an electron target are then described by two coupling constants g_L and g_R. We then obtain two relations between the four total cross sections which can be written as[21]

$$3\,\sigma(\bar\nu_\mu e^-) - \sigma(\nu_\mu e^-) = 3\,\sigma(\bar\nu_e e^-) - \sigma(\nu_e e^-) \qquad (46)$$

$$\left[3\sigma(\nu_\mu e^-) - \sigma(\bar\nu_\mu e^-) + 3\sigma(\nu_e e^-) - \sigma(\bar\nu_e e^-) - \tfrac{16}{3}\sigma_0\right]^2 =$$
$$= 4\left[3\sigma(\nu_\mu e^-) - \sigma(\bar\nu_\mu e^-)\right]\left[3\sigma(\nu_e e^-) - \sigma(\bar\nu_e e^-)\right] \qquad (47)$$

with $\sigma_0 = \dfrac{G^2 m_e E}{\pi}$

It must be noticed that formula (46) involving the right-handed amplitude is also valid when the final neutrino (antineutrino) is not identical to the initial ones. It constitutes a strict test of μ-e universality for pure (V,A) currents.

The compatibility of the (V,A) assumption with experiment can also be studied by looking at the g_L, g_R plane where a given value of the total cross section is represented by an ellipse centered at the origin $g_L = g_R = 0$ in the $\nu_\mu, \bar\nu_\mu$ case and at the point $g_L = -1, g_R = 0$ in the $\nu_e, \bar\nu_e$ case.

For the (V,A) case the model independent analysis of the experimental data is shown on Fig. 11 and the allowed domain is obviously non empty. The general constraints on the magnitude of g_L and g_R as given by the Gargamelle experiment are

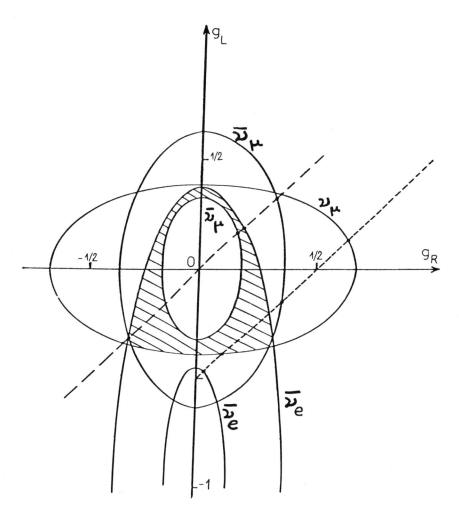

FIG. 11. Experimental data in the g_L, g_R plane.

$$|g_L| \leq 0.38$$
$$|g_R| \leq 0.36$$

Unfortunately the data are not accurate enough to put any restriction on the relative amount of vector and axial vector components for the leptonic neutral current.

6°) <u>Models</u>

In the (V,A) framework, various one-parameter models have been proposed for the leptonic neutral current. Let us give some example without discussing the basis of these models

PURE VECTOR $\qquad g_R = g_L$
Sakuraï[22]; Bég-Zee[16]

PURE AXIAL VECTOR $\qquad g_R = -g_L$
Cheng-Lo[42]

PURE V-A $\qquad g_R = 0$
Pakvasa-Tuan[28] ; Adler-Tuan[39]
Korke-Konuma[40]

(V,A) MIXTURE $\qquad g_R = g_L + \frac{1}{2}$
Salam-Weinberg[1,2,3,7]

For all these models an experimental fit can be done.

In the Salam-Weinberg[1,2,3,7] gauge theory of weak and electromagnetic interactions, neutral leptonic currents are predicted to be there and the g_L and g_R coupling constants are related to the mixing angle θ_W by

$$g_L = \sin^2 \theta_W - \frac{1}{2} \qquad g_R = \sin^2 \theta_W$$

It is easy to compute the lower bounds for the four elastic cross sections represented on Fig. 12 as function of $\sin^2 \theta_W$. The result is

$$\sigma(\nu_\mu e^-) \geqslant \tfrac{1}{8} \sigma_0 \qquad \sigma(\bar{\nu}_\mu e^-) \geqslant \tfrac{1}{8} \sigma_0$$

$$\sigma(\nu_e e^-) \geqslant \tfrac{1}{2} \sigma_0 \qquad \sigma(\bar{\nu}_e e^-) \geqslant \tfrac{1}{6} \sigma_0$$

with

$$\sigma_0 = \frac{G^2 m_e E}{\pi} = 0{,}848 \; E \; 10^{-41} \; cm^2 \; GeV^{-1}$$

The Gargamelle experiment using ν_μ and $\bar{\nu}_\mu$ beams has found results compatible with these constraints and the allowed range for the parameter $\sin^2 \theta_W$ is

$$0{,}11 \leqslant \sin^2 \theta_W \leqslant 0{,}35$$

The corresponding prediction for the elastic $\bar{\nu}_e - e^-$ scattering total cross section is

$$0{,}4 \leqslant \frac{\sigma(\bar{\nu}_e e^-)}{\sigma(\bar{\nu}_e e^-)_{V-A}} \leqslant 1{,}1$$

in agreement with the result of the reactor experiment.

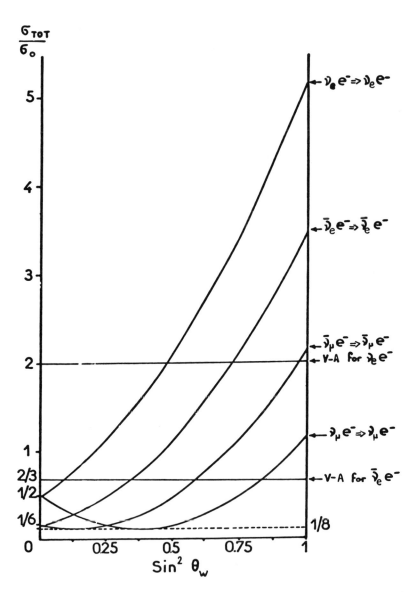

FIG. 12. Predictions of the Salam-Weinberg model for elastic cross sections on electrons.

THEORETICAL UNDERSTANDING OF NEUTRAL CURRENTS

/ COMMENT /

Other exclusive neutrino reactions where neutral currents have been observed

$$\nu + p \Rightarrow \nu + p + \pi^0 + n + \pi^+$$

$$\nu + n \Rightarrow \nu + n + \pi^0 + p + \pi^-$$

constitute an important piece of information. Methods analogous to those presented in PARTS I and II can be used but a satisfactory model independent treatment has not yet been given. The dynamics is usually implemented with a simple isobaric model where few refinements have been added. Comparison is then made for weak processes involving charged current and neutral current and reference is also given to π meson electroproduction where the data are accurate enough for testing the model. Let us simply emphasize the importance of the Δ (1236) region for the isospin properties of the neutral current.

/ SALAM-WEINBERG GAUGE THEORY /

RESUME

1. HADRONIC INCLUSIVE

Prediction

$$\sin^2 \theta_W = 0,40 \pm 0,05$$
$$\phi_c = 45° \pm$$

$$\frac{J_N}{J_\omega} = -0,16 \pm 0,05$$

2. HADRONIC ELASTIC - Predictions

R^p FORWARD = $0,19 \pm 0,02$

R^n FORWARD = $0,25$

R^p TOTAL = $0,11 \pm 0,01$

R^n TOTAL = $0,16 \pm 0,01$

3. LEPTONIC ELASTIC

$0,11 \leq \sin^2 \theta_W \leq 0,32$

THEORETICAL UNDERSTANDING OF NEUTRAL CURRENTS

APPENDIX I

KINEMATICS FOR INCLUSIVE REACTIONS

The nucleon target has a mass M and the notations are shown on Fig. 1. The Lorentz scalar variables are defined by

$$q^2 = (k - k')^2 \qquad W^2 = -(p+q)^2$$

$$M\nu = -p \cdot q$$

with the relation

$$W^2 = M^2 + 2M\nu - q^2$$

The lepton laboratory variables are

Incident (Final) lepton energy $\quad E \ (E')$
Lepton scattering angle $\quad \theta_\nu$

and we simply have

$$q^2 = 4EE' \sin^2 \frac{\theta_\nu}{2} \qquad \nu = E - E'$$

It is convenient to work with dimensionless variables

$$x = \frac{q^2}{2M\nu} \qquad \rho = \frac{p \cdot k'}{p \cdot k} = \frac{E'}{E} = 1 - y$$

The laboratory energy of the hadronic system X is given by

$$E_x = \nu + M$$

and the hadronic angle θ_x is related to scalar variables by

$$\cos \theta_x = \frac{1}{\sqrt{\nu^2 + q^2}} \left[\nu + \frac{q^2}{2E} \right]$$

APPENDIX II

NEUTRINO - ANTINEUTRINO ANNIHILATION CHANNEL

The states of the $\nu\bar{\nu}$ system can be classified according to the value of the angular momentum J, the parity P and the charge conjugation C. For $J \leq 1$ we obtain

STATE	J^{PC}	SYMBOL
$^1S_0$	0^{-+}	P
$^3P_0$	0^{++}	S
$^1P_1$	1^{+-}	T
$^3S_1 + {}^3D_1$	1^{--}	V
$^3P_1$	1^{++}	A

Let us now decompose the neutrino Dirac field ψ into a left-handed part ψ_L and a right-handed part ψ_R

$$\psi = \psi_L + \psi_R$$

where

$$\psi_L = \frac{1+\gamma_5}{2}\psi \qquad \psi_R = \frac{1-\gamma_5}{2}\psi$$

THEORETICAL UNDERSTANDING OF NEUTRAL CURRENTS

The Dirac covariants S, P, T, V, A take the explicit forms

$$S \Rightarrow \bar{\psi}\psi = [\bar{\psi}_R \psi_L + \bar{\psi}_L \psi_R]$$

$$P \Rightarrow i\bar{\psi}\gamma_5\psi = i[\bar{\psi}_R \psi_L - \bar{\psi}_L \psi_R]$$

$$T \Rightarrow \tfrac{i}{2}\bar{\psi}[\gamma_\mu, \gamma_\nu]\psi = \tfrac{i}{2}[\bar{\psi}_R[\gamma_\mu,\gamma_\nu]\psi_L + \bar{\psi}_L[\gamma_\mu,\gamma_\nu]\psi_R]$$

$$V \Rightarrow i\bar{\psi}\gamma_\mu\psi = i\bar{\psi}_L\gamma_\mu\psi_L + i\bar{\psi}_R\gamma_\mu\psi_R$$

$$A \Rightarrow i\bar{\psi}\gamma_\mu\gamma_5\psi = i\bar{\psi}_L\gamma_\mu\psi_L - i\bar{\psi}_R\gamma_\mu\psi_R$$

and it is easy to make the following observations

i) In the (S, P, T) case the initial and final neutrinos (or antineutrinos) have opposite helicities.

ii) In the (V, A) case the initial and final neutrinos (or antineutrinos) have the same helicity.

iii) The two (V, A) covariants

$$i\bar{\psi}_L\gamma_\mu\psi_L \quad\text{and}\quad i\bar{\psi}_R\gamma_\mu\psi_R$$

are separately Hermitian. It follows that they can be connected only to Hermitian leptonic or hadronic currents and a non Hermitian piece of the neutral current can contribute if and only if the initial and final neutrino are different particles.

iv) In the (S, P) case, starting with a left-handed neutrino beam, the only available covariant is $\bar{\psi}_R \psi_L$ which is not Hermitian. Therefore the initial and final neutrinos are distinct particles of opposite helicities and non Hermitian pieces of the neutral current can be present.

APPENDIX III

KINEMATICS FOR ELASTIC REACTIONS

The target of mass μ is in practice a nucleon $\mu = M$ or an electron $\mu = m_e$. Elastic scattering is then characterized by the equality $W^2 = \mu^2$

The laboratory variables are shown on Fig. 9.

    Incident neutrino (antineutrino) energy E
    Final neutrino (antineutrino) energy E'
    Neutrino (antineutrino) scattering angle θ_ν
    Recoil electron energy E_R
    Recoil electron angle θ_R

The conservation of energy-momentum implies the following relations between these variables

$$E + \mu = E' + E_R$$

$$\cos\theta_R = \left(1 + \frac{\mu}{E}\right)\sqrt{\frac{E_R - \mu}{E_R + \mu}}$$

$$q^2 = 2\mu(E - E') = 4EE'\sin^2\frac{\theta_\nu}{2}$$

We shall use again the dimensionless variable

$$\varrho = \frac{E'}{E}$$

and we have

$$q^2 = 2\mu E(1 - \varrho)$$

The kinematical constraint $0 \leq \cos\theta_R \leq 1$ implies

THEORETICAL UNDERSTANDING OF NEUTRAL CURRENTS

for ρ and q^2 the following limits

$$\frac{\mu}{2E+\mu} \leq \rho \leq 1$$

$$0 \leq q^2 \leq \frac{2\mu E}{1+\frac{\mu}{2E}}$$

APPENDIX IV

ELASTIC NEUTRAL FORM FACTORS

We now study the elastic neutral form factors for the two parameter models considered in PART II. The electromagnetic form factors of the nucleon are described by the isoscalar and isovector form factors $G^S_{E,M}(q^2)$ and $G^V_{E,M}(q^2)$ with the usual normalization

$$G^S_E(0) = \frac{1}{2} \qquad G^S_M(0) = \frac{\mu_p + \mu_n}{2}$$

$$G^V_E(0) = \frac{1}{2} \qquad G^V_M(0) = \frac{\mu_p - \mu_n}{2}$$

The axial vector form factor $G^V_A(q^2)$ can be measured in elastic neutrino scattering involving charged current

$$G^V_A(0) = \frac{-1}{2} g_A$$

For MODELS I, II, III, IV, the neutral elastic form factors are linear combinations of the previous quantities. The results are the following, the upper sign corresponding to the proton and the lower sign to the neutron.

MODEL I. Isovector neutral current
Proton and neutron neutral form factors are opposite

$$G^N_{E,M}(q^2) = \pm \alpha\, G^V_{E,M}(q^2)$$

$$G^V_A(q^2) = \pm \beta\, G^V_A(q^2)$$

MODEL II. Known currents
The three parameters α, β, γ are related by $\alpha = \beta + \gamma$

$$G^N_{E,M}(q^2) = \gamma G^S_{E,M}(q^2) \pm \alpha G^V_{E,M}(q^2)$$

$$G^N_A(q^2) = \pm \beta G^V_A(q^2)$$

For the Weinberg model $\beta = 1$

MODEL III. The quantum numbers of the neutral current are those of the electromagnetic current

$$G^N_{E,M}(q^2) = \alpha \left[G^S_{E,M}(q^2) \pm G^V_{E,M}(q^2) \right]$$

For the axial vector part we use $SU(3)$ symmetry. The electric charge operator Q belonging to the adjoint representation we have two reduced matrix elements $f_a(q^2)$ and $f_s(q^2)$ respectively associated to antisymmetric and symmetric isometries

$$G^V_A(q^2) = \frac{1}{2} \left[f_a(q^2) + f_s(q^2) \right]$$

$$G^N_A(q^2) = \frac{\beta}{2} \left\{ \left[f_a(q^2) - \frac{1}{3} f_s(q^2) \right] \pm \left[f_a(q^2) + f_s(q^2) \right] \right\}$$

The $SU(6)$ result $f_a/f_s = 2/3$ agrees with low q^2 experiments as for instance those of hyperon β decay. If it is extended to all q^2's we get

$$G^N_A(q^2) = \beta \left[\frac{1}{5} \pm 1 \right] G^V_A(q^2)$$

MODEL IV. Isoscalar neutral current (electromagnetic). Proton and neutron neutral form factors are the same. Proceeding for the axial vector part as in MODEL III we obtain

$$G^N_{E,M}(q^2) = \gamma G^S_{E,M}(q^2)$$

$$G_A^N(q^2) = \frac{5}{5} G_A^V(q^2)$$

MODEL V. Isoscalar baryonic current.

Proton and neutron neutral form factors are the same but they cannot be related to electromagnetic or weak charged elastic form factors.

The baryonic charge of the nucleon being unity we have the normalization condition

$$G_E^N(0) = 5$$

Let us remark that the axial vector baryonic charge f_1 can be obtained by looking at the correlation between beam and target polarizations in electron proton scattering if one believes in the Bjorken sum rule[+)]

$$G_A^N(0) = \xi f_1$$

+) J.D. BJORKEN Phys. Review D 1 1376 (1970)

For a quark parton model derivation of the sum rule see

M. GOURDIN Nuclear Physics B 38 418 1972

THEORETICAL UNDERSTANDING OF NEUTRAL CURRENTS

GARGAMELLE results used in the calculations

i) Hadronic Inclusive for neutral currents[+)]

$$R^\nu = \left(\frac{NC}{CC}\right)^\nu = 0.217 \pm 0.026$$

$$R^{\bar\nu} = \left(\frac{NC}{CC}\right)^{\bar\nu} = 0.56 \pm 0.08$$

$$\bar{r}_c = \frac{(CC)^{\bar\nu}}{(CC)^\nu} = 0.25$$

ii) Hadronic Inclusive for charged currents

$$\frac{\Delta_c}{\Sigma_c} = 0.466 \pm 0.021$$

$$\frac{\Sigma'_c}{\Sigma_c} = 0.298 \pm 0.045$$

iii) Leptonic Elastic for neutral currents

$$\sigma = \alpha E \; 10^{-41} \; cm^2 \; GeV^{-1}$$

$$\alpha(\nu_\mu e^-) \leq 0.26$$

$$\alpha(\bar\nu_\mu e^-) = 0.14 \pm 0.08$$

+) In order to take partially account of the effect due to the hadronic energy cuts we work with symmetrized ratios defined by

$$\frac{\Sigma_N}{\Sigma_c} = \frac{R^\nu + \bar{r}_c R^{\bar\nu}}{1 + \bar{r}_c} = 0.286 \pm 0.026$$

$$\frac{\Delta_N}{\Delta_c} = \frac{R^\nu - \bar{r}_c R^{\bar\nu}}{1 - \bar{r}_c} = 0.103 \pm 0.044$$

REFERENCES

GENERAL REVIEWS

M. Gourdin, 1974 Erice Lectures : Lepton and Hadron Structure, vol.12 A. Zichichi Editor, Academic Press (1975).

L. Wolfenstein, Talk given at the American Physical Society Meeting, Williamsburg (U.S.A.), September 1974.

K.V.L. Sarma, Talk given at the IInd High-Energy Physics Symposium, Visra-Bharati, Santiniketan (India), November 1974.

E.A. Paschos, Talk given at the Xe Rencontre de Moriond, Méribel (France), March 1975.

M. Gourdin, Rapporteur's Talk at the European Physical Society Conference, Palermo (Italy), June 1975.

---

1) A. Salam and J.C. Ward, Phys. Letters 13 (1964) 168.
2) S. Weinberg, Phys. Rev. Letters 19 (1967) 1264.
3) A. Salam, In Elementary Particle Theory 8th Nobel Symposium Stockholm (1968).
4) S.L. Glashow, J. Iliopoulos and L. Maiani, Phys. Rev. D 7 (1970) 1285.
5) D. Yu Bardin, S.M. Bilenky and B. Pontecorvo, Phys. Letters 32B (1970) 68.
6) T.P. Cheng and Wu-Ki Tung, Phys. Rev. D3 (1971) 733.
7) S. Weinberg, Phys. Rev. Letters 27 (1971) 1688.
8) G.'t Hooft, Phys. Letters 37B (1971) 195.
9) S. Weinberg, Phys. Rev. D 5 (1972) 1412.
10) H.H. Chen and B.W. Lee, Phys. Rev. D 5 (1972) 1874.
11) R. Budny, Phys. Lett. 39 B (1972) 553.
12) R. Budny and P.N. Scharbach, Phys. Rev. D 6 (1972) 3651.
13) Riazuddin and Fayyazuddin, Phys. Rev. D 6 (1972) 2032.
14) A. Pais and S.B. Treiman, Phys. Rev. D 6 (1972) 2700.
15) E.A. Paschos and L. Wolfenstein, Phys. Rev. D 7 (1973) 91.
16) M.A.B. Bég and A. Zee, Phys. Rev. Letters 30 (1973) 675.
17) H. Terazawa, Phys. Rev. D 8 (1973) 1817.
18) R.B. Palmer, Physics Letters 46 B (1973) 240.
19) L.M. Sehgal, Nuclear Physics B 65 (1973) 141.

20) C.H. Albright, Phys. Rev. D 8 (1973) 3162.

21) L.M. Sehgal, Phys. Lett. 48 B (1974) 60.

22) J.J. Sakurai, Phys. Rev. D 9 (1974) 250.

23) S.M. Bilenky and N.A. Dadajan, preprint Dubna 1975.

24) L.M. Sehgal, Physics Letters 48 B (1974) 133.

25) D.Z. Freedman, Phys. Rev. D 9 (1974) 1389.

26) A. Pais and S.B. Treiman, Phys. Rev. D 9 (1974) 1459.

27) G. Rajasekaran and K.V.L. Sarma, Pramana 2 (1974) 62.

28) S. Pakvasa and S.F. Tuan, Phys. Rev. D 9 (1974) 2698.

29) T.W. Donnelly, D. Hitlin, M. Schwartz, J.D. Walecka and S.J. Wiesner, Phys. Letters 49 B (1974) 8.

30) G. Rajasekaran and K.V.L. Sarma, Pramana 3 (1974) 44.

31) B.R. Kim and R. Rodenberg, Phys. Rev. D 10 (1974) 2234.

32) S.L. Adler, Talk given at the NAL Topical Conference on Neutrino Physics (March 1974).

33) S.P. Rosen, Talk at the Fourth International Conference on Neutrino Physics and Astrophysics, Downingtown (April 1974).

34) J.J. Sakurai, Talk at the Fourth International Conference on Neutrino Physics and Astrophysics, Downingtown (April 1974).

35) R.L. Kingsley, F. Wilczek and A. Zee, Phys. Rev. D 10 (1974) 2216.

36) V.S. Mathur, S. Okubo and J.E. Kim, Phys. Rev. D 10 (1974) 3648.

37) T.C. Yang, Phys. Rev. D 10 (1974) 3744.

38) G. Rajasekaran and K.V.L. Sarma, Phys. Letters 55 B (1975) 201.

39) S.L. Adler and S.F. Tuan, Phys. Review D 11 (1975) 129.

40) K. Koike and M. Konuma, preprint Kyoto 1974.

41) B. Kayser, G.T. Garvey, E. Fischbach and S.P. Rosen, Phys. Letters 52 B (1974) 385.

42) H. Cheng and C.Y. Lo, Phys. Letters 52 B (1974) 453.

43) J.J. Sakurai and L.F. Urrutia, Phys. Review D 11 (1975) 159.

44) M. Gourdin, Talk given at the Nordic Elementary Particle Meeting Copenhagen (November 1974).

45) R.L. Kingsley, R. Shrock, S.B. Treiman and F. Wilczek, Phys. Review D 11 (1975) 1043.

46) L.M. Sehgal, Phys. Letters 55 B (1975) 205.

47) S. Pakvasa and G. Rajasekaran, Preprint Hawaii (December 1974).

48) M. Gronau, Preprint Haifa.

49) L. Wolfenstein, Nucl. Physics B 91 (1975) 95.

50) G. Ecker and H. Pietschmann, Preprint Vienna (January 1975).

51) M. Gourdin. Lectures given at the XIV Internationale Universitäts wochen für Kernphysics, Schladming, Austria (February-March 1975).

52) C.M. Viallet, Thèse de 3ème Cycle, Paris (Avril 1975).

53) G.V. Dass, preprint CERN TH 2020, May 1975.

54) F. Martin, preprint Paris, PAR-LPTHE 75.12, June 1975.

EXPERIMENTAL REVIEWS

D.C. Cundy, Review Talk at the XVII Intern. Conf. on High Energy Physics, London (G.B.), July 1974.

Contributions of F.J. Hasert, B. Escoubes, F. Merritt at the International Colloquium "La Physique du Neutrino à Haute Energie", Paris, March 1975.

H. Faissner, Rapporteur's Talk at the European Physical Society Conference, Palermo (Italy), June 1975.

B.C. Barish, 1974 and 1975 Erice Lectures, A. Zichichi Editor.

DISCUSSION

CHAIRMAN: M. Gourdin

Scientific Secretary: W. Kozanecki

DISCUSSION No. 1

- *KOZANECKI:*

Your predictions for r_N, $\bar{\rho}$, etc., depend on the validity of the scaling assumption. Was the value $r_N = 0.6 \pm 0.2$ obtained by the Gargamelle group calculated from data in the scaling region only, i.e. for large neutrino energy, say, above 5 GeV, or for all energies? Inspection of CC y distributions show that scaling does not set in until around 5 GeV.

- *GOURDIN:*

It is true that it would have been better to use data from the scaling region only, but the statistics were too poor to do so. r_N was determined using the values of R_ν, $R_{\bar\nu}$, and $CC(\bar\nu)/CC(\nu)$ calculated for the whole data sample. However, while for the y distribution you should restrict yourself to the scaling region, you expect that the cross-section integrated over all x and all y will be less sensitive to this kind of approximation. Remember that we make the *a priori* assumption that we can compute the total cross-section using the scaling form of the differential cross-section in the whole physical region.

- *RÜCKL:*

You derived some of your bounds assuming the absence of second class currents. What are the experimental bounds on the existence of second class currents? And what are the main consequences of the potential existence of second class currents?

- *GOURDIN:*

To your first question, I will answer that a lot of work has been done to clarify this question, with no definite result. All I can say is that there is no evidence for the existence of second class currents.

These currents behave under crossing in a way opposite to the first class currents. Therefore, the interference between a 1st class matrix element and a 2nd class matrix element will behave in an anomalous way, just with a minus sign as compared to pure 1st class or pure 2nd class currents. It follows that the three functions a_0, a_1, a_2 for neutrinos and antineutrinos are no longer correlated. In particular, the difference $d\sigma^\nu - d\sigma^{\bar\nu}$ also contains three terms and is no longer proportional to $\cos\theta_t$. In principle, this is obviously a nice test. As an example of such a possibility in the V,A case, I will refer to Wolfenstein, who studies the possibility for the final neutrino to be different from the original one, as might be the case if heavy neutral leptons coupled to ordinary neutrinos exist.

- KOZANECKI:

Could you give a practical definition of scaling for an experimentalist working in the Gargamelle energy range? How can one be certain one is in the scaling region?

- GOURDIN:

Firstly, scaling is an asymptotic statement valid for q^2, $W^2 \to \infty$ with $x = q^2/2M\nu$ fixed. Experimentally, it was a surprise to find scaling to be valid at SLAC for $q^2 > 1$ $(GeV/c)^2$, $W > 2.6$ GeV which are not large compared to the mass of the target nucleon. We now have a better theoretical understanding of these facts; but it is obviously true that at low energy, for instance, in the Gargamelle experiment, the magnitude of the scaling region in the q^2, W^2 plane depends critically on two choices:

i) the scaling variable, and

ii) the scaling function.

For the first question, going from $x = q^2/2M\nu$ to $x' = q^2(2M\nu + M^2) = q^2/(W^2 + q^2)$ is an improvement in the W direction and allows, in the Bloom-Gilman duality sense, to include resonances. By the way, this fact is probably related to the observation that, for elastic form factors, universality in q^2-dependence is better realised when the quantity $q^2/W^2 = q^2/M_R^2$ is used.

For the second question, let me just remark that if the function $F(q^2, W^2)$ scales, the function $(\nu^2 + q^2)/\nu^2 \, F(q^2, W^2)$ also scales; but the low energy behaviour is obviously different in the two cases.

Finally, let me emphasize that these problems have not yet been investigated in neutrino experiments, where the three structure functions have not been separated and where scaling in the direct sense has not been established. We only have hopes that the situation will be similar to that at SLAC.

DISCUSSION No. 2

- DYDAK:

For $\bar{\nu}_e e^-$ scattering, is the reactor experiment you mentioned that of Reines et al.? How did you establish bounds for the Weinberg angle, despite the fact that this experiment has not yet proved the existence of neutral currents?

- GOURDIN:

The most recent data concerning the reactor experiment have been presented at the Balaton meeting last month. In my graph, I used only Reines's

DISCUSSION

old data; in the final table, the first lines are based on Gargamelle data only, while the subsequent lines contain information both from Gargamelle and from the reactor experiment. It is extremely difficult to establish the existence of neutral currents using reactor data only, as is obvious from looking at the ellipses: the latter are centred at R = 0, L = 2 (because of the CC contribution).

Therefore, the upper bound ellipse will almost certainly contain the origin, and the lower bound curve almost certainly will not -- because the bounds are not tight enough. You need a very accurate reactor experiment to reduce the gap between the two ellipses and to be able, for instance, to exclude L = 0, R = 0, and hence establish the existence of neutral currents. In the ν_μ case, by contrast, all you need is one clean event to exclude (0,0).

- *DYDAK:*

Since neutral currents exist, the de-excitation of nuclei may occur via processes like $N \to N + \nu + \bar{\nu}$, in competition with photon emission. Could such a process be detectable in the case of strongly hindered photon transitions?

- *GOURDIN:*

I would prefer to make a general comment here. In atomic and nuclear physics, there are two ways to evidence the necessary existence of neutral currents: first, interference effects between one-gamma and one-Z exchange; and secondly, one-gamma forbidden transitions which have to proceed via an alternate path.

Examples of one-photon-one-Z interference effects could occur in $e^+e^- \to \mu^+\mu^-$ and in electroproduction at very high energies; these effects would be competing with two-photon exchange, and therefore we must restrict ourselves to parity-violating effects which characterize weak interactions.

As an example of one-photon exchange forbidden transitions, consider the $0 \to 0$ transitions in nuclear physics. Again the detection would be very difficult, even using the highly accurate techniques developed in nuclear physics.

Finally, a technical remark in the context of the Weinberg-Salam model: neutral current processes are often times suppressed by one order of magnitude -- compared to estimates based on inclusive or elastic processes -- due to a factor of $\sin^2 \theta_W$. This makes the detection even more difficult.

- *MILLER:*

We are planning a neutrino experiment in BEBC with a sensitive deuterium target surrounded by neon. In principle, it should be possible to observe your suggested channels

$$\nu d \to \nu d, \quad \bar{\nu} d \to \bar{\nu} d,$$

$$\nu d \to \nu d \pi^0, \quad \bar{\nu} d \to \bar{\nu} d \pi^0.$$

In practice, I fear that the allowed range of momentum transfer to the deuteron will cut the cross-section to a very small value. Is this true?

- GOURDIN:

As usual in elastic deuteron processes, the deuteron wave function will kill the cross-section at large momentum transfers and, therefore, what we can hope is to measure something at small $q^2 [(\ll 1 \, (GeV/c^2)]$. Nevertheless, the elastic reaction provides a signature for isoscalar neutral currents, and π^0 production corresponds to an isovector neutral current. In this sense, a clean answer about the existnnce of a non-vanishing cross-section and about a difference between neutrinos and antineutrinos is of great interest. In particular, the necessity of an axial vector, ssoscalar neutral current has not been established yet. Of course, most of the time we expect a deuteron break-up reaction $\nu + d \rightarrow \nu + p + n$ to occur. These processes are very interesting if we can isolate in a clean way the nucleon pole contribution. But, in the case of a neutrino beam, the final lepton being also a neutrino is not observable, and it would be really very hard to be sure that the final spectator carries only very little kinetic energy.

- TRIANTIS:

We have seen that polarized target measurements would prove useful, in particular for sorting out the V,A fromtthe S,P,T contributions, but we know also that such measurements are very difficult experimentally. Are there any estimates concerning the importance of final state polaritieson, in particular in semi-inclusive reacions of the type $\nu + p \rightarrow \nu + h + X$, where one could measure the h polarization, as for instance, in the case $h = \Lambda^0$?

- GOURDIN:

In an elastic reaction, using a polarized target or measuring the tianl polarization of the recoiling target, yields equivalent information. But in inelastic reactions, with many particles in the final state, I do not think that measuring the polarization of only one of the products would help untangling the SPT-VA puzzle.

I have not investigated in detail the problem of polarization in the general SPTVA case; but I am sure one can derive the general formulae considering the annihilation channel as done for the angular distribution. You will get terms containing legendre functions up to second order, and the ϕ dependence will introduce terms like $\cos \phi$, $\cos 2\phi$, $\sin \phi$, $\sin 2\phi$, the angle ϕ being that of the incident plane -- beam direction, target polarization -- with the final particle plane. It is obvious that the final neutrino being unobservable, the lepton plane cannot be defined and the azimuthal integration is automatically performed.

THE STATUS OF NON-CHARMED HADRON SPECTROSCOPY

R.H. Dalitz
Department of Theoretical Physics
Oxford University

SUMMARY

Main Theme: (i) the effort required to do an experiment in old
(= non-charm) hadron resonance physics has been increasing - the
resonances still to be established are more elusive (due to low spin,
or small elasticity, etc.) and the standards of statistical accuracy
required for a significant result increase correspondingly.

(ii) the arena of deep and rapid progress in elementary
particle physics has moved to phenomena at higher energies and to
the search for new types of particle, with new quantum numbers, and
this movement is also the case for the physicists and for their
equipment.

(iii) yet there are still quite central questions
concerning hadron spectroscopy which are not settled but which
probably could be settled with exisiting accelerators and experimental
equipment.

SU(6) Symmetry for hadrons.   Do the existing analyses of their
spectra and decay modes constitute a theory?   The answer is no,
because (a) for the mass spectra, we must recognize that SU(6)
symmetry with the quark model required at least non-relativistic
internal motions, since it operates in the space of SU(3) x $SU(2)_\sigma$,
the last factor denoting the space of Pauli spin.

(b) there are well-known difficulties which bar the extension
of SU(6) symmetry to a complete Lorentz-covariant symmetry.
Extensive analyses have been made for hadron decay processes in terms
of the non-covariant symmetry $SU(6)_W$, but even here what we have
today is a set of prescriptions, different for different classes of
decay mode, which have been found not to conflict with the data.
For example, the data require that the transitions $\Delta L_z = 0$ and
$\Delta L_z = \pm 1$ should have unrelated strengths, and that the $SU(6)_W$

amplitudes must be "ℓ-broken", i.e. that when two ℓ-values are possible for the decay $H^* \to H_1^* + H_2^*$, the two amplitudes corresponding must be treated as being independent parameters.
<u>Mesonic States</u> are regarded as quark-antiquark states with internal orbital angular momentum L. The patterns of states are expected to correspond to the <u>35</u> and <u>1</u> representations of SU(6) symmetry. The states on the leading trajectory in the plot of L versus the number n of quanta of excitation energy (such that $M^2 = M_0^2 + n\epsilon$) are those with n = L; for a simple harmonic oscillator interaction, the quantum of radial excitation is 2ϵ and the N-th trajectory below the leading trajectory is given by displacing the leading trajectory to start at n = 2N for L = 0. This should hold separately for the <u>35</u> and the <u>1</u> representations. The SU(6) symmetry-breaking appears to be dominated by (1) a quark mass difference term, (2) a spin-spin interaction and (3) a spin-orbit interaction.

As is well known, the pseudoscalar and vector mesons fit quite well the (<u>35</u> + <u>1</u>) pattern for L = 0. There are also some candidates for assignment to the radial excitation with n = 2 and L = 0, the E(1420) meson and especially the ρ'(1600) state produced in e^+e^- annihilations at this energy; also possibly the state ω(1675).

For L = 1, four SU(3) octets are expected, with spin values J = 0, 1, 2 for S = 1 and J = 1 for S = 0. The $J = 2^{++}$ octet is well established and agrees with this expectation. For each of the other three octets, at least two meson states have been established, thus confirming the (<u>35</u>, L = 1) pattern expected. However, there are serious problems:

(a) there is no clear evidence for the A_1 meson with JPC = 1^{++}. The situation is confused, owing to the Deck mechanism which also leads to 3π final states peaking at low mass. It has been shown that the data on 3π final states is at least not necessarily inconsistent with the existence of an A_1 meson. The existence of a corresponding K_A meson has been established, and therefore SU(3) symmetry also requires the existence of an A_1 meson. The B-meson, with JPC = 1^{+-}, is well established, as is also the corresponding strange meson K_B.

(b) no evidence has yet been found for the I = 0 H and H' mesons expected, for which JPC = 1^{+-}. The H' meson ($\underline{1}$ representation for SU(6)) need not exist in this mass region, of course.

For L = 2, we have the well-established state g(1680) with JPC = 3^{--}, and there are two states, A_3(1640) with (2^{-+}) and L(1770) with (2^{-}?), which may correspond to the configuration S = 1, L = 2. For L = 3, the only established state is h(2050) with JPC = 4^{++}, deduced independently from its decays to $\pi^+\pi^-$ and $\pi^0\pi^0$.

The states ρ, A₂, g and h constitute our present knowledge of the leading ρ-trajectory for Δ^2 negative, where Δ denotes the momentum transfer. This part of the trajectory joins smoothly with the ρ-trajectory for Δ^2 positive, now known over a considerable range of Δ^2 from the Regge analysis of change-exchange reactions at very high energies.

There is also data which suggests that there may exist mesonic resonances which are "unstable bound states" of baryon-antibaryon systems, lying moderately close to the corresponding threshold. These resonances deserve further investigation, both experimentally and theoretically.

Baryonic States are regarded as three-quark states with internal orbital momenta summing to value L. These states may be classified by the permentation symmetry of their orbital wavefunction, which may be of Symmetric, Mixed symmetry, or Antisymmetric type. Empirically, the first two of these are found to correspond to SU(6) representations $\underline{56}$ and $\underline{70}$, respectively; no states of the third type have yet been identified (for good reasons) but they are expected to belong to $\underline{20}$ representations. In particular, the ground state supermultiplet is believed to have a symmetric space wavefunction (certainly so, with the quark shell model) and has the structure ($\underline{56}$, L = 0). This implies that the total wavefunction has even permutation symmetry, which violates the Spin-Statistics Theorem (and also the empirical fact that protons obey Fermi statistics). This difficulty can be avoided only if the quarks have a further degree of freedom, which must be mathematically three-dimensional. Gell Mann proposed the name "colour" for this variable, and its most fruitful form corresponds to the introduction of a three-dimensional colour space together with a symmetry SU(3)' in this space. With this formulation, Lipkin has pointed out that quark-quark forces mediated by the exchange of a colour octet of vector mesons would lead to strong binding only for quark systems which are SU(3)'-singlet, as is the empirical situation for all of the hadronic states we know. This formulation thus leads to a colour-gauge theory and such a theory has both asymptotic freedom and the possibility of leading to quark confinement.

When attention is confined to SU(3)'-singlet baryonic states, we are limited to the consideration of three-quark states which have even permulation symmetry for their wavefunction in space, spin and unitary-spin variables. The simplest and most natural model is the quark shell model, with harmonic orbitals, for it immediately leads to the expectation of a series of bands of states, with alternating parity $P = (-1)^n$ in sequence as the degree of excitation n (and therefore the mass) increases, as is empirically apparent for the lowest five bands of nucleonic resonances. Even with SU(6)-invariant quark-quark forces, this model leads to a supermultiplet structure even within one band. For example, the band with two quanta of excitation is expected to include five supermultiplets, namely the $(\underline{56},2^+)$, $(\underline{70},2^+)$, $(\underline{20},1^+)$, $(\underline{56},0^+)^*$ and $(\underline{70},0^+)$ supermultiplets; in this approximation, there are two equal-spacing rules, between the states $(\underline{20},1^+)$, $(\underline{70},L^+)$ and $(\underline{56},L^+)$, in that order, for the two cases $L = 0$ and 2. In the next approximation, for comparison with experiment, quark-quark forces are introduced which break the SU(6)-symmetry in various ways, for example spin-spin forces and spin-orbit forces which may or may not break (SU3) symmetry as well, together with the quark mass difference terms. The strengths of all these SU(6)-breaking terms are obtained by fitting the calculated mass spectrum for the n = 0, 1 and 2 supermultiplets to the well-established baryonic resonance states. The parameter values obtained do not indicate a simple situation. All of the SU(6)-breaking terms have caomparable orders of magnitude; no one effect is dominant. The major spin-orbit terms are also SU(3)-breaking in character; although the various spin-orbit terms almost cancel together for N* and Δ* states, this is not at all the case for the

Λ^*, Σ^* and Ξ^* states.

The SU(6) pattern ($\underline{56}$, L=0$^+$) is well established for the lowest baryonic octet and decuplet. For the first excited supermultiplet, ($\underline{70}$, L = 1$^-$), all the nucleonic members are now known and well established; also the two unitary singlet states $\Lambda(1405)$ and $\Lambda(1520)$ with JP = $\frac{1}{2}^-$ and $\frac{3}{2}^-$, respectively. Three of the other five predicted Λ^* states are well established; the missing states are ΛD03 and ΛS01, which are both predicted to be broad and strongly inelastic. The Σ^* situation is less clear, but four out of the seven predicted Σ^* states appear well-established. There is still doubt concerning two ΣD13 states and the uppermost ΣS11 state. Of the seven Ξ^* states predicted, only two have been observed; neither of the two Ω^* states are known.

For the next band of positive parity states (n = 2), the SU(6) supermultiplets ($\underline{56}$,2$^+$) and ($\underline{56}$,0$^+$)* are clearly required by the data, for there are now established N* states corresponding to each octet, and Δ^* states corresponding to each decuplet, for these two supermultiplets. For the supermultiplets ($\underline{70}$,2$^+$) and ($\underline{70}$,0$^+$), the situation is controversial. The three states NF17(1990), ΛF07(2020), and ΛP03(1860) which can be conveniently assigned to ($\underline{70}$,2$^+$) are still open to question; also, Cashmore, Hey and Litchfield (CHL, hereafter) have proposed assignment of the first two states to the n = 4 supermultiplet ($\underline{56}$,4$^+$). The state NP11(1750) assigned to ($\underline{70}$,0$^+$) is quite a well-established state, but CHL propose its assignment to the n = 4 radial excitation ($\underline{56}$,0$^+$)**. No states belonging to the ($\underline{20}$,1$^+$) supermultiplet are known, but it is predicted that their formation and their decay should each involve relatively complicated processes.

In the SU(6) limit, each supermultiplet may be conveniently represented by one point, labelled S, M or A, on a plot of L versus the number n of excitation quanta (the mass being given by $M^2=M_0^2+n\epsilon_B$). With the quark shell model, we have (1 x S) for n = 0, (1 x M) for n = 1, (2 x S, 2 x M, 1 x A) for n = 2, (2 x S, 4 x M, 2 x A) for n = 3, and so on, the numbers of supermultiplets for excitation n increasing roughly like n^3 with increasing n. Even along the leading trajectory,

the supermultiplets soon become multiple. At least one 56 supermultiplet is predicted on the leading trajectory for every value of
n except n = 1. At n = 6, there are two 56 supermultiplets predicted
with L = 6, at n = 12 there are three 56 supermultiplets predicted
with L = 12, and so on. This complexity is a consequence of the fact
that L is the sum of the two internal orbital angular momenta
corresponding to the two independent harmonic oscillators for the
internal motion.

The reassignments of N and Λ states proposed by CHL are motivated
by the question whether the SU(6) supermultiplets required by the data
might correspond to a simpler (L, n) pattern, such that the only
SU(6) supermultiplets are $(\underline{56}, L^+)$ for n = L = even, and $(\underline{70}, L^-)$ for n =
L = odd, together with the attached sequence of radial excitations
with n = L + 2, L + 4,..., but with no multiple occurrence for any
supermultiplet. The decision whether or not the data require the
n = 2 supermultiplets $(\underline{70}, 2^+)$ and $(\underline{70}, 0^+)$ is clearly crucial for
this question. The simpler (L, n) pattern is precisely that which
results when there is only one internal oscillator, as for the case of
a diatomic molecule, where each Regge trajectory consists of the
sequence of rotational excitations for a given vibrational state,
there being a different trajectory for each vibrational state. Quark
theories which lead to such an (L, n) pattern involve essentially the
suppression or freezing of one of the internal oscillators, by one
mechanism or another, for example by the assumption that there is an
$\ell = 0$ di-quark unit which retains its identity in the motion.

With the full quark shell model, there is still relative
simplicity along the leading trajectory, and there is therefore much
interest in the determination of the characteristics of the resonance
states with highest spin in each band n, and their confrontation with
theoretical calculations which have essentially no free parameter.
For Δ resonances with $J = L + \frac{3}{2}$ and positive parity, the sequence
L = 0, 2, 4 is well established, the partial width for $\Delta^{++} \to p\pi^+$
being known for each resonance from the measured width and elasticity;
bumps are also observed in the $\pi^+ p$ total cross-section at the correct
mass values for L = 6 and 8, and it has been natural to interpret these
as due to the higher resonances in this rotational sequence. These

data are at present in quite satisfactory agreement with the theoretical calculations, and it is desirable that the empirical situation be established with more certainty and extended in mass range, if possible. A similar analysis could be carried out for the high spin states in the sequence of supermultiplets ($\underline{70}$, L^-) for $L = 1, 3, 5...$

DISCUSSION

CHAIRMAN: R.H. Dalitz

DISCUSSION No. 1

- CALDI:

If the A_1 continues to be non-existent, do you have any explanation in the context of the quark model or, if not, how serious a blow to the quark model would you consider it to be?

- DALITZ:

I have no suggestion to explain it. If it continues to be non-existent I would consider this to be a very serious problem. This is why many people have spent so much time trying to decide whether an A_1 resonance is necessarily excluded by the data. It is very hard to lose a quark-antiquark state altogether.

- SUNDERMEYER:

What happened to the ε?

- DALITZ:

The ε was always a rather theoretical object and the theorists have changed their minds. Also, we now have much more detailed data.

- SUNDERMEYER:

What about the Berkeley K-matrix analysis?

- DALITZ:

ε is a name which has been used for many different phenomena. In this case, poles were introduced to fit the data. The ε pole is almost on the imaginary axis and very far from the physical region. Such distant resonance poles are always suspect, and it is difficult to extract them from the experimental data with any certainty. For this reason, there has always been a good deal of unhappiness about the interpretation of this pole as a meson.

- SUNDERMEYER:

What does this mean for the ε'?

- *DALITZ:*

The ε' has nothing to do with the old ε. This is just bad nomenclature. The ε' should be called the $S^{*'}$.

- *KURODA:*

What kind of experiments would be needed to establish whether the A_1 is there or not? How can a kinematic enhancement be distinguished from a resonance?

- *DALITZ:*

The most promising situation would be one where there is no diffractive background. That is why people have looked for the A_1 in charge-exchange processes such as $\pi^-p \to \pi^+\pi^-\pi^0 n$. The background is indeed reduced, but so is the whole peak so nothing could be concluded; indeed, the data were still well fitted by the Reggeized Deck model, according to Berger. In the somewhat similar "Q" region for the $\bar{K}\pi\pi$ state, experiments on $K^-p \to (K^{*-}\pi^+)n$ are again well-fitted by this Deck model.

We are accustomed to the idea that, if a two-body system $(a + A)$ has an elastic resonance M_ℓ^* in partial wave ℓ, then production of the system $(a + A)$ with internal orbital angular momentum ℓ necessarily shows the effect of the resonance. In the one-channel system $(a + A)$, non-resonant background is necessarily intimately related with the resonant scattering, through unitarity. In this situation, the resonance M_ℓ^* must appear. However, this is not the case for two channel interactions. If these channels are $(a - A)$ and $(b - B)$, the resonant amplitude will have the form

$$\frac{1}{k} \frac{\sqrt{\gamma_a \gamma_b}}{E_R - E - i\gamma/2} . \qquad (1)$$

The resonant state then involves the $(a - A)$ and $(b - B)$ systems in definite ratio, the state being

$$\psi_{res} = \sqrt{\gamma_a}\, \psi(a - A) + \sqrt{\gamma_b}\, \psi(b - B) . \qquad (2)$$

There is then an orthogonal non-resonant state

$$\psi_{non-res} = -\sqrt{\gamma_b}\, \psi(a - A) + \sqrt{\gamma_a}\, \psi(b - B) . \qquad (3)$$

If the production process leads naturally to the latter state, then the resonance M_ℓ^* will not be excited and will not show in the final state. However, one would not expect that the state (3) should never be produced; generally, the production process will lead to a linear superposition of (2) and (3), and the resonance will be seen more or less strongly. Our intuition based on experience with single two-body interactions is misleading; even if the system $(a - A)$ is produced with internal angular momentum ℓ, the resonance M_ℓ^* is not necessarily excited. Three-particle and more complicated states may be regarded as more elaborate multichannel situations in this respect.

DISCUSSION

- *MILLER:*

An alternative technique for producing the A_1 was discussed by Stodolsky and others some years ago. This would use axial-meson dominance of the weak axial current in neutrino interactions -- a very similar process to rho-meson electroproduction. The cross-section should be small, but not negligible.

- *COLEMAN:*

I am fully aware that no-one has succeeded in deriving $SU(6)_W$ from a Lagrangian theory, but is the application of $SU(6)_W$ to 3-particle vertices successful in describing experimental data?

- *DALITZ:*

It does appear to have a close connection with the experimental data. Sets of 3-particle vertices can be classified as "$SU(6)_W$-like", but other sets of vertices appear to have a systematic reversal of sign, in which case we speak of "anti-$SU(6)_W$" being effective. These sign reversals do have empirical consequences, since there can often be interferences between two matrix elements, one being of the class "$SU(6)_W$-like" and the other of the class "anti-$SU(6)_W$". The empirical situation does not correspond directly and simply to $SU(6)_W$, even when this is imposed only on three-particle interaction vertices.

- *TRIANTIS:*

The $\bar{p}N$ bound state at 1796 MeV/c^2 which you mentioned this morning is quite interesting, due to its large binding energy of around 80 MeV. Could you comment on the 2 order of magnitude difference in this binding energy from that of the deuteron? Should these two-nucleon systems not have comparable binding energies?

- *DALITZ:*

Let us consider first the NN potential. We believe that this consists of a very strong attraction outside a strongly repulsive core. The attractive potential calculated from pion (and vector) meson theory had depth of order 1 GeV just outside the hard core. For the deuteron, we adjust the parameters finely to give the observed binding energy of 2.26 MeV; if these parameters (e.g. the hard-core radius) are changed slightly, the binding energy will quickly become very large, say 20 to 100 MeV, or the bound state will quickly cease to exist. In the deuteron there is a delicate balance between the internal kinetic energy and the potential energy; if the latter is changed, this balance quickly changes in one direction or the other.

For the $\bar{N}N$ system, the situation is quite similar, except that the core consists of a strongly absorptive interaction. With the forces due to pion and vector meson exchanges, a prediction of a binding energy of order 50 to 100 MeV does not seem unreasonable. If the binding energy were predicted to be much higher than this, one would need to think about the inclusion of relativistic effects for the N and $\bar{N}$, but that is another question.

DISCUSSION No. 2

- *GOTTLIEB:*

Could you comment on the situation regarding the so-called exotic states?

- *DALITZ:*

Most of the positive evidence has come from the analysis of K^+p and K^+n scattering. There is a bump in the K^+p total cross-section, near the inelastic threshold, about which there has been much discussion. If this corresponds to a resonance, then it appears that it can only have a small amplitude. The general opinion, at present, is that the data do not call for a resonance in the $I = 1$ channel.

The more serious evidence is that obtained for the $I = 0$ KN interaction. There are several acceptable phase-shift analyses, and one of these solutions does indicate the occurrence of an elastic resonance in the partial wave ZP11. Quite recently, data have also been obtained on the process $K_L p \to K_S p$, which also bears on this question. I have not seen these data, but my impression is that they favour the Z_0^* resonance interpretation.

It is not necessary that all resonance states should correspond to the simple three-quark states we have discussed. For example, for baryon number two, we know of one state, the deuteron, which we do not believe to be a six-quark bound state. We believe the deuteron to be a bound state of two nucleons (each consisting of a three-quark bound state), formed by long-range forces between the two nucleons due to the exchange of mesons. Other states of this general kind may also exist.

- *DE GRAND:*

What is the status of the $N^*(1420)$, the Roper resonance, in the $SU(6)_W$ classification?

- *DALITZ:*

The decay data for $N^*(1420) \to \pi \Delta$ and those data on the photoproduction process $\gamma N \to N^*(1420)$, both analysed by Cashmore et al., indicate that this N^* could only be a member of a $\{56\}$. It is therefore natural to assign it to the $N = 2$ radial excitation $\{56, L = 0^{+*}\}$. Other members of this multiplet may include the states $\Delta P33(1690)$, $\Sigma P11(1670)$, and $\Lambda P01(1570)$.

- *JOSHI:*

Do you support the use of a linear or a quadratic mass formula?

- *DALITZ:*

I believe the quadratic form to be correct. It is hard to give a theoretical justification for this, since we do not know what relativistic equation should be used to discuss these states. From a purely empirical point of view, the quadratic form does seem to work best.

DISCUSSION

- *DE GRAND:*

As an aside, the MIT Bag model mass formula, derived from the virial theorem, involves mass to the power 4/3.

DISCUSSION No. 3

- *MILLER:*

I have a wild suggestion. Could it be possible that in high energy pp collisions a kind of "optical-pumping" may take place, in which higher {70} multiplet states are excited and fall back to give relatively copious production of the {20} states which you mentioned? If the {20} states are sufficiently long-lived, then their electromagnetic or weak decays might account for the direct lepton production in hadron collisions which has been reported by Cronin and others.

- *DALITZ:*

A {20} state is excited in both quark variables $\left[\text{i.e. } (1p)_\rho (1p)_\lambda\right]$, so it would be difficult to produce directly, either electromagnetically or weakly, from a proton target, since these currents couple to only one quark at a time. Such states could certainly be produced via pionic decay, following the excitation of a higher {70}. They might well decay very slowly if their mass is low compared with the {70,1-} states to which they would be expected to decay by single-pion emission. The double quark excitation would also make difficult any electromagnetic or weak decays of the {20} states to the ground state {56} states. More complicated decay modes may be possible to the ground state {56}, such as two-pion emission processes.

I do not know if any work has been done on the electromagnetic or weak decays of the {20} states. This may be worth looking into further. However, I am inclined to think that if there were any {20} states lying so low in mass that they suffered strong-decay hindrance due to their "two-quark-excitation" character, then we would already know about them.

- *CANNING:*

Satisfaction of dual sum rules demands hadron states of ever increasing number of quarks and antiquarks. Are these distinguishable from your "unstable bound states"? For example, a $qqq\bar{q}$ state is required in $b\bar{b}$ scattering. A b bound to $\bar{b}$ retaining its $qqq\bar{q}\bar{q}\bar{q}$ identity, would nevertheless in drawing a dual diagram for a meson decay involve the $q\bar{q}q\bar{q}$ intermediate state. Is it so easy to distinguish when one has quark-gluon or hadron-meson forces?

- *DALITZ:*

Does duality demand resonances or would continuum channels be sufficient?

— *CANNING:*

The predictive power of the theory is less if resonance dominance is abandoned.

— *DALITZ:*

I have never regarded these duality arguments for these states as compelling.

— *CANNING:*

You gave an analysis of how to distinguish "unstable bound states" from resonances. Is there some simpler way to distinguish them -- say, by comparison of widths with those of ordinary resonances of the same mass?

— *DALITZ:*

The phenomenological identification of such states is not always straightforward. It involves the extrapolation of (K-matrix) amplitudes obtained from the analysis of physical data, into unphysical sheets of the energy plane. Of course, we are accustomed to do this already for the identification of a resonance pole. In both cases, the identification is more convincing when there is a theoretical model used to calculate the energy-dependence of the amplitude, whose parameters can be adjusted to give the best fit to the physical data.

The possibility of an "unstable bound state" is the more likely, when the scattering is particularly strong in the neighbourhood of a threshold. The interactions in the neighbourhood of the $K\bar{K}$ threshold were a natural candidate for this interpretation but the detailed analysis of the pole situation apparently shows it to be otherwise. On the other hand, for the $\bar{N}N$ system we can calculate the $\bar{N}N$ forces from meson theory. If these forces suggest the existence of a bound state (which would be unstable for decay to pions), in some energy region for some spin and parity, and a resonance with the right spin and parity were found in this neighbourhood, then it would be natural to consider this interpretation seriously, and to adjust the parameters of these forces to find whether the best fit to this resonance data is an acceptable fit.

It is not really possible to give a general prescription for the recognition of this class of resonance. Besides the detailed empirical data, some theoretical guidance about the possibilities is also desirable.

VERY HIGH ENERGY HADRONIC INTERACTIONS

A.N. Diddens
CERN, Geneva, Switzerland

INTRODUCTION

The first part of these lecture notes reviews the current experimental situation in the field of high energy interactions of hadrons resulting in a 2-body final state. The second part is mainly concerned with two subjects, diffraction dissociation and correlation studies, from the field of many body final states.

1. Two-Body Final States

    1.1 Total cross sections

A summary of total cross sections of $\pi^\pm$, $K^\pm$ and $p^\pm$ on protons, above 10 GeV/c, is shown in Fig. 1. The ANL-FNAL-Rockefeller University measurements for most particles extend up to 280 GeV/c.[1] The rise of the total cross sections with energy for all particles - except antiprotons, for which higher energies are needed, presumably - is confirmed. Fig. 2a shows the K^+p data on an expanded scale, Fig. 2b similarly the pp data.

The same group also performed measurements on deuterium, from which particle-neutron total cross sections were derived, exhibiting a similar trend as the data on protons. An independent np total cross section measurement in this energy range was performed by a Michigan University group[2]; their data is shown in Fig. 3, top half, and compared with a fit to the pp data (full line), shown in the bottom half. Thus the np and pp total cross sections are very similar in this energy range, in agreement with the Okun-Pomeranchuk rule.[3]

The differences of particle and antiparticle total cross sections are plotted as a function of the energy in Fig. 4 and apparently approach each other as a power of the energy:

$$\Delta\sigma_{tot} \propto s^{\alpha-1}$$

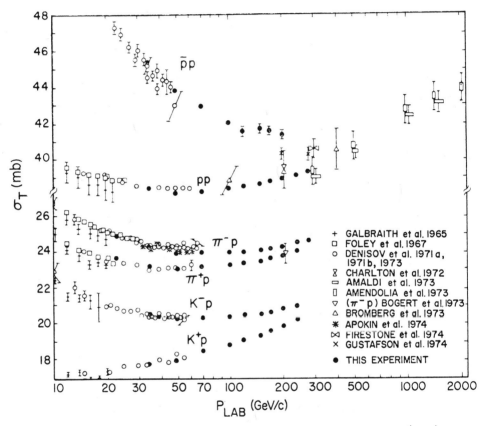

Fig.1. The energy dependence of the total cross sections of $\pi^{\pm}$, $K^{\pm}$ and $p^{\pm}$ on protons.

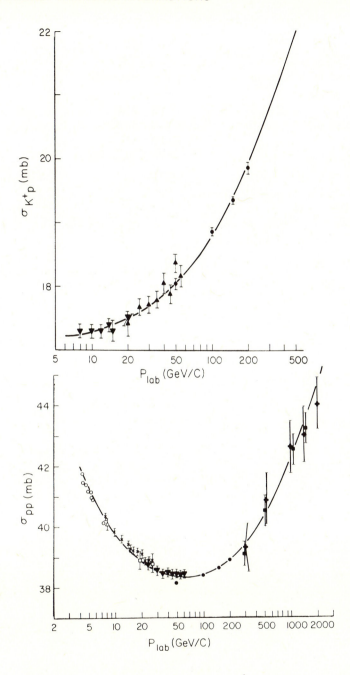

Fig.2.a. The high energy behaviour of the K^+p total cross section.
b. The high energy behaviour of the pp total cross section.

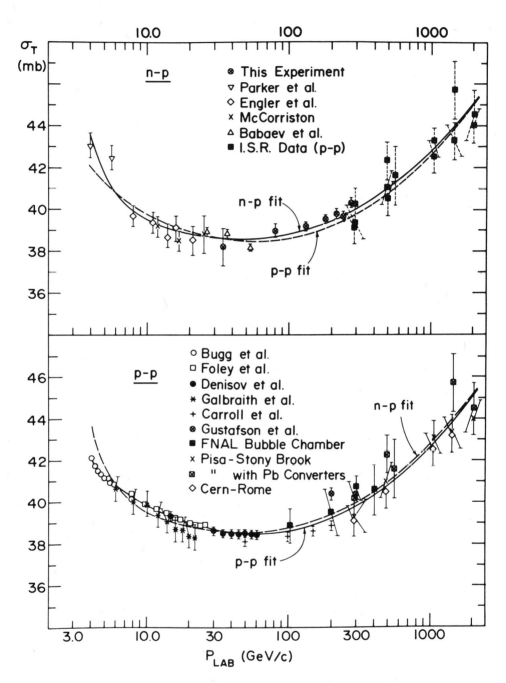

Fig.3. A fit to the pp total cross section, the lower half, is compared to the np total cross section in the top half.

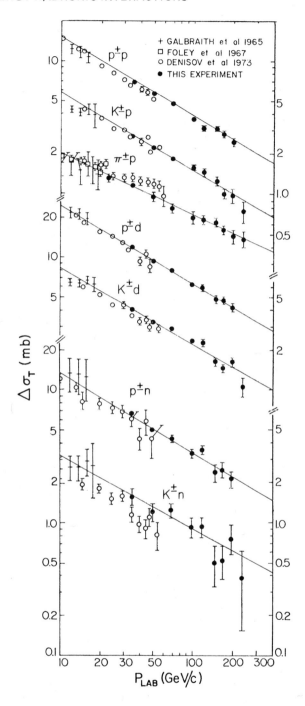

Fig.4. The difference of the antiparticle and particle total cross section as function of the energy.

with $\alpha = 0.39 \pm 0.02$, 0.40 ± 0.04 and 0.55 ± 0.03 for the $p^{\pm}p$, $K^{\pm}K$ and $\pi^{\pm}p$ differences, respectively. Similar values of α are found for the other cross section differences. Thus the Pomeranchuk theorem[4] seems well satisfied by the data.

A recent review of the validity of various sum rules has been given by Wetherell.[5]

Using the ISR data on pp total cross sections and the power law behaviour thus found, the $\bar{p}p$ total cross section can be predicted at higher energies than measured, Fig. 5. According to this picture[1] it will have increased by 1 mb at 600 GeV incident $\bar{p}$ momentum.

Having thus described the experimental situation, the theoretical situation may be summarised as follows. A clear understanding of why total cross sections rise with energy does not exist. Models exist that predict a continuous rise with energy, for instance the model of Cheng-Walker-Wu.[6] Others, based on the Gribov Reggeon calculations, consider the rise as a transient effect (Frautschi and Margolis[7], Dean[8]). Complex Regge pole theories[9] predict oscillating cross sections. Although the data points have small errors (typically 0.5% systematic error, and statistical errors between a few per mille and a few per cent), a distinction between a rise $\propto \log s$, $\propto \log^2 s$, and $\propto s^{\varepsilon}$ cannot not be made.[10]

Some remarks relevant to the subject can nevertheless be made.

(i) Lipkin[11] formalizes the usual concept of the total cross sections being made up of two components, a Regge component R and a Pomeron component $f(s)$, the latter with at the most a mild energy variation, as

$$\sigma_{tot}(s) = Rs^{\frac{1}{2}} + f(s) \tag{1a}$$

which can be rewritten as

$$\sqrt{P_{LAB}}\,\sigma_{tot}(s) = R + \sqrt{P_{LAB}}\,f(s) \tag{1b}$$

leading to the plot shown in Fig. 6. The nucleon total cross sections have been multiplied by 2/3 to take account of the difference in quark content of the mesons and the nucleons. Without increasing our insight very much, the plot shows that total cross sections (or scattering amplitudes at $t = 0$) follow an orderly pattern, and behave very similarly for the various particles.

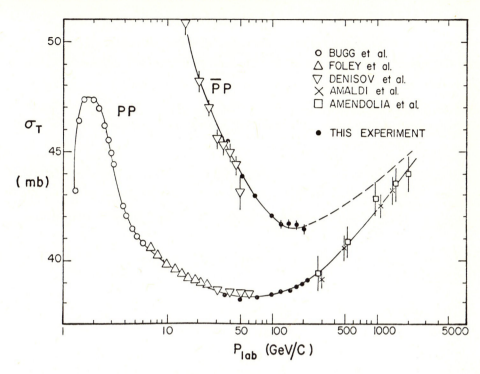

Fig.5. Prediction for the energy dependence of the $\bar{p}p$ total cross section.

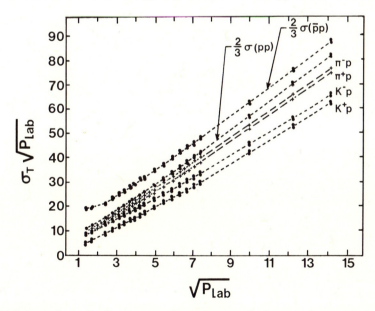

Fig.6. Lipkins plot of the energy dependence of the total cross sections, emphasizing the similarity between the behaviour of the various particles.

(ii) Morrison[12] analysed the inelastic cross sections, σ_{INEL}. A summary of the energy dependence of σ_{INEL} for six incident particles, $\pi^\pm$, $K^\pm$ and $p^\pm$ on protons, is shown in Fig. 7. The two exotic combinations K^+p and pp show a continuous rise with energy, and are remarkably similar in their behaviour. The other four cross sections initially decrease but subsequently also start to rise. It may be concluded that σ_{INEL} has a rising component, from the lowest energies onward.

(iii) The impact parameter representation of the scattering amplitude and the representation in $\sqrt{|t|}$ -space are related by a Fourier transform. Because of the Heisenberg uncertainty principle a physical significance is assigned to this mathematical relation. The increase in the total cross section is thus found to be a peripheral phenomenon: it is not the opaqueness of the nucleon that increases but the radius.[13]

(iv) The two foregoing arguments lead to an analogy between the rise of the strong interaction cross section and the relativistic rise of the ionisation. The latter is a result of the Lorentz contraction of the electro-magnetic field in the direction of motion, which is compensated by an extension of the range of the interaction to keep the field-energy constant. In the case of the strong interactions the corresponding Maxwell-equations are lacking so that the comparison can be called only an analogy.

(v) Total cross sections of gamma's on protons and deuterons[14] are shown in Fig. 8. They are flat up to the highest energy available, about 25 GeV.

The cross sections of ρ^0 and ϕ mesons on protons can be derived from photoproduction measurements, assuming vector-dominance and diffractive elastic scattering for the vector particles.[15]

$$\frac{d\sigma}{dt}(\gamma p \to Vp) = \frac{3\Gamma}{\alpha m_V}(Vp \to Vp) = \frac{3\Gamma}{16\alpha\pi m_V}\left[\sigma_{TOT}(Vp)\right]^2 e^{-B|t|} \quad (2)$$

where Γ is the branching ratio for the decay into e^+e^- and m_V is the mass of the vector particle. Obviously the energies reached are too low to decide whether vector particles will also show rising total cross sections, Fig. 9.

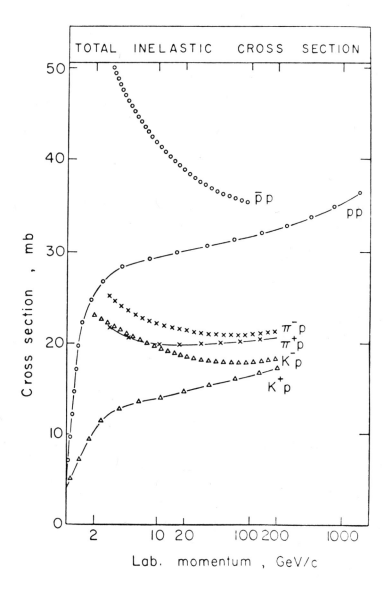

Fig.7. Summary of the energy dependence of the inelastic cross sections of $\pi^{\pm}$, $K^{\pm}$ and $p^{\pm}$ on protons.

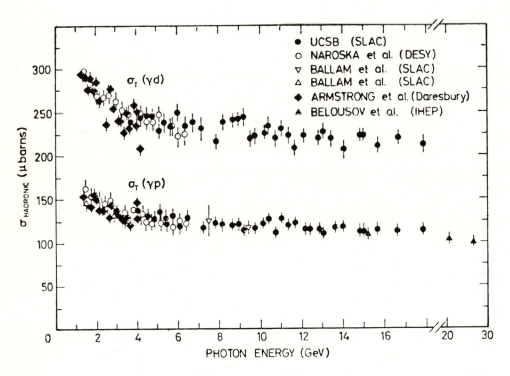

Fig.8. The energy dependence of the γp and γd total hadronic cross section.

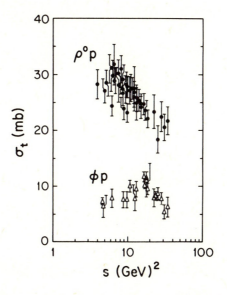

Fig.9. The ρp and the φp total cross section, derived from photoproduction data.

1.2 Elastic Scattering

1.2.1 The region near t = 0

At around $|t| \simeq 10^{-2}$ GeV/c² the interference between the Coulomb amplitude and the real part of the strong scattering amplitude f is maximal. The shape of the cross section thus allows a determination of ρ = Ref/Imf in the forward direction, with Imf(o) determined by the total cross section via the Optical theorem. Since the elastic scattering at 10 GeV/c and above is mainly diffractive, spin dependent amplitudes are assumed to be negligible. In case this simplifying assumption is not made, one may consult, for instance, Ref. 16.

Khuri and Kinoshita[17] proved that total cross sections rising like a power of the logarithm

$$\sigma_{TOT} \propto (\log E)^{\nu} \qquad \text{for } E \to \infty \qquad (3)$$

imply that ρ is approaching zero from above

$$\rho = +\frac{\pi \nu}{2} \frac{1}{\log E} \qquad \text{for } E \to \infty \qquad (4)$$

The formal proof is in fact for the even signature amplitude (sum of particle and antiparticle amplitudes) and to apply the result to the two amplitudes separately the subsidiary condition is needed that the odd signature amplitude vanishes sufficiently fast, which seems to be the case (Fig. 4).

In the energy range of the PS and AGS accelerators the results of the best measured reactions ($\pi^{\pm}$p and pp) show that ρ is still negative. It is expected, therefore, that ρ will cross zero at a higher energy somewhere, in order to satisfy (4) finally.

Fig. 10 shows a compilation of measurements of ρ, including a preliminary result for a FNAL-Yale group at 70 GeV[18]; the results are compared with a dispersion relation calculation by Hendrick and Lautrup.[19] The agreement is found to be reasonable, except in the case of $\bar{K}$p scattering, where the experimental points have a very large scatter, although their errors are also sizeable.

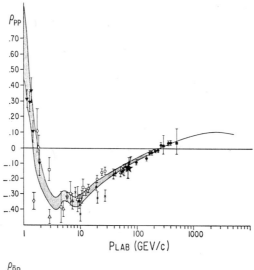
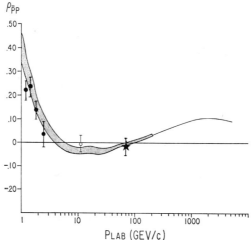

Fig.10 The energy dependence of the ratio of the real and imaginary parts of the scattering amplitude. The point at 70 GeV/c is preliminary. a) pp, b) p̄p, c) π^-p, d) π^+p, e) K^-p; the point at 14 GeV/c has recently been the subject of a change f) K^+p. The points with the large errors are derived from a comparison between σ_{TOT} and $d\sigma/dt$ (t = 0).

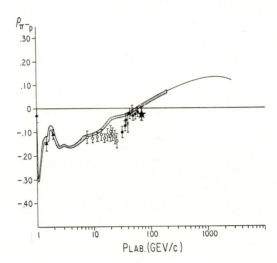

Fig.10c

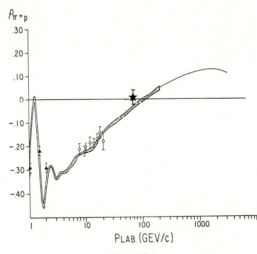

Fig.10d

VERY HIGH ENERGY HADRONIC INTERACTIONS 1081

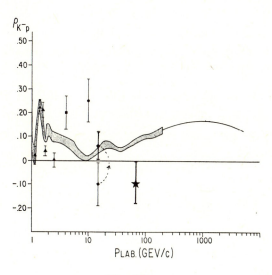

Fig.10e

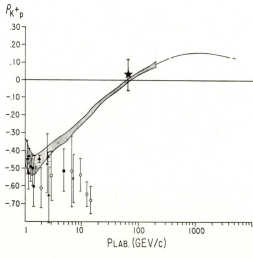

Fig.10f

1.2.1 The region around $|t| \simeq 0.2$ GeV/c^2

The t-dependence of the differential cross section is usually parameterised as $\exp(-B|t|)$. The energy-dependence of the slope parameter B for pp scattering is shown in Fig. 11; it is determined over the t-range $|t| \leqslant 0.15$ GeV^2. At about this value of $|t|$ the single parameterisation with one slope is found to break down[20], Fig. 12, and often a second slope B is introduced for $|t| \geqslant 0.15$ GeV^2. Also this second slope B is found to increase with energy, as shown in Fig. 13. Both energy dependences could be proportional to logs.

This two-slope phenomenon is also shown by π^-p scattering data[21] at 8 GeV/c, where the break occurs at $|t| \simeq 0.2$ GeV^2, (Fig. 14). For K^-p scattering at this energy the effect is smaller, and in $\bar{p}p$ scattering indiscernable.

The single arm spectrometer group at NAL performed a set of measurements[22] on elastic scattering on protons between 50 and 175 GeV/c, for all six charged particles $\pi^\pm$, $K^\pm$, $p^\pm$. An example of the data is shown in Fig. 15 and the slopes at $|t| \simeq 0.2$ GeV/c^2 are plotted against energy in Fig. 16, together with data at lower energies. K^+p and pp scattering are found to have a continuously increasing slope from 10 GeV/c onwards (shrinking diffraction pattern); $\bar{p}p$ shows a decreasing slope; the other three particles give a small effect of B increasing with energy. The correlation with the behaviour of σ_{tot} or σ_{INEL} is obvious.

1.2.3 The region around $|t| \simeq 1$ GeV/c^2

Some typical pp data covering the region from PS up to the ISR energies is shown in Fig. 17, demonstrating a nice dip, presumably of diffractive origin. The CHOV collaboration[23] made a detailed comparison of pp elastic scattering at the lowest and the highest ISR energy (Fig. 18), demonstrating conclusively that the dip moves to smaller values of t with increasing energy, in agreement with a naive geometrical model according to which the total cross section, slope, the position of the dip and the differential cross section itself have an energy dependence given by an energy dependent radius (R(s)) (Geometrical Scaling)[24]. The prediction and the experimental values of the CHOV group are compared in Table 1. The agreement is very good.

VERY HIGH ENERGY HADRONIC INTERACTIONS

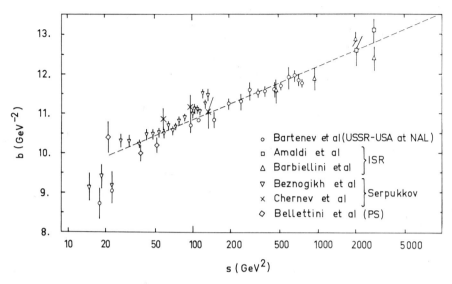

Fig.11 The energy dependence of the slope parameter of the exponential t-dependence of elastic pp scattering for $|t| \leqslant 0.15$ GeV/c^2.

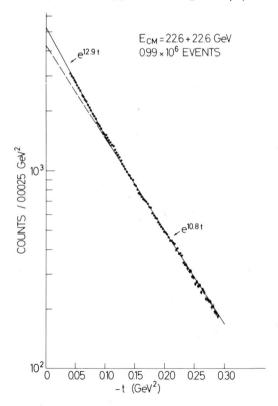

Fig.12. The t-dependence of elastic pp scattering at ISR energies.

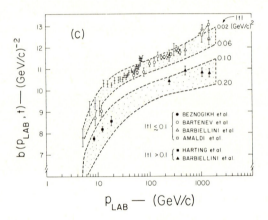

Fig.13. The energy dependence of the exponential slope parameter of the differential cross section for elastic pp scattering in two different ranges of t. The figure is from reference 56.

VERY HIGH ENERGY HADRONIC INTERACTIONS

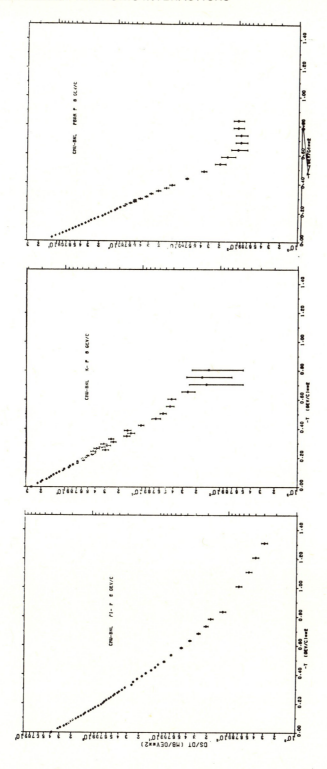

Fig.14. Diffraction scattering of π^-p, K^-p and $\bar{p}p$ at 8 GeV/c incident momentum.

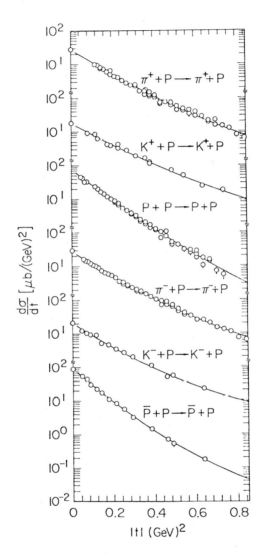

Fig.15. Elastic scattering at 100 GeV/c incident momentum for $\pi^{\pm}p$, $K^{\pm}p$ and $p^{\pm}p$.

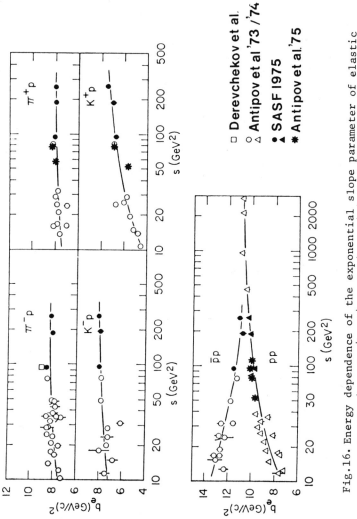

Fig.16. Energy dependence of the exponential slope parameter of elastic scattering of $\pi^{\pm}p$, $K^{\pm}p$ and $p^{\pm}p$ at $|t| = 0.2$ GeV/c^2.

Table 1

Test of geometrical scaling

Variable	Prediction	Experiment
σ_{tot}	$\propto R^2(s)$	1.14 ± 0.03
$B(0.2 \leq \|t\| \leq 0.8)$	$\propto R^2(s)$	1.13 ± 0.03
t_{DIP}	$\propto R^{-2}(s)$	$(1.14 \pm 0.03)^{-1}$
$d\sigma/dt$ at 2nd maximum	$\propto R^4 f(R^2 t)$	$(1.27 \pm 0.11)^2$

Figs. 17 and 18 give an indication that the depth of the dip is greatest at 200 GeV/c, in the measurements of the Michigan-ANL-FNAL-Indiana collaboration.[25] If this were true, the dip might be a transient phenomenon, as function of energy. A candidate for the uncovering and filling of the dip is the real scattering amplitude, which, in the very forward direction, passes through zero in this energy region.

The diffraction dip has been found only in pp scattering. FNAL data[25] of 100 GeV/c K^+p scattering and 200 GeV/c $\bar{\pi}p$ scattering (Fig.19a,b) show some irregularities in their t-behaviour, but not a spectacular dip. Whether higher energies, higher precision or larger t-values are needed, is not clear. It is also conceivable that these other particles have a profile function in impact parameter space that will never give a pronounced diffraction pattern.

1.2.4 The region around 90° (C.M.)

A compilation of proton-proton data in this angular range is shown in Fig. 20, showing that only below 30 GeV incident momentum experiments have been covering this large t-region, up to now.

In the parton interchange model[26] the energy dependence of the cross section at fixed angle is described by

$$\frac{d\sigma}{dt}(\theta) \propto s^{-n} \tag{5}$$

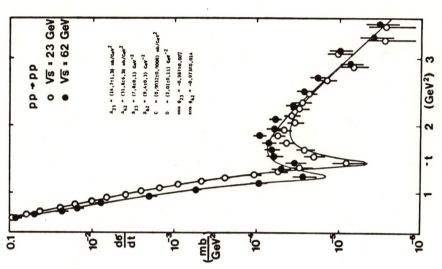

Fig.18. The differential cross section for elastic pp scattering at two ISR energies.

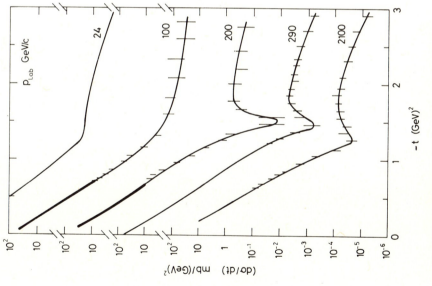

Fig.17. Summary of the energy behaviour of elastic pp scattering.

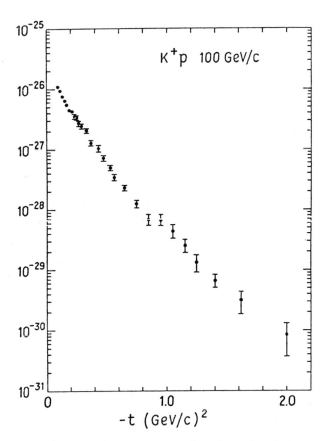

Fig.19a. The differential cross section for elastic scattering for a) K^+p at 100 GeV/c, and b) $\bar{\pi}p$ at 200 GeV/c.

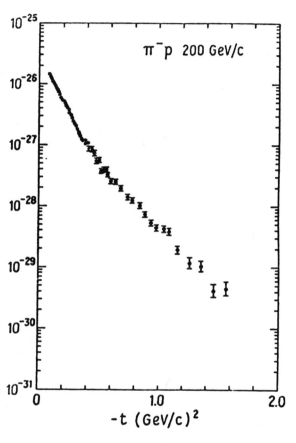

Fig.19b

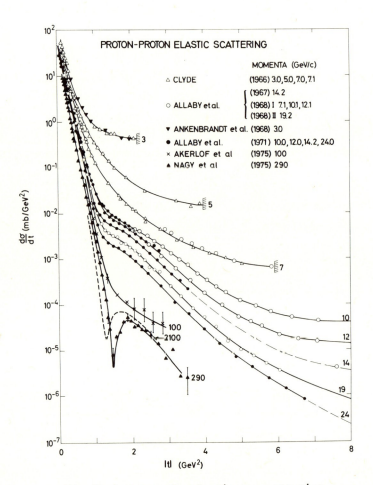

Fig.20. Large angle elastic pp scattering.

where the power n is related to the sum of the partons in the initial and final state according to

$$n = \Sigma(\text{partons}) - 2 \qquad (6)$$

Thus for large angle pp or $\bar{p}p$ scattering n = 10, for meson-proton scattering n = 8.

A comparison of relation (5) with proton-proton data is shown in Fig. 21, demonstrating a qualitative agreement with the data. On closer inspection, however, the points show a systematic deviation. This led Schrempp and Schrempp[27] to a parameterisation in two components, a smoothly decreasing term and an oscillating term:

$$\frac{d\sigma}{dt}(\theta) \propto e^{-Rg(\theta)p_T}(1 + C(\theta)\cos(Rp_T + \phi)) \qquad (7)$$

where p_T is the transverse momentum, $p_T \simeq \sqrt{tu/s}$, $g(\theta)$ and $C(\theta)$, R and ϕ are four adjustable constants, the first two depending on the angle. The fit to the oscillating part is shown in Fig. 22 a,b for $\theta = 90°$ and $\theta = 68°$, respectively. The pronounced effect leads to R = 0.73f, a typically strong interaction radius. It can be concluded that diffractive effects are still felt at $t \simeq 10$ GeV/c^2. Whether scattering at 90° is all diffraction scattering or whether it is just a modulation of a more fundamental parton behaviour, is a question of conviction at this moment.

A comparison of equ (5) with data for other than pp reactions was made by Barger[28].

Fig. 23 shows a comparison of $\bar{p}p$ and pp scattering at large angles, at 5 and 6 GeV/c[29]. The 90° cross section of $\bar{p}p$ is two orders of magnitude lower than that for pp scattering. A convincing argument for this fact has not been given yet.

1.3 <u>Polarisation in elastic scattering</u>

The operation of the polarised proton beam at ANL has made possible the experiments shown in Fig. 24, where the polarisation in pp and pn scattering is compared, at energies from 2 to 6 GeV/c.[30] A pure

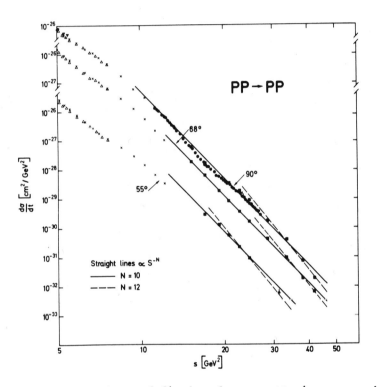

Fig.21. The energy dependence of fixed angle pp scattering compared with a power law.

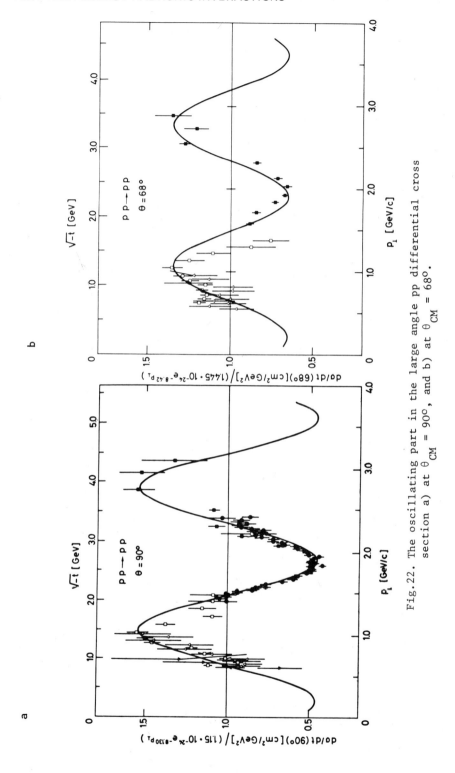

Fig. 22. The oscillating part in the large angle pp differential cross section a) at $\theta_{CM} = 90°$, and b) at $\theta_{CM} = 68°$.

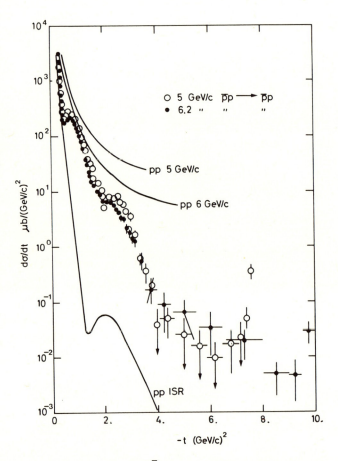

Fig.23. Comparison between pp and p̄p large angle scattering at 5 and 6 GeV/c incident momentum.

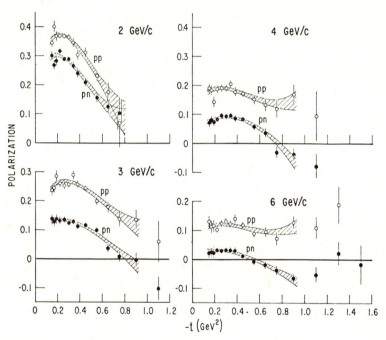

Fig.24. Comparison between the polarization in np and pp scattering at 2, 3, 4 and 6 GeV/c.

isovector exchange would imply that the polarisation in these two reactions is equal but opposite in sign, like the polarisation in $\pi^{\pm} p$ scattering. Isoscalar exchange would have them identical. The data show that neither of these two extreme cases is valid. The data also show that the polarisation in absolute magnitude decreases with energy.

A Saclay, IHEP, Dubna, Moscow collaboration[31] has measured the polarisation parameter P for $\pi^{\pm}$, $K^{\pm}$ and $p^{\pm}$ scattering in protons at 40 or 45 GeV/c. The data is shown in Fig. 25. The $\pi^{\pm} p$ data shows a maximum polarisation of about 5% in magnitude, smaller than at PS energies. The proton data is compared with data at lower energies in Fig. 26. The double zero in the polarisation at 12 GeV/c has changed character at 45 GeV/c, being a single zero now. $K^+ p$ and $\bar{p} p$ have non-zero polarisation but the errors are rather big. In $K^- p$ scattering the polarisation seems to be small. The collaboration is also involved in the measurement of the rotation parameter R.

The polarised proton beam at ANL in combination with a polarised target, has provided results on the Wolfenstein parameter R, D and C_{nn} in pp scattering[32]. Fig. 27 presents another result[33] of this group on the difference between the total cross section for the spins parallel and antiparallel (perpendicular to the beam direction). Whereas at 2 GeV/c the spin-spin amplitudes in the forward direction are seen to be relatively important, their effect seems to decrease rather fast with energy.

Summarizing, polarisation effects are still with us in the 40 GeV/c region, but their magnitude is diminishing with energy.

1.4 Charge exchange scattering

The $\pi^- p \to \pi^0 n$ charge exchange scattering has been studied by a Berkeley-Caltech collaboration in the 20 to 100 GeV/c region.[34] Data at 100 GeV/c are shown in Fig. 28; the total charge exchange cross section is shown in Fig. 29, together with data at lower energies, demonstrating a behaviour

$$\sigma^{CH.EX} \propto P_{LAB}^{-1.15} \qquad (8)$$

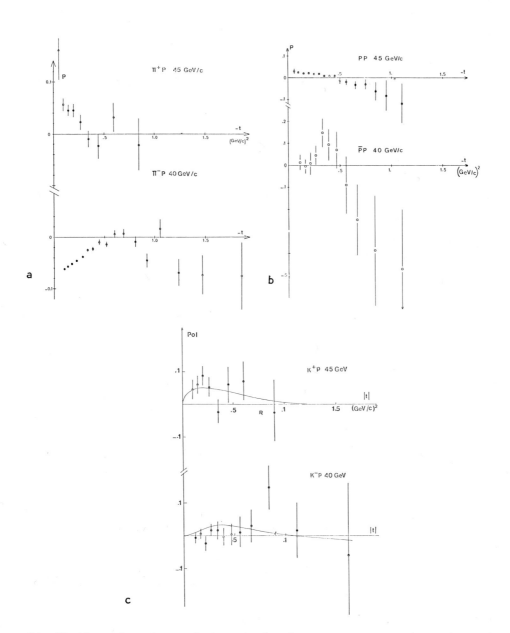

Fig.25. The t-dependence of the polarization parameter P at 40 or 45 GeV/c for a) $p^{\pm}p$, b) $\pi^{\pm}p$ and c) $K^{\pm}p$.

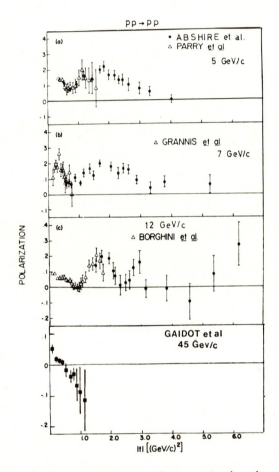

Fig.26. Comparison of the t-dependence of the polarization parameter P at 4 different energies in elastic pp scattering.

VERY HIGH ENERGY HADRONIC INTERACTIONS

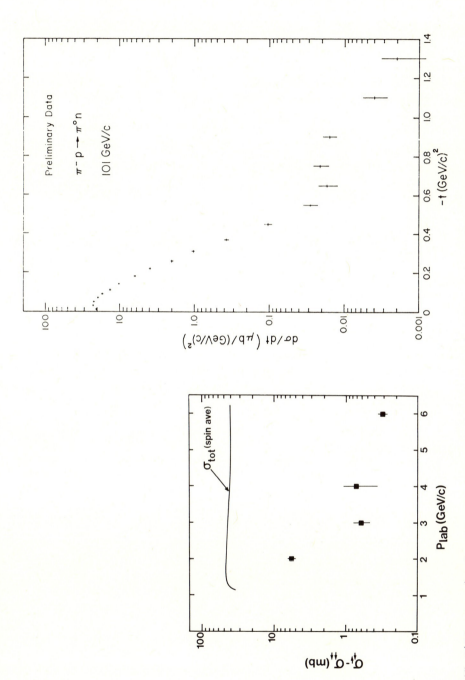

Fig. 27. Difference in the proton-proton total cross section for antiparallel and parallel initial spin states.
Fig. 28. $\pi^- p$ charge exchange scattering at 100 GeV/c incident momentum.

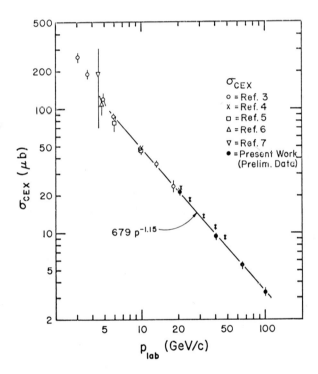

Fig.29. The energy dependence of the total cross section for $\pi^- p$ charge exchange scattering.

A sketch of the behaviour of the charge-exchange process at various energies is shown in Fig. 30; the dip at t = 0 is seen to be rather energy-independent, implying that the ratio of the spin flip and spin-nonflip amplitudes is energy independent in the forward direction. The energy dependence of the cross section at t = 0 and large t is, however, very different, and can be described by the Regge formula

$$\frac{d\sigma}{dt} = \beta |t| \left(\frac{s}{s_0}\right)^{2\alpha(t)-2} \tag{9}$$

with $\alpha(t)$ being the parameterisation of the rho-trajectory, which also explains the dip in the cross section at $|t| \simeq 0.6$ GeV/c². The rho-trajectory thus determined is shown in Fig. 31; the experimental points are near to the line drawn through the ρ and g poles.

The charge exchange cross section at t = 0 and the difference of the $\pi^\pm p$ total cross sections, $\Delta\sigma$, are related by[35]

$$\frac{d\sigma^{CH.EX}}{dt}(t=0) = \frac{\pi}{p_{LAB}^2} (1+\rho^2) (\Delta\sigma)^2 \tag{10}$$

with ρ being the ratio of the real and the imaginary scattering amplitude for charge exchange scattering. If the rho-intercept at t = 0 is 0.5, then $\rho = 1$. The comparison of the left hand side (full line) and the right hand side of equ (10) is shown in Fig. 32, demonstrating a reasonable agreement, especially in view of the difficulties provided by the charge exchange dip at t = 0.

The regeneration amplitude for K_s by K_L, f_{21}/k and the regeneration phase ϕ, are related to the difference of the $K^\pm p$ total cross sections by

$$\Delta\sigma = -4\pi \frac{2|f_{21}|}{k} \sin\phi \tag{11}$$

The data for a comparison are provided by Albrecht et al.[36] and are shown in Fig. 33. Obviously the agreement is very good.

In conclusion, it seems that the single Regge-pole theory works very well for these exchange reactions.

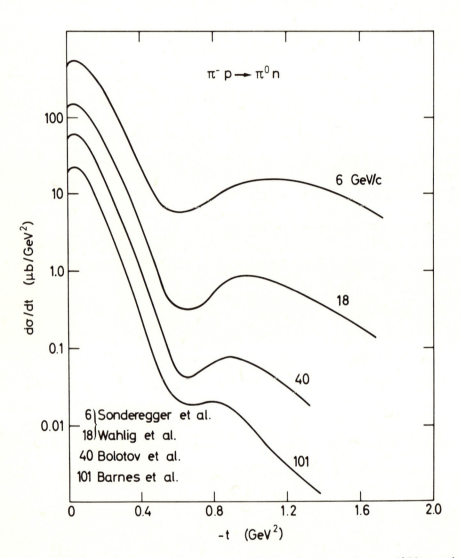

Fig.30. Sketch of the energy dependence of the charge exchange differential cross section.

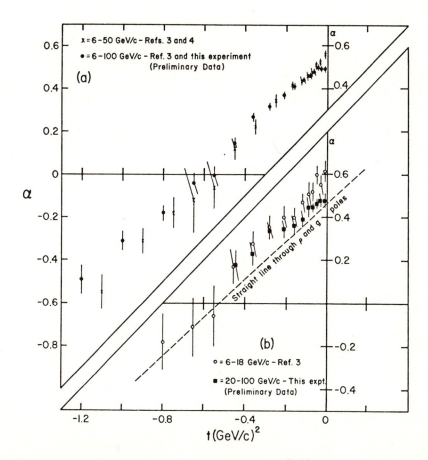

Fig.31. The effective trajectory derived from $\pi^- p \to \pi^0 n$. Data from the energy ranges of the CERN PS, Serpukhov and FNAL are combined in various ways.

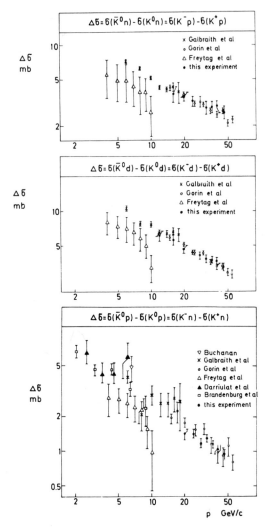

Fig.32. Comparison of the left and right hand side of equ. (10).

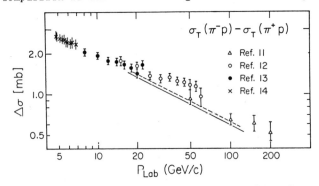

Fig.33. Comparison of the left and right hand side of equ. (11).

2. MANY BODY FINAL STATES

2.1 Diffraction Dissociation

Diffraction dissociation is the process

$$A + B = A' + X \qquad (12)$$

under restrictive conditions, postulated as such by Good and Walker and by Feinberg and Pomeranchuk.[37] The condition is that the incident wave A in exciting the target B to X changes its four momentum and quantum numbers so little that the outgoing wave A' is coherent with A. Introducing the Feynman variable x, this condition, viz, that the frequency change of the incident wave times the time to traverse the target is much smaller than one, can be written in natural units as

$$(1-x) M_T \frac{1}{m_\pi} \ll 1 \qquad (13)$$

with M_T and m_π being the mass of the target and the pion, respectively. For a proton or a deuteron target, equ (13) thus reads

$$(1-x) \ll 0.14 \qquad (14)$$

The relation between M_X, the mass of X, and x is given by

$$M_X^2 \simeq s(1-x) \qquad (15)$$

Assuming that diffraction dissociation takes place up to a maximum value of 1-x, satisfying equ (14), it is clear that the maximum value of M_X, that can be produced by this process, increases linearly with s.

At PS and AGS energies much effort[38,39] has been devoted to the reaction

$$P + P \rightarrow P + X \qquad (16)$$

where it was found that a number of nucleon isobars are excited in the mass spectrum of X. As an example[38] Fig. 34 is shown, which demonstrates also the mild energy dependence of this process.

Two experiments have been performed at FNAL, one by a Stoneybrook-Columbia collaboration[40], on a proton target, and the other by a Dubna-FNAL-Rockefeller-Rochester collaboration[41], on a deuteron target. In

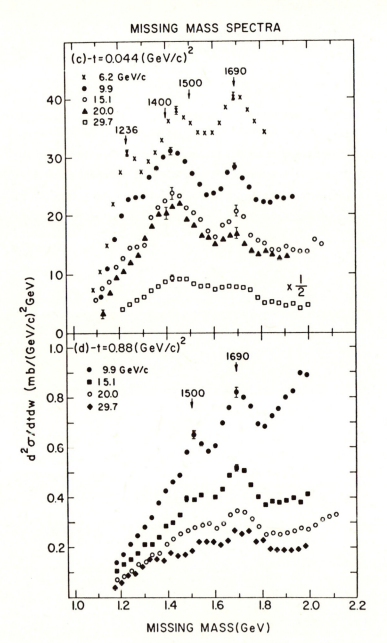

Fig.34. Comparison of the differential cross section for the reaction pp → pX, with X in the region of the nucleon resonances, at various energies.

both cases the recoil particle is detected and the projectile proton is excited to X, the reaction thus being

$$p + p \to X + p$$
$$p + d \to X + d \quad (17)$$

The excitation spectrum in the resonance region is shown in Fig. 35, at 100 and 275 GeV/c, and $|t| = 0.025$ GeV^2. The data from the proton and the deuteron target coincide, once the deuteron data are corrected by the deuteron coherence factor $F_D(t)$, which takes care of the chance that the deuteron dissociates in receiving a kick $\sqrt{|t|}$. This factor $F_D(t)$ is essentially the square of the deuteron form factor as measured in elastic pd scattering. Fig. 35 demonstrates that also in the FNAL energy range nucleon resonances are excited. The energy dependence of the nucleon isobar excitation, shown in Fig. 36, is seen to remain rather weak; the 20 GeV/c data in this figure is from an AGS experiment.[38]

A larger range of the excitation spectrum is shown in Fig. 37, at 275 GeV/c and two values of t; the curves drawn through the points are $\propto M_X^{-2}$. Theoretical models based on the triple-Regge analysis expect such a behaviour;[42] integration of a differential spectrum $\propto M_X^{-2}$ up to a fixed value of s would result in a cross section rising logarithmically with s, if no other energy dependences are present.

The facts about the large M_X^2 part of the spectrum are the following: the Stoneybrook-Columbia group[40] performed their experiment in the momentum range 150 to 400 GeV/c, $1-x \leqslant 0.13$ and $0.02 \leqslant |t| \leqslant 0.2$ GeV^2. The results are i) the double differential cross section $d^2\sigma/dtdM_X^2 \propto M^{-\alpha}$ with $\alpha \simeq 3$ to 4; ii) the integrated cross section for excitation of a single nucleon, σ_{SD}, is constant to within about $\pm 5\%$; for $0 \leqslant M_X^2 \leqslant 0.06S$ one finds $\sigma_{SD} \simeq 2.5$ mb and for $0 \leqslant M_X^2 \propto 0.1S$ σ_{SD} 3.2 mb.

The Dubna-FNAL-Rockefeller-Rochester collaboration[41], on the d-target, also measured between 150 and 400 GeV/c, with $M_X^2 \leqslant 0.03S$ and $M_X^2 \geqslant 5$ GeV^2, $0.03 \leqslant |t| \leqslant 0.12$ GeV/c^2. Their data can be parameterised as

$$\frac{d\sigma}{dtdM_X^2} = a(s,M_X^2) \exp\{-b(s,M_X^2)|t| + ct^2\} \quad (18)$$

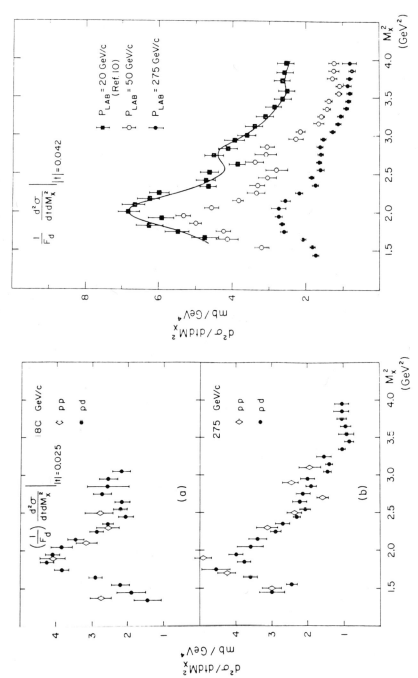

Fig. 35. Comparison of the nuclear resonance production in pp → Xp and pd → Xd at FNAL energies.
Fig. 36. Comparison of the differential cross section for the reaction pd → Xd, with X in the region of the nucleon resonances, at various energies.

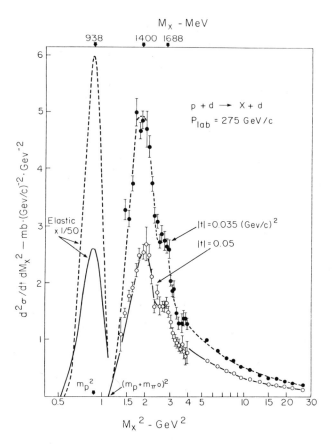

Fig.37. Dependence on M_X^2 of the reaction pd → Xd at 275 GeV/c and two values of the four-momentum transfer. The line through the data points at large M_X^2 is $\propto M_X^{-2}$.

The parameter c is taken from the deuteron form factor; $b(s,M_X^2)$ is shown in Fig. 38a, and is thus found to be a constant, independent of s and M_X^2; the parameter $a(s,M_X^2)$ is with high accuracy $\propto M_X^{-2}$ and in addition has a slight energy dependence (Fig. 38b). Thus equ (18) is transformed into

$$\frac{d^2\sigma}{dt dM_X^2} = \{\frac{A}{M_X^2}(1+\frac{C}{P_{LAB}})\exp(-B|t|)\} F_d(t) \qquad (19)$$

The term between brackets is the proton dissociation term. The values of the constants are $A = 3.5 \pm 0.2$, $C = 54 \pm 16$, $B = 6.5 \pm 0.3$ in the usual units. Integration leads to a constant cross section in the NAL energy range, $\sigma_{SD} \simeq 3$ mb.

The M_X dependences of the two groups are not necessarily in contradiction because of the very different ranges of M_X^2 used in the analyses.

Future experiments at the ISR should give additional information on the energy dependence of the quasi-elastic process of diffraction dissociation.

2.2 Low multiplicity exclusive channels

Diffraction dissociation being a low-multiplicity phenomenon, the question can be asked whether reactions such as

$$pp \to pp\pi^+\pi^-$$
$$\pi^-p \to \pi^-p\pi^+\pi^- \qquad (20)$$
$$pp \to p(n\pi^+)$$

have diffractive characteristics, like a constant cross section, or instead have an asymptotically vanishing cross section, like $S^{-\alpha}$.

A summary of what is known[43-46] in this respect is shown in Fig. 39. The first two reactions seem to give evidence for a levelling off; the last one not yet, within its rather large errors.

VERY HIGH ENERGY HADRONIC INTERACTIONS

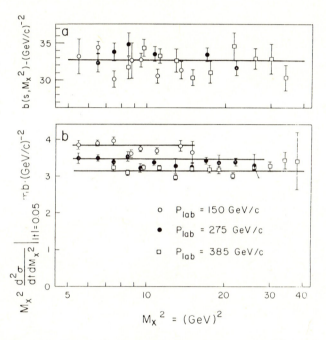

Fig.38a. Dependence of the slope parameter b on s and M_X^2 (equ. 18).
b. Demonstration that the parameter $a(s,M_X^2)$ in equ. (18) is $\propto M_X^{-2}$.

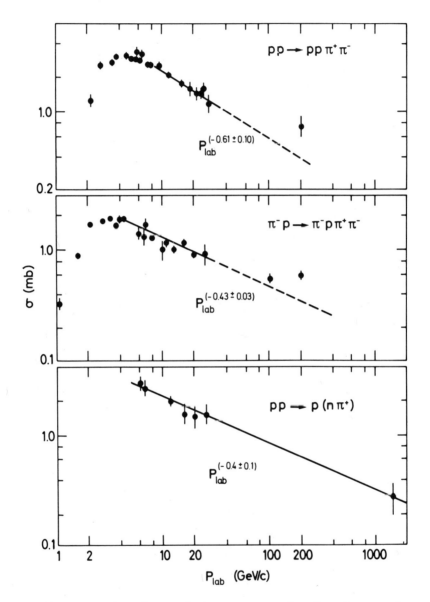

Fig.39. Energy dependence of the cross section for the reactions
a) pp → ppπ⁺π⁻, b) π⁻p → π⁻pπ⁺π⁻, and c) pp → pnπ⁺.

2.3 Characteristics of many-particle production

Diligent studies by many people have taught us a number of properties of many-particle final states. These can be summarised as follows:[47]

i) Pions are the most abundantely produced particles, making up a fraction of about 90% of the produced particles.

ii) The multiplicity grows slowly with energy, perhaps logarithmically. The multiplicity distribution follows some bell-shaped curve.

iii) The average transverse momentum is limited, $<P_T> \leqslant 0.4$ GeV/c.

iv) The transverse momentum distribution has a tail with high momentum components, increasing in importance with energy.

v) A leading particle effect exists, the two incident particles not losing their identity in a collision process but on the average preserving about 50% of their energy.

vi) The produced particles form a plateau in rapidity space, the length of which increases logarithmically with energy.

vii) The more abundantly produced particles show approximately Feynman sealing in the fragmentation region; on the central plateau the violation of the rule could be larger than 10% in the ISR range of energies.

viii) Correlation studies have shown the existence of short range order: particles are produced in clusters.

Some recent results of correlation studies will be treated in the next sections.

2.4 Some results from recent correlation studies

2.4.1 Correlations in the central region

An experiment has been performed in the 30-inch hydrogen bubble chamber at FNAL, studying $\pi^-\pi^-$, $\pi^+\pi^+$ and $\pi^-\pi^+$ correlations, in 200 and 300 GeV/c pp collisions and 100 GeV/c π^-p collisions.[48] The multiplicity of the events studied is limited to $6 \leqslant n \leqslant 14$, in order to avoid complications due to diffraction dissociation in the low multiplicity events and due to phase space limitations in the high multiplicity events. The semi-inclusive correlation formation between a charged particle at rapidity y_1 and another at y_2 in an event with n charged particles is defined as

$$C_n(1,2) = \frac{1}{n(n-1)\sigma_n} \frac{d^2\sigma_n}{dy_1 dy_2} - \frac{1}{n\sigma_n} \frac{d\sigma_n}{dy_1} \cdot \frac{1}{n\sigma_n} \frac{d\sigma_n}{dy_2} \qquad (21)$$

and similarly, although different in detail, for the (--), (++) and (-+) combinations. This correlation function is zero in the absence of correlations. According to the independent cluster model $C_n(y_1 = y_2 = 0)$ behaves as $\simeq n^{-1}$; this seems supported by the data. The data of different multiplicities can therefore be averaged by considering instead of C_n the quantity

$$\langle J_n C_n(\Delta y) \rangle = \frac{\Sigma J_n \cdot C_n(\Delta y)\sigma_n}{\Sigma \sigma_n} \qquad (22)$$

with J_n = n-1; analogous definitions apply for the other three particular charge combinations.

This averaged correlation function has been plotted in Fig. 40 as a function of the rapidity difference $\Delta y = y_1 - y_2$ with $|y_2| \leqslant 0.5$, and of the azimuthal difference $\Delta\phi$. For the first time one may note the difference in the behaviour for the various charge combinations; the like pairs of pions have a strong short range correlation at $\Delta\phi = 0$, the unlike pions have a broader and smaller correlation at all ϕ, but especially at $\phi = \pi$. The effective mass plots do not show evidence of ρ or any other resonance production. The authors contribute the strong correlation of like pion pairs partly to Bose-Einstein effects, but mention as additional possibilities the production of A_2-type clusters, with high transverse momentum, and (--++) neutral cluster production.

2.4.2 Correlation studies with one particle in the fragmentation region

The CERN-Roma-Pisa-Stoneybrook collaboration[49] studied the correlation between an identified negative particle with measured momentum,

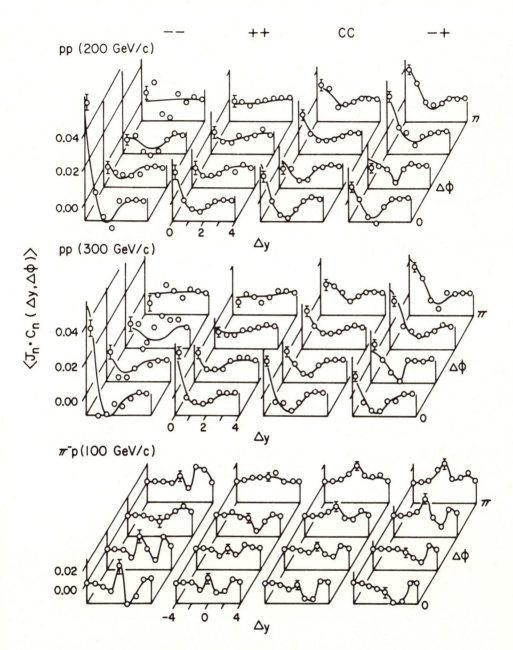

Fig.40. Dependence of the average correlation, as defined in equ. (22), for $\pi^+\pi^+$, $\pi^-\pi^-$ and $\pi^+\pi^-$ on the difference in rapidity and azimuthal angle between the two particles; $|y_1| \leqslant 0.5$.

produced at $0°$, and the other n-1 charged particles, produced at ISR energies:

$$p + p \rightarrow \begin{Bmatrix} \pi^- \\ K^- \\ \bar{p} \end{Bmatrix} + \text{n-1 charged particles} \qquad (23)$$

The negative particle has a relatively high momentum, $x \geqslant 0.4$, thus being in the fragmentation region of one of the incident protons. The charged particles have only their angle measured, and are thus defined by the pseudo-rapidity $\eta = \log \text{tg} \tfrac{1}{2}\theta$; they are accepted over a solid angle of nearly 4π. The correlation function is defined as

$$R(\eta,y) = (\sigma_{IN} \frac{d\sigma}{d\eta dy} / \frac{d\sigma}{d\eta} \frac{d\sigma}{dy}) - 1 \qquad (24)$$

where y is the rapidity of the detected negative particle and σ_{IN} the inelastic cross section.

The correlation function for the detected particle being a π^- or a K^- is shown in Fig. 41 for various values of its momentum; the detected particle is on the right side ($\eta = +\infty$). It is clear from this figure that a cluster of particles accompanies the negative detected particles; in addition the π^- case shows also an enhancement on the opposite side. Fig. 42 is for $x = 0.4$ only and demonstrates that also antiprotons are accompanied by a cluster. In this figure the data is further subdivided into various multiplicity classes. Concentrating first on the π^- particle, it is obvious that the cluster formation for π^- is a low multiplicity phenomenon. Monte-Carlo simulation studies and investigation of the $pp\pi^+\pi^-$ final state show that the cluster mass in about 2 GeV, and that the cluster angular distribution has a slope and a slope dependence on the cluster mass that characterise it as a diffractive phenomenon; Fig. 43 shows that π^- detection enhances the fraction of low multiplicity events. In conclusion, a π^- in the fragmentation region signals a diffractively produced leading cluster (baryon number + 1).

Although the kaons and antiprotons are also accompanied by a cluster, closer inspection nevertheless points to a number of differences: these

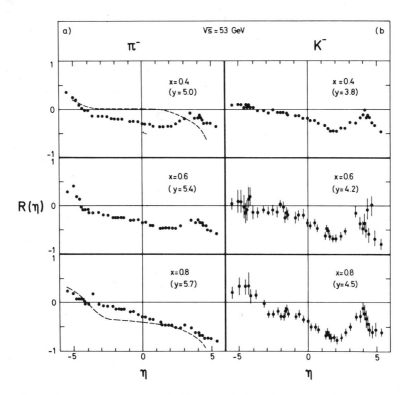

Fig.41. Dependence of the correlation as defined in equ. (24), between a fast forward π^- or K^- and the other charge particles, on the pseudo rapidity of the latter ones. The fast particle is detected on the right hand side at $0°$ ($\eta = +\infty$) and is produced in pp collision at ISR energies.

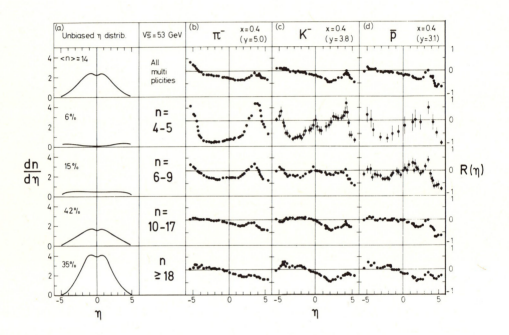

Fig.42. The figures in the top row of columns b) and c) are the same as in the top row of Fig. 41; column d, top row, shows similar data for a fast forward $\bar{p}$. The following row shows the same data for various subclasses of multiplicity. Column a) shows the uncorrelated spectrum.

VERY HIGH ENERGY HADRONIC INTERACTIONS

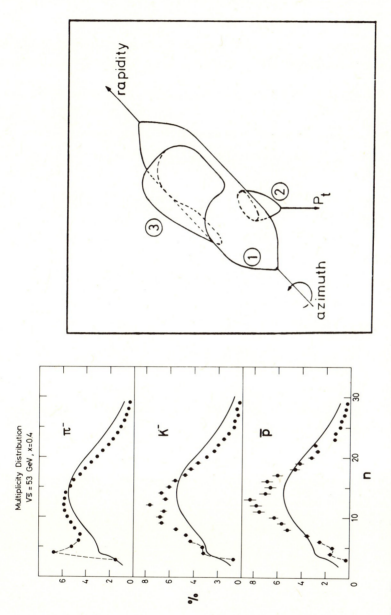

Fig.43. Comparison of the unbiased multiplicity distribution (full line) with the distribution when a fast forward π^-, K^- or $\bar{p}$ is selected, in pp collisions at ISR energies.

Fig.44. Pictorial representation of the correlations when a π^0 with high transverse momentum at 90° CM is selected in pp collisions at ISR energies.

clusters are not a low-multiplicity phenomenon; particle production in the central region remains strong, whereas there is a depletion in the case of the $\bar{\pi}$; the enhancement in the opposite hemisphere is absent. Thus kaons and antiprotons seem much less accompanied by leading clusters with diffractive properties. Instead they show characteristics of more inelastic, central collisions.

2.4.3 Correlations in events with a large transverse momentum particle

This field, recently reviewed by Darriulat[50], has been pioneered at the ISR by several groups, the Pisa-Stoneybrook[51] collaboration with their 4π charged particle detector, the Aachen-CERN-Heidelberg-München[52] collaboration with a streamer chamber (also 4π coverage), the CERN-Rockefeller[53] and CERN-Columbia-Rockefeller-Saclay combination[54] with π^0 detection (direction and energy) over two opposite fractional solid angles, with added coverage for charged particles. Usually a high transverse momentum π^0 ($P_T \geqslant 3$ GeV/c), produced at $90°$, served as trigger. This effort has led to distinguishing three components in the reaction products of such an interaction, shown pictorially in Fig. 44. The experimental evidence for this picture is shown in Fig. 45, where the correlated charged particle density is shown in six intervals of the azimuthal angle, as a function of the pseudo rapidity; the π^0 ($p_T \geqslant 2$ GeV/c) is detected at $90°$; the full line shows the particle density without the requirement of the π^0.

The three components have the following characteristics:

i) the first component has all the properties of a normal non-diffractive interaction. The multiplicity in this component is, however, reduced and is equal to that of an interaction at an effective energy

$$\sqrt{s}_{eff} \leqslant \sqrt{s} - 2p_T$$

Correlation studies of two pions in this first component do not show any influence of the fact that a high p_T π^0 has been produced as well in the interaction.

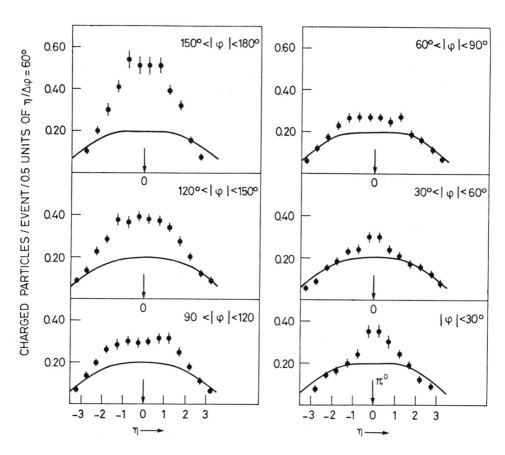

Fig.45. Dependence of the production of charged particles in pp collisions at ISR energies on their pseudo-rapidity, for various intervals of the azimuth. A π^0 with high transverse momentum is detected at 90° CM ($\eta = 0$); the full line shows the unbiased distribution.

ii) the second component is formed by a cluster of particles accompanying the high p_T π^0. It contains on the average about one charged particle and one gamma ray, spread over an internal $\Delta y = \pm 1.5$, $\Delta \phi = \pm 60°$. The probability of finding another high p_T pion in this cluster is enhanced and increases with s; the enhancement is about 2 orders of magnitude at $p_T = 3$ GeV/c at the highest ISR energies.

iii) the third component is opposite the high p_T π^0. It covers about $\Delta y = \pm 2$, $\Delta \phi = \pm 90°$. The multiplicity in this component increases with the transverse momentum of the trigger. The probability of finding a high p_T particle in it is also enhanced by a comparable factor, as in the second component, but independent of s this time.

In parton models[55] a high transverse momentum particle is thought to signal the fundamental parton-parton interaction and two opposite jets of particles are supposed to emerge. Before identifying the second and third components with these two jets it should be realised that simple arguments of momentum conservation can explain a large part of the characteristics of the third component. The cluster of particles in the second component that accompanies the particle with high transverse momentum might contain information of more physical interest. Perhaps nature is giving us a signal here, waiting to be deciphered.

References

1) A.S. Carroll et al., to be published. For previous work see A.S. Carroll et al., Phys. Rev. Letters 33 (1974) 928 and 932.

2) M.J. Longo et al., Phys. Rev. Letters 33 (1974) 725.
   P.V. Ramana Murthy et al., Nucl. Phys. B92 (1975) 269.

3) L.B. Okun and I.Ia. Pomeranchuk, Zh. Eksperim i Teoret Fíz 30 (1956) 307; Soviet Phys. JETP (English translation) 3 (1956) 307.

4) I.Ia. Pomeranchuk, Zh. Eksperim. i Teoret. Fiz. 34 (1958) 725; Soviet Phys. JETP (English transl.) 7 (1958) 499.

5) A.M. Wetherell, Rapporteur talk at the International Conference on High Energy Physics, Palermo (Italy), 1975.

6) H. Cheng, J.K. Walker and T.T. Wu, Phys. Letters 44B (1973) 97 and 283.

7) S. Frautschi and G. Margolis, Nuovo Cim. 56A (1968) 1155.

8) N.W. Dean, Phys. Rev. 182 (1969) 1695.

9) G.F. Chew and D.R. Snider, Phys. Letters 31B (1970) 75.

10) A.N. Diddens, Proc. 17th Int. Conf. on High Energy Physics, London (1974), p. I-41.

11) H. Lipkin, "New Plots for Energy Dependence of Hadron Total Cross Sections".

12) D.R.O. Morrison, preprint CERN/D.Ph.II/Phys. 74-30 (1974), submitted to Nuclear Physics B.

13) U. Amaldi, Proc. II International Conference on Elementary Particles (Aix-en-Provence), Journal de Physique, Tome 34, C1-1973, p.241.

14) G. Giacomelli, Proc.16th Int. Conference on High Energy Physics, Chicago-Batavia (1972), Vol. 3, p. 219.

15) V. Barger and R.J.N. Phillips, preprint Univ. of Wisconsin COO-881-445 (1975).

16) J. Soffer and D. Wray, Phys. Letters 43B (1973) 514.

17) N.N. Khuri and T. Kinoshita, Phys. Rev. 137B (1965) 720 and 140B (1965) 706.

18) C. Ankenbrandt et al., to be published (Fermilab-Conf-75/61-EXP).

19) R.E. Hendrick and B. Lautrup, Phys. Rev. $\underline{11D}$ (1975) 529.

20) G. Barbiellini et al., Phys. Letters $\underline{39B}$ (1972) 663.

21) D. Birnbaum et al., to be published. See also Ref. 10.

22) Fermilab Single Arm Spectrometer Group, submitted to Phys. Rev. Lett. (Fermilab-PUB-75/48-EXP).

23) N. Kwak et al., Physics Letters $\underline{50B}$ (1975) 233.

24) A.J. Buras and J. Dias de Deus, Nucl. Phys. $\underline{B71}$ (1974) 401.

25) C.W. Akerlof et al., to be published (Fermilab-Pub-75/69-EXP; ANL-HEP pp. 74-43).

26) V.A. Matveev, R.M. Muradyan and A.N. Tavkelidze, Nuovo Cimento Letters $\underline{5}$ (1972) 907.
S.J. Brodsky and G.R. Farrar, Phys. Rev. Letters $\underline{31}$ (1973) 1153; Phys. Rev. $\underline{11D}$ (1975) 1309.

27) B. Schrempp and F. Schrempp, Phys. Letters $\underline{55B}$ (1975) 303.

28) V. Barger, Proc. 17th Int. Conf. on High Energy Physics, London (1974), p. I-193.

29) A. Lundby, Symposium on Antinucleon-Nucleon Interactions, Liblice-Prague (1974) Cern 74-18, p. 383

30) R. Diebold et al., Physics Rev. Letters $\underline{35}$ (1975) 632.

31) C. Bruneton et al., Phys. Letters $\underline{B44}$ (1973) 471; F. Lehar et al., Proc. 17th Int. Conf. on High Energy Physics, London (1974), P.I-27; A. Gaidot et al., paper submitted to International Conference on High Energy Physics, Palermo (Italy) 1975. See also Ref. 5. A. Gaidot et al., Physics Letters $\underline{57B}$ (1975) 389.

32) J.R. O'Fallon et al., Phys. Rev. Letters $\underline{32}$ (1974) 77. See also Ref. 10.

33) W. de Boer et al., Phys. Rev. Letters $\underline{34}$ (1975) 558.

34) A.V. Barnes et al., Proc. 17th Int. Conf. on High Energy Physics, London (1974) P.I-37. See also Ref. 10 and Ref. 28.

35) R.J. Eden, "High Energy Collisions of Elementary Particles" Cambridge University Press, 1967.

36) V.K. Birulev et al., Phys. Lett. $\underline{38B}$ (1972) 452; K.F. Albrecht et al., Phys. Lett. $\underline{48B}$ (1974) 257 and Ref. 10.

37) L. Good and W.D. Walker, Phys. Rev. 120 (1960) 1857.
    E.L. Feinberg and I. Pomeranchuk, Suppl. Nuovo Cim. 3 (1956) 652.

38) R.M. Edelstein et al., Phys. Rev. D5 (1972) 1073.

39) J.V. Allaby et al., Nucl. Phys. B52 (1973) 316.

40) R.D. Schamberger et al., Phys. Rev. Letters 34 (1975) 1121.

41) Y. Akimov et al., Phys. Rev. Letters 35 (1975) 763, 766.

42) See for instance:
    D.P. Roy and R.C. Roberts, Nucl. Physics B77 (1974) 240.
    R.D. Field and G.C. Fox, Nucl. Phys. B80 (1974) 367.

43) M. Derrick et al., Phys. Rev. 9D (1974) 1215.

44) E.L. Berger et al., Preprint CERN/D.PH.II/Phys. 74-26 (1974).

45) H.H. Bingham et al., Preprint LBL-2460 (1974).

46) E. Nagy et al., Proc. 17th Int. Conf. on High Energy Physics, London (1974) p.I-13. See also Ref. 10.

47) See for instance:
    K. Zalewski, Proc. 17th Int. Conf. on High Energy Physics, London (1974) p. I-93.
    P.V. Landshoff, idem p. V-57.

48) B.Y. Oh et al., Physics Letters 56B (1975) 400. See also J. Ranft and G. Ranft, Physics Letters 57B (1975) 373.

49) U. Amaldi et al., Physics Letters 58B (1975) 206, 213

50) P. Darriulat, Rapporteur talk at the International Conference on High Energy Physics, Palermo (Italy) 1975.

51) G. Finocchiaro et al., Physics Letters 50B (1974) 396.
    R. Kephart et al., to be published. See also P.V. Landshoff, Proc. 17th Int. Conf. on High Energy Physics, London (1974), p.V-57.

52) K. Eggert et al., submitted to Nuclear Physics B. See also Ref. 50.

53) F.W. Büsser et al., Physics Letters 51B (1974) 306, 311.

54) F.W. Büsser et al., Physics Letters 55B (1975) 232.

55) S.M. Berman, J.D. Bjorken, J.B. Kogut, Phys. Rev. $\underline{D4}$ (1971) 3388
R. Blankenbeckler, S. Brodsky and J. Gunion, Phys. Letters $\underline{42B}$ (1972) 461, and Phys. Rev. $\underline{8D}$ (1973) 287.
For a general review, see: S.D. Ellis, Proc. 17th Int. Conf. on High Energy Physics, London (1974), p.V-23, R. Savit, SLAC-PUB-1324 (T/E) 1973.

56) K. Goulianos, Preprint Rockefeller University COO-2232A-21 (1975).

DISCUSSION

CHAIRMAN: A.N. Diddens

Scientific Secretaries: D. Johnson and H. Kirk

DISCUSSION No. 1

- *RÜCKL:*

What was the conclusion from the pp experiment using a polarized beam and polarized target? What can one generally learn from such experiments?

- *DIDDENS:*

The conclusion is that the relative importance of spin-spin effects at t = 0 is diminishing with energy. In general, such experiments are needed to make a complete amplitude analysis possible.

- *WEISSKOPF:*

I would be interested in seeing the data again for $d\sigma/dt$ plotted for fixed angles versus s in connection with the parton prescription offered by Brodsky and Farrar.

- *DIDDENS:*

The data I showed were for $d\sigma/dt$ in pp elastic scattering at fixed angles around $90°$ versus s, and comparing the resulting distribution with a power law s^{-10} fall-off after the ideas of Brodsky and Farrar. In their model, the cross-section should show a power-law behaviour like s^{-n} or

$$d\sigma/dt\,(\theta) \alpha\, s^{-n},$$

where $n = \Sigma$ partons $- 2$. In this formula, n represents the sum of partons in the initial and final state. Hence, for example

Reaction	n
$pp \rightarrow pp$, $\bar{p}p \rightarrow \bar{p}p$	12 − 2 = 10
$\pi p \rightarrow \pi p$	8
$\gamma p \rightarrow \pi^+ n$	7
$pp \rightarrow pN^*$	10 .

Barger, in his review talk at the 1974 London Conference, made a compilation for many reactions, showing that often the agreement with the model is quite good.

- *WEISSKOPF:*

Does not $\gamma p \to \gamma p$ with $n = 6$ violate meson dominance?

- *DIDDENS:*

Yes, I agree. I do not know how to answer that.

- *WEISSKOPF:*

I noticed that there is not such a good agreement for this power law behaviour in the plots you showed for

$$\gamma p \to \pi^+ n$$
$$\to \pi^- \Delta^{++}$$
$$\to \pi^0 p \ .$$

If you multiply $d\sigma/dt$ by s^8 instead of s^7, what would have happened? Is the agreement better or worse?

- *DIDDENS:*

I have not tried another power, but judging from Barger's figure, I would guess the agreement would become worse. One further remark should be made about the Brodsky-Farrar approach. The form factors for baryons and pions are predicted to follow a power law in t. For baryons, it is predicted to go as t^{-2}, for the pion, as t^{-1}. I remind you that the experimental data on the proton form factor can be very well fitted with the dipole formula, the pion data with a monopole formula, thus in agreement with the expectation.

- *SUNDERMEYER:*

Do you know if Regge phenomenologists have been able to describe the polarization data for πp and, in particular, whether the question of weak or strong cuts has been decided?

- *DIDDENS:*

As far as I know, this question is still open. I know of no particular progress in this direction.

- *TRIANTIS:*

You showed that the geometrical picture could be used to explain the shift of the dip in the pp differential cross-section around $t = -1.2$ GeV/c^2 and that the same picture is also useful in relating cross-section, slopes, etc. It is not obvious that one cannot also use the same picture for the charge exchange cross-sections. Is there any special reason why this geometrical picture should be better for pp than for charge exchange?

DISCUSSION

- *DIDDENS:*

If our present interpretation is correct in that the dip in charge exchange is due to the ρ-trajectory crossing α = 0 at t = -0.6, then this dip should remain fixed. To my knowledge, it is quite well fixed. The geometrical scaling picture, in a sense, has not much theoretical basis. It is purely heuristic; it seems to work, so we have used it. In that sense, therefore, there is no contradiction with the πp charge exchange data.

- *ZICHICHI:*

One should add that in the previous case, when you speak about optics, the exchanged object is most likely the Pomeron. Hence, when you exchange a Pomeron, you get optical diffraction scattering. In πp charge exchange, you exchange the ρ meson trajectory.

- *JENSEN:*

I noticed a bump, or at least a non-zero polarization, for the $\bar{p}p$ elastic cross-section around 40 GeV/c at approximately the same position in -t, for which the pp elastic polarization passed through zero. Could you comment on the significance of this result?

- *DIDDENS:*

I think all one can really say about the $\bar{p}p$ polarization data is that there is about a 3 standard deviation non-zero value for the region of -t you speak of. One should note the very large errors and the fact that even at low energy, the $\bar{p}p$ polarization data is relatively of poor quality.

- *KURODA:*

What is the method used to measure ρ, the real to imaginary ratio of scattering amplitude, away from the forward direction? I believe this measurement is important because the dip in large angle elastic pp scattering is deepest around 200 GeV/c, while the real part of the forward scattering amplitude does not vanish until around 300 GeV/c, so there is some discrepancy. One might expect that the dip is deepest at the energy where the real part vanishes. Hence, the measurement of the real part at non-forward angles, if possible, is very useful in clarifying the situation.

- *DIDDENS:*

About the method, let us consider a simpler example, say, πp scattering. One performs a complete set of measurements of dσ/dt, polarization, of all angles, including zero, so that one can do a complete amplitude analysis. You find from Coulomb interference near the forward direction, the absolute phase at t = 0, and you follow this out to larger values of t in the amplitude analyses. This program has been performed in the 5 GeV/c region [   ]. In pp, of course, you have many more amplitudes to measure, so it is not as simple any more.

— KURODA:

Are there any theoretical predictions of ρ from dispersion relations at $t \neq 0$?

— DIDDENS:

Only at $t = 0$, the total cross-section measurements and the optical theorem give us a pure imaginary scattering amplitude to use as input to the dispersion relation to calculate the real part. Away from $t = 0$, a reasonable part of the differential cross-section can come from other spin dependent amplitudes.

DISCUSSION No. 2

— JOHNSON:

In the pp experiment at 300 GeV/c, why is it that the $\pi\pi$ correlation function for $\pi^-\pi^-$ is about twice as large as for $\pi^+\pi^+$ at $\Delta y = \Delta \phi = 0$?

— DIDDENS:

It could be a statistics effect. In any case, the authors do not assign any special importance to this effect.

— JOHNSON:

This effect is seen also at 200 GeV/c, where it is even more striking. It seems to me that there should be no difference between a $\pi^-\pi^-$ and $\pi^+\pi^+$ correlation in the region.

— WEISSKOPF:

It can be understood perhaps. In pp collisions you have a surplus of positive charges. An excess of positive charges could tend to smear out the correlation in the $\pi^+\pi^+$ case.

— DIDDENS:

Perhaps as well, low energy protons have been misidentified as π^+'s.

— ZICHICHI:

If anything, this would produce a more enhanced correlation, not less.

— WEISSKOPF:

There is a surplus of positive charge; therefore, there will be a lot of pionization of π^+'s that are statistically distributed, but to make two π^-'s, you really need special processes, at least to a greater extent, and hence there will be more correlation for the $\pi^-\pi^-$.

DISCUSSION

- *ZICHICHI:*

It would be interesting to know which of the two correlations is strongest.

- *WEISSKOPF:*

They do not give us absolute intensities; they give us correlations.

- *MUREI:*

You mentioned three possibilities to explain the difference in the correlation functions between $\pi^+\pi^-$ and $\pi^+\pi^+$. Would you explain the last possibility of cluster formation effects in more detail? In particular, did you assume that the peak in the $\pi^\pm\pi^\pm$ correlation at $\Delta y = \Delta\phi = 0$ can be reproduced without Bose-Einstein statistics, only by postulating the existence of clusters?

- *DIDDENS:*

The authors mention, in addition to Bose-Einstein effects, the production of resonances decaying into 3 charged particles and the production of (--++) neutral clusters. Such clusters would give a contribution to (--), (++), and (-+) charge correlations. Low mass neutral (-+) clusters, which might be more abundant, would contribute mainly to the $\Delta\phi = \pi$ correlation function of (-+).

- *RÜCKL:*

There seems to be no difference at large Δy between $\pi^+\pi^-$, $\pi^-\pi^-$, and $\pi^+\pi^+$ correlations. Is this a kinematic effect?

- *DIDDENS:*

At large Δy, there should be no correlations any more if long range correlations are absent in the sample selected. For large Δy, all correlations should be zero, then.

- *RÜCKL:*

How does this difference in Δy vary as a function of energy?

- *DIDDENS:*

The only data existing of a reasonable accuracy at these energies are what I have shown; namely 200 and 300 GeV/c pp data and 100 GeV/c π^-p data. Hence, any energy dependence must be inferred from this graph.

- *RÜCKL:*

You said that the plateau in the rapidity distribution becomes broader at higher energy like log s. I have heard that the height of the plateau increases at ISR energies. Has this effect been established, or is it only a rumour?

— *DIDDENS:*

To my knowledge, it is established. Various people have seen it, but the errors are still rather large. It is probably not more than a 3 standard deviation effect at present. One must make good luminosity measurements and this is difficult at the ISR, but could be done. To my knowledge, nothing has happened in the past year and the state is still the same as at the London Conference.

— *CREUTZ:*

With regard to the high p_T-triggered π^0 data, how clear is the separation between the same-side bump? It seems that there is an enhancement at every $\Delta\phi$ angle.

— *DIDDENS:*

That is generally true. The peak is quite sharp at small $\Delta\phi$, then seems first to diminish, then to increase again as $\Delta\phi$ increases.

— *TRIANTIS:*

Do the measurements of diffraction dissociation at the ISR involve experimental difficulties similar to elastic measurements at small angles?

— *DIDDENS:*

Elastic scattering measurements at the ISR have been performed down to momenta transfers $|t| \geq 10^{-3}$ $(GeV/c)^2$. Diffraction dissociation measurements have been performed typically down to $|t| \gtrsim 0.2$ $(GeV/c)^2$. In the future, it is hoped to perform diffraction dissociation measurements at the ISR down to $|t| \gtrsim 0.05$ $(GeV/c)^2$.

RESULTS FROM STUDIES OF HIGH ENERGY COSMIC RADIATION

C.B.A. McCusker

Cornell-Sydney University Astronomy Centre,
School of Physics, University of Sydney, 2006. Australia

1. INTRODUCTION

In this paper I will discuss the observations (up to primary energies of $\sim 10^8$ GeV) on high transverse momenta: the evidence for massive long lived particles in cosmic radiation obtained in deep underground experiments, in mountain and airplane altitudes and using cloud and emulsion chambers, in air shower cores and in 'time of flight' experiments at sea level; and the evidence for continuing changes in reaction characteristics even at energies as high as 10^{18} and 10^{19} eV.

2. HIGH TRANSVERSE MOMENTA

Primary cosmic ray particles of energies between say 10^5 and 10^8 GeV interact in the atmosphere to produce what are called extensive air showers [often shortened to EAS]. The total number of particles (N) in any one shower at sea level depends on the primary energy, the nature of the primary particle (to some extent) and on the chance fluctuations in the development of the cascade. For primaries of these energies N would lie in the range 10^4 to 10^8. Some of these particles may be as much as 2.5 kms from the axis of the shower. However, it is the region within a few metres of the axis which yields useful results on high p_T. The useful results come from studies of both the electronic and hadronic components in the cores of showers.

2.1 Electron distributions in shower cores

In some showers the electron density (i.e. the number of electrons per square metre in the shower plane) increases approximately as $\frac{1}{r}$ as one approaches the axis. This increase may continue up to distances as small as can be measured, < 0.5 cms in some cases. These are called single cored showers. In other cases the electron density distribution may have two or more maxima distributed over a few square metres in the centre of the shower. Some times only two maxima occur[1]; in other cases there are so many overlapping maxima that several square metres in the centre of the shower have an almost constant density.

2.2 A comparison of single cored and multi cored showers of the same total size

Single cored and multi cored showers of the same size, N, have markedly different properties quite apart from the shape of their electron distributions near the core. Table 1 shows some of these differences for a sample of 115 showers with $10^5 \leq N \leq 3.3 \times 10^5$ particles which fell on the Sydney 64-scintillator experiment. The single cored showers have a much higher (3.1 times) mean central electron density than the multi cored showers. They have a much higher ratio of Δ_1 to Δ_2; a much higher maximum hadron energy (4.2 times) on average and a much higher ratio of E_1 to E_2. These differences are all explained if the single cored showers are produced by primary protons and the multi cored showers by primary nuclei with $A > 1$.

Table 1

Core type	$<N> \times 10^{-5}$	$<\Delta_1>$	$<\frac{\Delta_1}{\Delta_2}>$	$<E_1(had)>$	$<\frac{E_1(had)}{E_2(had)}>$
Single	2.02	786	2.22	1290 GeV	10.3
Multiple	2.10	254	1.15	309 GeV	2.5

Mean properties of single and multi cored air showers of size around 2×10^5 particles. Δ_1 is maximum number of electrons on any one scintillator. Δ_2 is the second highest number. $E_1(had)$ is the maximum observed hadron energy. E_2 is the second highest.

2.3 Explanation of the multi cored events

If, say, a helium nucleus hits the atmosphere, it breaks up, generally at the first collision with an air nucleus, into its four

STUDIES OF HIGH ENERGY COSMIC RADIATION

constituent nucleons. These then make four independent cascades. In each collision, the nucleon acquires a transverse momentum component so as they move down the atmosphere, the nucleons, on average, separate. At a height generally between 2 and 6 kms above sea level they will produce one or more π^o mesons whose decay γ-rays initiate a cascade which produces one of the peaks in the electron distribution at sea level. The separation of the peaks will be determined by the mean transverse momentum and the p_T distribution which holds for collisions in the energy region from E_o (where E_o is the primary energy) down to the energy of the nucleon at the level at which it produces the π^o meson which gives the electron peak at sea level.

2.4 Comparison of Monte Carlo simulations of EAS and experimental distributions

One can simulate air showers (having a primary particle of any given A and a total primary energy E_o GeV) using the Monte Carlo technique. One needs to put in the characteristics of particle interactions, of course, and this includes both $<p_T>$ and the p_T distribution. When this is done using a Cocconi-Koester-Perkins distribution and a mean of 1 GeV/c for the forward going nucleon or isobar, one gets good agreement with the experimental observations for the smaller showers $[10^4 < N < 10^5$ particles$]^{1)}$ using $E_o = 10^6$ GeV . However, when the simulation is carried out for showers with $E_o = 10^7$ GeV , even showers started by primaries with $A = 64$ produce only single cored showers on an array of 0.5×0.5 m scintillators. Experimentally the situation is exactly the opposite. Instead of all the showers being single cored at the largest size, they are all multi cored. The experimental result is given in Table 2 . The fact that the core separation stays large is very good evidence that the mean transverse momentum is still increasing in the energy region around $10^6 - 10^7$ GeV .

2.5 Determination of p_T for individual showers

Two observations can be made on the electromagnetic cascade in any one core. These are the central density Δ_1 , over some defined area, and the rate at which the density falls off as distance from the axis increases. If $\Delta(r) \propto \dfrac{1}{r^\alpha}$, then this means finding α . Once Δ_1 and α are known, the theory of the electromagnetic cascade gives

Table 2

Shower size in particles	10^5	3.3×10^5	10^6	3.3×10^6
Number of events	132	333	234	59	32
Fraction of single cored showers	0.49	0.32	0.31	0.12	0

Fraction of single cored showers seen on the Sydney 64 scintillator array for five different ranges of shower size.

both the energy of the γ-rays starting it <u>and</u> their height of production above the apparatus (h). We can also find r the distance of this particular peak from the centroid of the electron distribution. Thus we have

$$p_T = \frac{r\, p_L}{h}$$

If r was due only to the transverse momentum of the produced π^o then p_T would be this transverse momentum. However, the distance from the shower axis of the nucleon (which produces the π^o) is also caused by the previous collisions of the nucleon with air nuclei. Hence, p_T measured in this way is some function of the mean transverse momentum at energies from E_o (the primary energy) down to the nuclear energy at the π^o production level. Figure 1 shows the value of $\frac{r\, p_L}{h}$ obtained for the major subcore of a number of showers plotted against the size of the shower. One sees that for shower sizes up to 10^5 particles $\frac{r\, p_L}{h}$ has values around 0.5 GeV/c, but after that it increases and the very large showers give very large values of $\frac{r\, p_L}{h}$.

2.6 <u>The distribution of p_T at very high energies</u>

During a run of the 64 scintillator array from April 1968 to January 1969, 51 showers of size $> 10^6$ particles struck the array. 45 of these

STUDIES OF HIGH ENERGY COSMIC RADIATION

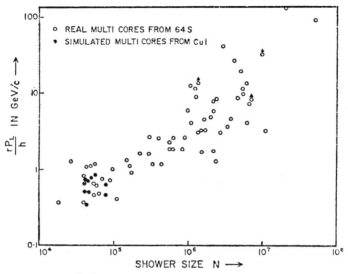

Figure 1 Values of $\frac{r\,p_L}{h}$ (in GeV/c) of the largest subcore plotted against the total number of particles in the shower. Open circles are for real showers. Solid circles are simulated showers of primary energy 10^6 GeV simulated using the 'normal' transverse momenta distribution and mean value.

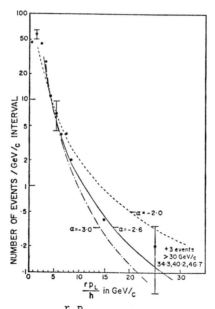

Figure 2 The distribution in $\frac{r\,p_L}{h}$ (in GeV/c) for 215 subcores in 45 large multi cored showers observed with the Sydney 64 scintillator array in 1968. The absence of small values of $\frac{r\,p_L}{h}$ is an instrumental effect. The three curves are power laws with the exponents as shown.

events were multi cored showers and these had 215 measurable subcores. Of these, 168 had $\frac{r \, p_L}{h} > 1$ GeV/c. Three types of subcore were not included in this sample:

a) cores so small that it was difficult to distinguish them from fluctuations of the background,

b) 84 cores so large that they saturated the central scintillator (at $\sim 10,000$ particles/scintillator) and

c) an unknown number of cores that fell off the array. A differential $\frac{r \, p_L}{h}$ spectrum for the measurable subcores is shown in figure 2.

One can also use this distribution to estimate the fraction of collisions producing high p_T (say > 2 GeV/c ; at 1500 GeV this is about 6×10^{-4}). To do this we consider the two extreme assumptions namely that the primary particles which caused the multi cored showers were either all helium nuclei or were all iron nuclei. In the former case we are dealing with a beam of 180 nucleons; in the latter with a beam of 2520 nucleons. The fraction of collisions producing $p_T > 2$ GeV/c then lies between 0.17 and 0.012. Up to the maximum energy at which it has been possible to make measurements this fraction is increasing.

2.7 Discussion

The very large transverse momenta imply a very strong force acting at high energies. One candidate for this force is the quark-quark binding force [2]. Another possibility, which is not necessarily incompatible with the first, is that very massive particles are produced carrying very high transverse momentum which is balanced by a very large p_T imparted to the incident nucleon. I will, in a moment, review the evidence for the production of such massive particles.

3. HADRONS IN AIR SHOWER CORES

The experiments on the electromagnetic component in EAS cores which gave the evidence for high transverse momenta were carried out with a variety of instruments:- scintillators, ionisation chambers, neon hodoscopes, glass fronted spark chambers, large Wilson cloud chambers and

STUDIES OF HIGH ENERGY COSMIC RADIATION

emulsion chambers. Evidence for high p_T from a study of the energetic hadrons close to the core of showers similarly used a variety of techniques.

3.2 Hadrons in EAS at 800 g/cm² atmospheric depth

The cosmic ray group of the Tata Institute[1], Bombay operated an air shower array at Ootacamund in Southern India, altitude 800 g/cm². The array included a large ionisation calorimeter. It also determined the core position and direction of showers reasonably accurately. Table 3 gives the distribution in distance from the axis for hadrons of energy > 1 TeV for two ranges of shower size.

Table 3

Shower size \ Distance from core in metres	0 to 1.4	1.4 to 2	2 to 2.8	2.8 to 4	4 to 5.6
<3×10⁵ particles	$\frac{10}{157}$	$\frac{0}{362}$	$\frac{0}{143}$	$\frac{0}{496}$	$\frac{0}{493}$
>3×10⁵ particles	$\frac{10}{176}$	$\frac{4}{182}$	$\frac{3}{275}$	$\frac{2}{490}$	$\frac{1}{788}$

The lateral distribution of hadrons of > 1 TeV in showers of size less than and greater than 3×10^5 particles. The numerator gives the number of high energy hadrons and the denominator the total number of showers observed in that distance range.

The effect is the same as in the electromagnetic component. In the larger showers, in which the longitudinal momentum is higher, the distance from the axis of the high energy hadrons is greater, obviously implying a much increased mean transverse momentum.

3.3 The Tien Shan experiment

This was a complicated array[1]. Its central detector was a 36 m² ionisation calorimeter, of depth 1440 g/cm² and having 15 layers of ionisation chambers. Above the calorimeter, an array of 64 scintillators gave the electromagnetic core structure of the shower. The calorimeter gave the energy of the hadrons and their position with respect to the shower axis. The height of production of the hadrons was deduced from a study of the cascades resulting from the decay of π^o mesons. They found that 5% (31 events) of showers with $N_e > 10^5$ (at 3340 m above sea level) had $E.r \geq 2.10^{13}$ eV for at least one hadron. The proportion of high p_T events <u>increases</u> with shower size. 15 hadrons (in 190 showers) had $p_T.h > 10^4$ GeV/c.m . In these showers the accompanying cascades showed that the π^o mesons had originated at a mean height $<h>$ = 0.6 ± 0.1 kms . If the 15 hadrons came from the same height, they had

$$<p_T> = 15 \pm 3 \text{ GeV/c}$$

3.4 'Centauro'

This event was detected in emulsion chamber number 15 exposed by the Japanese Brazilian Emulsion Collaboration on Mt. Chacaltaya. The chamber had a surface area of 40 m² and was exposed for one year. Figure 3 is a diagrammatic sketch of the event. In the upper chamber there are only 14 showers with a total energy in γ-rays (ΣE_γ) of 27.1 TeV . In the lower chamber however, there are 140 showers within a radius of 0.75 cms with ΣE_γ = 230 TeV . Most of these showers are due to γ's from π^o mesons produced in very high energy nuclear interactions in the pitch layer. It was possible to trace two γ-ray cascades in the top chamber to the lower chamber. This gave their height of production as 50 ± 10 m above the chamber. A check on the production height of the whole beam of hadrons can be made by measuring the increase in spread of the 140 γ-ray cascades as they go through the 7 cm deep lower chamber. This gives the height as 50 ± 20 m above the chamber. 42 hadronic interactions were observed. Another 13 hadrons are expected to have traversed the chamber without interacting. Some hadrons are expected to have energies below the detection threshold. Altogether 90 ± 16 hadrons (+ anti hadrons) are estimated to have originated in the interaction occurring 50 m above the chamber. The number of π^o mesons is ≤ 3 and might well be zero. The investigators interpret the event as the production of a short lived particle in a 2×10^{15} eV interaction. The mass of the particle is (210 ± 80) GeV .

Figure 4 shows the distribution in transverse momentum (with respect to their own centre of momentum) of the observed hadrons from the decay.

STUDIES OF HIGH ENERGY COSMIC RADIATION

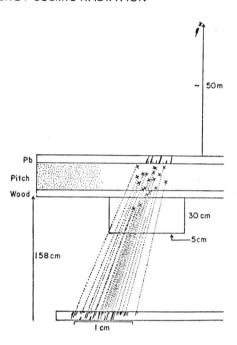

Figure 3 A schematic diagram of the event 'Centauro'.

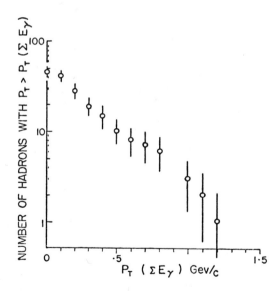

Figure 4 The distribution in transverse momentum of the event 'Centauro'. Only the measured γ-ray energy has been used in deriving p_T and hence the real p_T of the hadrons will be 4 to 6 times larger. Also the transverse momentum is only with respect to the fireball CMS.

The transverse momentum measured with respect to the direction of the incident particle will, of course, be larger.

4. EVIDENCE FOR MASSIVE LONG LIVED PARTICLES IN THE COSMIC RADIATION

The suspected particle causing the Centauro event could well be short lived, i.e. with $\tau \simeq 10^{-23}$ sec . A number of other events observed have lifetimes in the range 10^{-13} to $\geq 10^{-8}$ sec and masses in excess of 3 GeV and, in one case at least, probably in excess of 12 GeV .

4.1 Niu's X particle

In 1971 K. Niu and his co-workers published evidence [1] suggesting the existence of a particle of mass around 3.0 GeV and lifetime $\sim 3 \times 10^{-13}$ sec . They followed this up two years later with further evidence from their own exposures and an analysis of certain other events observed by others. Very recently another Japanese group have published details of an event observed in a balloon flight at an altitude of 10 g/cm^2 which has similar characteristics. A sketch of this event is shown in Figure 5. The primary particle of energy ~ 18 TeV interacted at point α . Amongst the charged products are two particles X_1 and X_2 . After 3.04 cms track, X_1 changes direction by 3.6×10^{-3} radians, 3 mms from the scattering point a pair of electrons materialised from a γ-ray. A second γ-ray from the scattering point produced a cascade in the emulsion chamber. The second particle X_2 travelled 6.34 cms before scattering through 1.2×10^{-2} radians. Two γ-rays give electron pairs at 4.1 and 8.2 mms respectively from the scattering point. The masses of the two particles depend on the nature of the charged secondary. The possible values are given in Table 4.

4.2 The Yunnan event

The Cosmic Ray Research Group of the Yunnan Institute of Atomic Energy [1,3] used a large cloud chamber [1.5 × 1.5 × .3 m^3] in a magnetic field of 7000 gauss to investigate interactions of high energy cosmic ray particles in a block of material above the chamber. The maximum detectable momentum of the arrangement was 48 GeV/c . The

Table 4

Assumed nature of charged particle	Mass in GeV	
	X_1	X_2
Σ	2.3 ± .6	2.4 ± .6
P	2.0 ± .5	2.1 ± .5
K	1.7 ± .4	1.7 ± .4
π	1.5 ± .4	1.6 ± .4

The masses of two X particles calculated on four assumptions as to the nature of the charged secondary. The lifetimes are 5×10^{-13} seconds for X_1 and 4×10^{-12} seconds for X_2.

exposure was slightly delayed so that the specific ionisation of the particles could be measured. One photograph showed three tracks a, b and c tracing back to one point in the producing layer and having the same width. Track a had a momentum of $6.6 ^{+ 1.0}_{- 0.8}$ GeV/c . The other two had momenta > 48 GeV/c . Tracks a and b had ionisations equal with the errors, namely 32.7 ± 1.3 and 33.7 ± 1.0 drops per cm . These measurements identify a as a fast negative pion, well on the plateau of the ionisation curve. Track b is also on the plateau and can be a pion, kaon or proton. Track c however, with momentum > 48 GeV/c , has an ionisation of only 21.3 ± 1.0 drops/cm . This is less than the <u>minimum</u> of the ionisation curve [0.88 ± 0.11 × minimum ionisation]. Track c then is due to a fractionally charged particle or a singly charged particle of mass > 12 GeV . Its lifetime is $\gtrsim 10^{-8}$ seconds.

4.3 The Kolar Gold Field experiment

A number of experiments on neutrino interactions have been carried out at great depths in the Kolar Gold Mines by a collaboration between

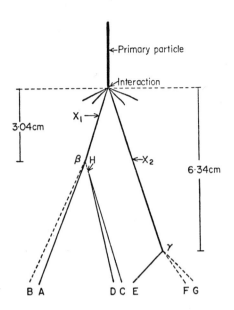

Figure 5 The primary particle interacted at point α to give, amongst other charged secondaries the two particles X_1 and X_2. After 3.04 cms X_1 scattered at point β. Two γ-rays trace back to this point. X_2 scattered at point γ and again two γ-rays trace back to this point. The γ-rays are detected by the electron pairs and subsequent electromagnetic cascades they induce.

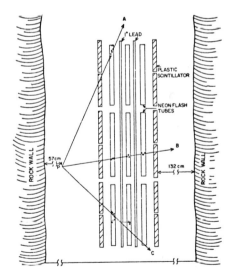

Figure 6 An event, seen in plan, recorded during the Kolar Gold Fields experiment at a depth of 7×10^5 gm/cm^2.

the Tata Institute of Fundamental Research and the Osaka City University[4]).
In one arrangement two vertical walls of plastic scintillator two metres
apart horizontally triggered the array when there was a coincidence
between them. In between the two walls of plastic scintillator were
three banks of neon flash tubes separated by two one inch thick lead walls.
Figure 6 shows one of the events. Three particles come from a point in
the air 57 cms <u>outside</u> the rock wall of the tunnel. Two of the
particles pass through 2" of lead and without interacting and therefore
are not electrons. Out of 20 events produced by neutrino interactions,
5 were of this type. In four cases, three decay particles were seen;
in the fifth, two only but a third track could well have missed the
detector. In all, the particles traversed $\sim$ 350 g/cm^2 of lead and
400 g/cm^2 of iron; no nuclear interactions were observed and one
particle was certainly identified as a muon. Very possibly, all are muons.
The likely interpretation of these events is that a massive [2 to 5 GeV]
particle is produced in the rock by neutrino interaction decays in the
air in a 3-charged particle mode. The lifetime needs to be around
10^{-9} seconds so the particle is not the hypothesised intermediate vector
boson. Also, since the only particles seen are coming from the decay
vortex (except in one event, the particle going into the vortex was also
observed), it seems likely that the process is <u>not</u> ν + target $\rightarrow$ target
+ μ + new particle so the interaction may be of the neutral current
type.

4.4 'Time of flight' experiments

Yock [1]) and his co-workers operated a scintillator telescope under
600 g/cm^2 of concrete for 1900 hours. The telescope timed the flight
of a particle over two 1.524 m paths and made five measurements of its
energy loss $\frac{dE}{dx}$. Between scintillators 4 and 5 there was an absorber
of 15.7 g/cm^2 and <u>below</u> scintillator 5, a further absorber above the
spark chamber which was the lowest detector. Table 5 gives the details
of the two anomalous particles found in the 1900 hours. The masses are
consistent with the Yunnan event.

Table 5

$\beta = v/c$	dE/dx in scintillators 1 to 4 in MeV/gm/cm^2	Charge/e	Range	dE/dx in scintillator 5	90% confidence limits on the Mass in GeV/c^2
0.34±0.1	10.1 ± .9	0.98±0.05	>Absorber 1	15.0 ± 1.8	4.2 to 38
0.35±0.1	10.3 ± .9	0.97±0.05	<Absorber 1+2	10.1 ± 1.8	> 8.5

Details of the two anomalous events seen with the Auckland telescope.

A time of flight telescope using a path length of 3.0 metres and 8 measurements of $\frac{dE}{dx}$ has recently been in operation in the University of Sydney.

4.5 Particles delayed behind the air shower front

Close to the core of an air shower the disc of charged particles is < 2 m thick in direction of the shower axis. The particles have velocities very close to c. Any very heavy particle produced a kilometre or so above the array might lag behind this front by a detectable amount. The Tata group used their fast timing air shower array and ionisation calorimeter at Ootacamund to search for such particles. In 1280 showers, they found 42 particles which deposited > 20 GeV in the ionisation calorimeter and lagged behind the shower front by > 28 n sec. A difficulty of the arrangement is that of chance coincidence of a fast particle hitting the ionisation calorimeter just after a shower front has passed through it. To get over this, the Tata group carried out a second experiment using a large multiplate cloud chamber instead of the ionisation calorimeter. A scintillator was placed in the middle of the cloud chamber. The cloud chamber also showed the direction of the hadron and this could be compared with the direction of the shower given by the fast timing array. So one can be certain that the candidate is associated with the shower and is delayed. In 2800 hours they found two good candidates. One gave a visible release of energy of 36 GeV and was delayed by 41 n sec behind the front. The other gave 28 GeV and was 25 n sec behind the front.

5. INTERACTIONS AT $> 10^{17}$ eV

There are at present three large air shower arrays operating. The Haverah Park array near Leeds in Northern England has a collecting area of ~ 11 km^2. The Yakutsk array in Siberia covers 25 km^2 and the Sydney array in the Pilliga State Forest near Narrabri, N.S.W. at present covers 40 km^2 and is being extended to 90 km^2.

5.1 The rapid development of large showers

The Sydney array has detected $\sim 15,000$ showers of energy $\gtrsim 10^{17}$ eV and a few of energy $> 10^{20}$ eV. Below 10^{19} eV the radiation comes equally from all directions and its intensity increases steeply with increasing energy. Above 10^{17} eV showers can be observed at all zenith angles from 0 to 90° if one uses (as we do in Sydney) the muon component for detection. If one releases a particular intensity (say I showers per m^2 per sec per steradian) and one determines the muon shower size that gives this intensity looking at different zenith angles, then one gets the development curve from 1000 g/cm^2 [zenith angle 0°] to 2000 g/cm^2 [zenith angle 60°] for showers of that particular energy. If showers of that energy have not reached their maximum development by the time they reach sea level, then as θ is increased from 0, N_μ will also increase. The experimental situation is shown in Figure 7. The surprising thing is that even the largest showers which it has been possible to measure are past their maximum by the time they reach sea level. At very high energies the particles are losing energy much more quickly than one would expect if say 2000 GeV were the beginning of Asymptotia.

Acknowledgements

I am greatly indebted to my colleagues in the Sydney group for much discussion and to Professor H. Messel for his encouragement. The Sydney experiments have been supported by the U.S.A.F. Office of Scientific Research, the Australian Research Grants Committee and the Science Foundation for Physics in the University of Sydney.

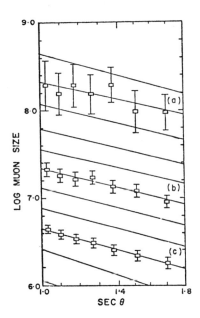

<u>Figure 7</u> Each line represents the number of muons at different atmospheric depth from a shower of a given primary energy. θ is the zenith angle at which the observation was made. All showers are obviously past their maximum.

REFERENCES

1) C.B.A. McCusker, Physics Reports (in the press).
   This is a review article giving references to the original work.

2) A.M. Bakich, D. Melley, C.B.A. McCusker, D. Nelson, L.S. Peak, M.H. Rathgeber & M.M. Winn, Canad. J. Phys. <u>46</u> No.10 Part 2 30 (1968).

3) The Cosmic Ray Research Group of the Yunnan Institute of Atomic Energy, Scientia Sinica XVI No. 1 123 (1972).

4) M.R. Krishnaswany, M.G.K. Menon, V.S. Narasimhan, N. Ito, S. Kawakami, & S. Miyake, University of Tokyo, C.R.L. Report - 21 (1975).

FERMION SYSTEMS IN DIFFERENT DIMENSIONS[*]

Leon N. Cooper

Department of Physics, Brown University, Providence, Rhode Island, USA

1. INTRODUCTION

Fermion systems in one, two and three space dimensions provide explicit models of some of the abstract fields discussed in particle physics. One dimensional fermion systems have been constructed experimentally. Theoretical analysis of such systems, in particular the Luttinger model developed by Mattis, Lieb, Luther, Peschel and Emery, has progressed recently. In two and three dimensions the singlet spin superconductor shows striking macroscopic coherence effects analogous to those that occur in gauge field theories and microscopic interference effects which display the time reversal properties and the fermion character of the underlying field.

2. ONE DIMENSION

Systems of interacting fermions provide examples of many of the concepts discussed in this school. Since it is possible to actually construct such systems, their properties can be studied experimentally. It may seem surprising that one can fabricate one dimensional systems in a three dimensional world. In quantum mechanics this is possible since the ratio of the width to length of a sample can be made small enough so that the energy required to excite a level associated with the small dimension is large compared to that required for excita-

---

[*] Work supported in part by the U. S. National Science Foundation and the Advanced Research Projects Agency, USA.

tions of levels associated with the large dimensions and large compared to k_BT. The sample thus behaves as though it were one dimensional.

There now exist several examples of what are believed to be quasi one dimensional electron systems. These include the organic charge transfer salt TTF-TCNQ[1], the platinum salts $K_2[Pt(CN)_4]$ $Br_{0.30}$ · $3(H_2O)$ [2] and plastically deformed cadmium sulfide in which screw dislocations run in one dimension[3]. Recently[4], the polymer $(SN)_x$ has been shown to undergo a superconducting transition at 0.3 K and to exhibit precurser loss of resistivity characteristic of quantum fluctuations which indicate a small dimension of the order of 10^{-6} cm [5].

These are remarkable systems in several ways. In TTF-TCNQ we have an organic molecule with metallic properties. The deformed cadmium sulfide shows metallic conductivity in the direction parallel to the dislocations and a striking anisotropy in electrical conductivity of the order of

$$\frac{\sigma_\parallel}{\sigma_T} \approx 10^8.$$

All of these one dimensional systems display an increasing electrical conductivity which peaks at some "critical" temperature and then diminishes rapidly (possibly exponentially).

Not unreasonably there has been a great deal of theoretical interest in these systems. It is generally believed that the conductivity peaks are associated with a lattice distortion at $k \approx 2k_F$ due to the appearance of a soft phonon mode at this frequency known as the Peierls transition[6]. This arises due to the instability of the one dimensional lattice electron system due to processes near the Fermi surface of the form:

These are divergent in one dimension and the degree of divergence is the same as that of the pair scattering process:

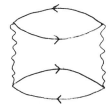

The Peierls transition may be regarded as arising from electron transitions from one side of the Fermi surface (line) to the other. This requires a momentum transfer of $q \simeq 2k_F$ which is provided by the lattice distortion associated with the soft phonon mode. Since the lattice distortion has a direction, it is only in this direction that the momentum transfer can occur.

In electron pair transitions, the momentum is transferred from one pair of electrons to the other so that the entire Fermi surface is available for transitions. In two or three dimensions, therefore, the electron pair process dominates. In one dimension, however, single electron transitions across the Fermi surface, mentioned above, which lead to the Peierls transition are of the same order as the electron pair transitions--thus giving the one dimensional system its special character.

There have been interesting theoretical developments of a one dimensional many fermion model introduced by Luttinger[7]. This has been analyzed by Mattis and Lieb[8], Luther and Peschel[9] and Luther and Emery[10], the last having found a solution for a single value of the coupling constant which displays superconductor like behavior (divergence of the pair correlation function and an energy gap in the fermion part of the spectrum) at the absolute zero.

In this model one begins with spinless fermions which fill a "Fermi line" and separates electrons into two classes of spinless fermions with creation (annihilation) operators, $a_{1k}^*(a_{1k})$ and $a_{2k}^*(a_{2k})$. The kinetic energy is linearized, in an approximation introduced by Tomonaga[11], and becomes

$$K = v_F' \sum_k k(a_{1k}^* a_{1k} - a_{2k}^* a_{2k}) .$$

In Luttinger's model[7] interactions are introduced between electrons on opposite sides of the Fermi surface in such a way that only small momentum transfers occur:

This Hamiltonian was diagonalized by Mattis and Lieb[8] and they obtained a plasmon spectrum. Luther and Emery[10] have generalized this model to

FERMION SYSTEMS IN DIFFERENT DIMENSIONS

include both spin and large momentum scattering (scattering across the Fermi line). The additional scattering terms are characterized by two coupling constants $U_\parallel$ and $U_\perp$

$U_\parallel \ (s = s')$

$U_\perp \ (s = -s)$

Luther and Emery found that this problem is exactly soluble for a single value of $U_\parallel$

$$\frac{U_\parallel}{2\pi \, v_F} = -3/5$$

In this remarkable solution a fermion excitation spectrum appears

$$E = v_F' \, k_F + \sqrt{v_F'^2 |k - k_F|^2 + \Delta^2}$$

with the energy gap $\Delta = U_\perp/2\pi\alpha$, where α is a cut-off parameter, equivalent to a band width of v_F'/α. The singlet spin pair correlation function diverges at $T = 0$, thus indicating a superconducting state (at least at $T = 0$).

This solution has been obtained only for a single value of $U_\parallel$ so it is of great interest to generalize to other values. Ertugrul Berkcan and I have been investigating this question; we believe we have such a generalization valid for certain discrete values of $U_\parallel$ between

$$-1 < \frac{U_\parallel}{2\pi \, v_F} < -3/5 \; .$$

Since this one dimensional system is closely related to the massive

Thirring model and the Sine-Gordon equation, discussed by S. Coleman[12], it seems reasonable to believe that results and insights in one area can be applied to the other. We believe further that the one dimensional solution can be connected to three dimensional systems using techniques introduced by Stölan and Cooper[13].

3. TWO DIMENSIONS

The singlet spin superconductor provides a concrete example of the field discussed by Coleman[12], which corresponds to the mapping $G = SO(2)$:

$$\psi = e^{i\chi}$$

and whose homotopy classes are labelled by the 'winding number'

$$n = \frac{1}{2\pi} \int_0^{2\pi} \left(\frac{d\chi}{d\theta}\right) d\theta .$$

For the superconductor, the Ginzburg-Landau equation determines a field

$$\psi = \psi_0 \, e^{i\chi}$$

which may be interpreted as a superfluid 'wave function', where $|\psi_0|^2$ is proportional to the density of superfluid 'carriers' and χ is their common phase. The full many-body wave function is the properly symmetrized product of these 'carrier' wave functions.

The requirement that the covariant derivative go to zero at infinity becomes the requirement that the current go to zero in the interior of the superconductor (Meissner effect). For a 'charge carrier' of mass m^* and charge q^* the current operator is

$$J = \frac{q^*}{m^*} \left(P - \frac{q^*}{c} A\right) .$$

FERMION SYSTEMS IN DIFFERENT DIMENSIONS

In a state of current, I, this has the expectation value

$$<J>_I = \frac{q}{m^*}(<P>_I^* - \frac{q}{c}<A>_I)$$

Consider a curve, c, deep inside the superconductor where the current flow is zero (Meissner effect). Integrating around this curve we obtain

$$0 = \oint <P>_I^* \cdot d\ell - \frac{q}{c} \oint <A>_I \cdot d\ell .$$

The last term is proportional to the flux enclosed inside the curve which need not be zero if there is a hole in the material.

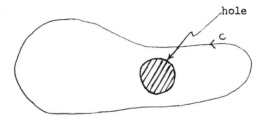

If the superfluid is characterized by the 'wave function', $\psi_0 e^{i\chi}$

$$<P>_I = |\psi_0|^2 p$$

where

$$p = \hbar \nabla \chi$$

and

$$<A>_I = |\psi_0|^2 A .$$

The flux through the hole is

$$\Phi = \oint A \cdot d\ell = \frac{c}{q^*} \oint p \cdot d\ell$$

The requirement that ψ (the carrier wave function) be single-valued gives the quantization condition:

$$\oint p \cdot d\ell = \frac{h}{2\pi} \int_0^{2\pi} \left(\frac{d\chi}{d\theta}\right) d\theta = nh \qquad n = 0, \pm 1, \ldots$$

so that

$$\Phi = n\left(\frac{ch}{q^*}\right).$$

The enclosed flux thus comes in units of

$$\Phi_0 = \frac{ch}{q^*}.$$

Experimental observation in superconductors has established that

$$q^* = -2e$$

and this is one of the verifications of the paired ground state wave function. Thus each 'carrier' has a common momentum and a charge $q^* = -2e$ [14].

Type II superconductors allow penetration of the magnetic field and therefore break up naturally into superconducting and normal regions. Each normal region allows penetration of the magnetic field in units of the fundamental flux quantum $\frac{ch}{2e}$.

4. THREE DIMENSIONS

The ground state superconductor wave function is a symmetrized product of the Ginzburg-Landau functions, $\psi = \psi_0 e^{i\chi}$. The common phase, χ, for all of the pairs is responsible for macroscopic interference effects--quantum mechanical behavior on a large scale. The symmetrized product of pairs which makes up the condensed phase has many Bose properties. The fermion character of the underlying fields of which the 'boson' ground state field is composed is shown most clearly in what might be called microscopic interference effects.

These 'coherence effects' are in a sense manifestations of interference in spin and momentum space on a microscopic scale, analogous to the macroscopic quantum effects due to interference in ordinary space. They depend on the behavior under time reversal of the perturbing fields[15]. It is intriguing to speculate that if one could somehow amplify them properly, the time-reversal symmetry of a fundamental interaction might be tested.

Near the transition temperature the coherence effects produce quite dramatic contrasts in the behavior of coefficients which measure interactions with the conduction electrons. Historically, the comparison with theory of the behavior of the relaxation rate of nuclear spins[16] and the attenuation of longitudinal ultrasonic waves in clean samples[17] as the temperature is decreased through T_c provided an early test of the detailed structure of the theory.

To illustrate how such effects come about we consider the transition probability per unit time of a process involving electronic transitions from the excited state k to the state k' with the emission or absorption of energy to or from the interacting field. What is to be calculated is the rate of transition between an initial state $|i\rangle$ and a final state $|f\rangle$ with the absorption or emission of the energy $\hbar\omega|k' - k|$ (a phonon for example in the interaction of sound waves with the superconductor). All of this properly summed over final states and averaged with statistical factors over initial states may be written:

$$\omega = \frac{2\pi}{\hbar} \times \frac{\sum_{i,f} \exp(-W_i/k_B T)|\langle f|H_{int}|i\rangle|^2 \delta(W_f - W_i)}{\sum_i \exp(-W_i/k_B T)}$$

We focus our attention on the matrix element $<f|H_{int}|i>$. This typically contains as one of its factors matrix elements between excited states of the superconductor of operators such as the current operator, $<\kappa'|j(x)|\kappa>$, the charge density operator, $<\kappa'|\rho(x)|\kappa>$ etc. Call such matrix elements $B_{\kappa'\kappa}$

$$B = \sum_{\kappa\kappa'} B_{\kappa'\kappa} c^*_{\kappa'} c_\kappa$$

where $c^*_{\kappa'}$ and c_κ are the creation and annihilation operators for electrons in the states κ' and κ. We use the symbol κ to designate momentum and spin $\kappa \equiv k,\uparrow - \kappa \equiv -k,\downarrow$.

To display the dependence on the character of the operators under time reversal write (for operators which do not flip spins)

$$B = \sum_{\kappa\kappa'} (B_{\kappa'\kappa} c^*_{\kappa'} c_\kappa + B_{-\kappa-\kappa'} c^*_{-\kappa} c_{-\kappa'}) .$$

Many of the operators, B, we encounter (e.g., the electric current, charge density or the spin operator) have a well-defined behavior under the operation of time reversal so that

$$B_{\kappa'\kappa} = \pm B_{-\kappa-\kappa'} \equiv B_{k'k}$$

Then B becomes

$$B = \sum_{kk'} B_{k'k}(c^*_{k'\uparrow} c_{k\uparrow} \pm c^*_{-k\downarrow} c_{-k'\downarrow})$$

where the upper (lower) sign results for operators even (odd) under time reversal.

The matrix element of B between the initial state, $\Psi_{...k\uparrow...}$, and the final state $\Psi_{...k'\uparrow...}$ contain contributions from $c^*_{k'\uparrow} c_{k\uparrow}$, shown in Figure 1 and unexpectedly from $c^*_{-k\downarrow} c_{-k'\downarrow}$, shown in Figure 2. As a

FERMION SYSTEMS IN DIFFERENT DIMENSIONS

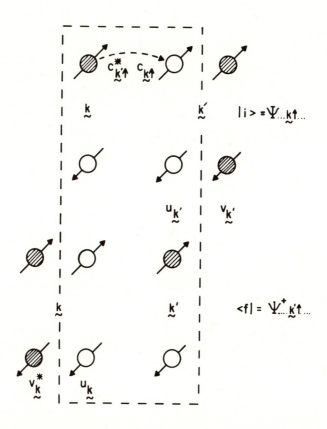

Fig. 1 The two states $|i\rangle$ and $\langle f|$ shown are connected by $c^*_{k'\uparrow} c_{k\uparrow}$ with the amplitude $u_k u_{k'}$.

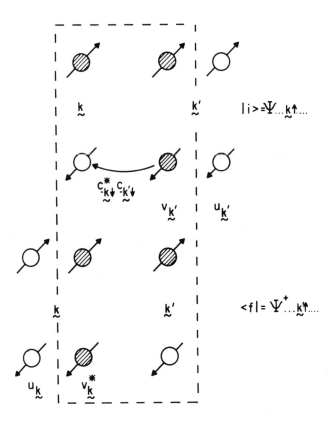

Fig. 2 The two states $|i\rangle$ and $\langle f|$ are also connected by $c^*_{-k\downarrow}c_{-k'\downarrow}$ with the amplitude $-v_{k'}v^*_k$.

FERMION SYSTEMS IN DIFFERENT DIMENSIONS

result the matrix element squared $|<f|B|i>|^2$ contains terms of the form

$$|B_{k'k}|^2 (u_{k'} u_k \mp v_{k'} v_k^*)^2,$$

where the sign is determed by the behavior of B under time reversal:

upper sign   B even under time reversal

lower sign   B odd under time reversal

Here $v_k (u_k)$ is the amplitude for occupation (non-occupation) of the paired state $(-k\downarrow, k\uparrow)$ in the ground-state wave function:

$$u_k = [\tfrac{1}{2}(1 + \epsilon/E)]^{1/2}$$

$$v_k = [\tfrac{1}{2}(1 - \epsilon/E)]^{1/2} e^{i\chi}$$

$$E = \sqrt{\epsilon^2 + |\Delta|^2}$$

where ϵ is the energy of an excitation measured from the Fermi surface

$$\epsilon = \frac{\hbar^2}{2m}(k^2 - k_F^2)$$

and Δ is proportional to the Ginzburg-Landau field

$$\Delta = |\Delta| e^{i\chi} \sim \psi_0 e^{i\chi}.$$

Applied to process involving the emission or absorption of boson quanta such as phonons or photons, the squared matrix element above is averaged with the appropriate statistical factors over initial and summed over final states; substracting emission from absorption probability per unit time, we obtain typically

$$\alpha = \frac{4\pi}{\hbar} |m|^2 \sum_{kk'} |(u_{k'} u_k \mp v_{k'} v_k^*)|^2 (f_{k'} - f_k) \delta(E_{k'} - E_k - \hbar\omega_{|k-k'|})$$

where f_k is the occupation probability in the superconductor for the excitation k↑ or k↓. (In the expression above we have considered only quasi-particle or quasi-hole scattering processes--not including processes in which a pair of excitations is created or annihilated from the coherent part of the wave function--since $\hbar\omega_{|k'-k|} < \Delta$ is the usual region of interest for the ultrasonic attenuation and nuclear spin relaxation we shall contrast.)

For the ideal superconductor, there is isotropy around the Fermi surface and symmetry between particles and holes; therefore sums of the form $\sum_k$ can be converted to integrals over the superconducting excitation energy, E:

$$\sum_k \rightarrow 2N(0) \int_\Delta^\infty \frac{E}{(E^2 - \Delta^2)^{1/2}} \, dE$$

where $N(0) E(E^2 - \Delta^2)^{-1/2} = N(0) E/\epsilon$ is the density of excitations in the superconductor. The appearance of this density of excitations is a surprise. Contrary to our intuitive expectations, the onset of superconductivity seems initially to enhance rather than diminish electronic transitions, as might be anticipated in a reasonable two-fluid model.

But the coherence factors $|(u'u \mp v'v^*)|^2$ are even more surprising; they behave in such a way as to sometimes completely negate the effect of the increased density of states. This can be seen using the expressions above for u and v for the ideal superconductor to obtain

$$(u'u \mp v'v)^2 = \frac{1}{2}\left(1 + \frac{\epsilon \epsilon' \mp \Delta^2}{EE'}\right).$$

In the integration over k and k', the $\epsilon \epsilon'$ term vanishes. We thus define $(u'u \mp v'v)_s^2$. In the usual limit where $\hbar\omega_{|k'-k|} \ll \Delta$, $\epsilon \approx \epsilon'$ and $E \approx E'$, this becomes

$$(u^2 - v^2)^2_s \to \frac{1}{2} \frac{\epsilon^2}{E^2}$$

operators even under time reversal

$$(u^2 + v^2)^2_s \to \frac{1}{2} (1 + \frac{\Delta^2}{E^2})$$

operators odd under time reversal

For operators even under time reversal, therefore, the decrease of the coherence factors near $\epsilon = 0$ just cancels the increase due to the density of states. For the operators odd under time reversal the effect of the increase of the density of states is not cancelled and should be observed as an increase in the rate of the corresponding process.

In general the interaction Hamiltonian for a field interacting with the superconductor (being basically an electromagnetic interaction) is invariant under the operation of time reversal. However, the operator B might be the electric current $j(r)$ (for electromagnetic interactions) the electric charge density $\rho(r)$ (for the electron-phonon interaction) or the z component of the electron spin operator, σ_z (for the nuclear spin relaxation interaction). Since under time-reversal

$$j(r, t) \to - j(r, - t)$$

(electromagnetic interaction)

$$\rho(r, t) \to + \rho(r, - t)$$

(electron-phonon interaction)

$$\sigma_z(t) \to - \sigma_z(- t)$$

(nuclear spin relaxation interaction)

these show strikingly different interference effects.

Ultrasonic attenuation in the ideal pure superconductor for $q\ell \gg 1$ (the product of the phonon wave number and the electron mean free path) depends in a fundamental way on the absorption and emission of phonons. Since the matrix elements have a very weak dependence on changes near the Fermi surface in occupation of states other than k or k' that occur in the normal to superconducting transition, calculations within the quasi-particle model can be compared in a very direct manner with similar calculations for the normal metal, as $B_{k'k}$ is the same in both states. The ratio of the attenuation in the normal and superconducting states becomes:

$$\frac{\alpha_s}{\alpha_n} = -4 \int_\Delta^\infty dE (u^2 - v^2)_s^2 \left(\frac{E}{\epsilon}\right)^2 \frac{df(E)}{dE}$$

Since $(u^2 - v^2)_s^2 = (1/2)(\epsilon/E)^2$, the coherence factors cancel the density of states giving

$$\frac{\alpha_s}{\alpha_n} = 2f(\Delta[T]) = \frac{2}{1 + \exp(\frac{\Delta(T)}{k_B T})}.$$

Morse and Bohm[17b)] used this result to obtain a direct experimental determination of the variation of Δ with T. Comparison of their attenuation data with the theoretical curve is shown in Figure 3.

In contrast, the relaxation of nuclear spins which have been aligned in a magnetic field proceeds through their interaction with the magnetic moment of the conduction electrons. In an isotropic superconductor this can be shown to depend upon the z component of the electron spin operator

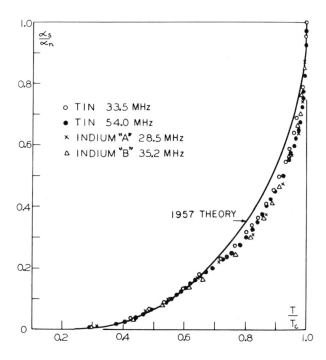

Fig. 3 Comparison of observed ultrasonic attenuation with theory. The data are due to Morse and Bohm[17b].

$$B_{\kappa'\kappa} = B(c^*_{k'\uparrow} c_{k\uparrow} - c^*_{-k\downarrow} c_{-k'\downarrow})$$

so that

$$B_{\kappa'\kappa} = -B_{-\kappa-\kappa'}.$$

This follows in general from the property of the spin operator under time reversal

$$\sigma_z(t) = -\sigma_z(-t).$$

The calculation of the nuclear-spin relaxation rate proceeds in a manner not too different from that for ultrasonic attenuation, resulting finally in a ratio of nuclear-spin relaxation rates in superconducting and normal states in the same sample:

$$\frac{R_s}{R_n} = -4 \int_\Delta^\infty dE (u^2 + v^2)_s^2 \left(\frac{E}{\epsilon}\right)^2 \frac{df(E)}{dE}$$

But $(u^2 + v^2)_s^2$ does not go to zero at the lower limit so that the full effect of the increase in density of states at $E = \Delta$ is felt. Taken literally, in fact, this expression diverges logarithmically at the lower limit due to the infinite density of states. When the Zeeman energy difference between the spin-up and spin-down states is included, the integral is no longer divergent but the integrand is much too large. Hebel and Slichter, by putting in a broadening of levels phenomenologically, could produce agreement between theory and experiment. More recently Fibich[18] by including the effect of thermal phonons has obtained the agreement between theory and experiment shown in Figure 4.

This striking result, depending on the behavior of operators under time reversal, might possibly be used to test the character of an elementary interaction. This would be an important new use of the superconducting wave function but not, as Sir James Frazer has shown,

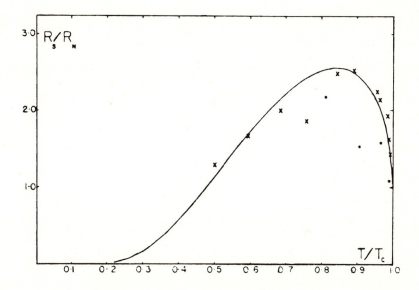

Fig. 4 Comparison of the observed nuclear spin relaxation rate with theory. The circles are the experimental data of Hebel and Slichter[16a]; the crosses are the data of Redfield and Anderson[16b].

the first--or the most dramatic use of time reversal.

".... it appears from the testimony of an anonymous Christian, who wrote in the fourth century of our era, that Christians and pagans alike were struck by the remarkable coincidence between the death and resurrection of their respective deities, and that the coincidence formed a theme of bitter controversy between the adherents of the rival religions, the pagans contending that the resurrection of Christ was a spurious imitation of the resurrection of Attis, and the Christians asserting with equal warmth that the resurrection of Attis was a diabolical counterfeit of the resurrection of Christ. In these unseemly bickerings the heathen took what to a superficial observer might seem strong ground by arguing that their god was the older and therefore presumably the original, not the counterfeit, since as a general rule an original is older than its copy. This feeble argument the Christians easily rebutted. They admitted, indeed, that in point of time Christ was the junior deity, but they triumphantly demonstrated his real seniority by falling back on the subtlety of Satan, who on so important an occasion had surpassed himself by inverting the usual order of nature[19]."

REFERENCES

1) For example, L. B. Coleman, et al., S. S. Commun. 12, 1125 (1973); L. B. Coleman, et al., Phys. Rev. B 10, 1298 (1974).

2) C. G. Kuper, Proc. R. Soc. Lond. A 227, 214 (1955).

3) C. Elbaum, Phys. Rev. Lett. 32, 376 (1974).

4) V. V. Walatka, et al., Phys. Rev. Lett. 31, 1139 (1973); R. L. Greene, et al., Phys. Rev. Lett. 34, 89 (1975); A. N. Bloch, et al., Phys. Rev. Lett. 34, 1561 (1975).

5) R. L. Civiak, C. Elbaum, W. Junker, C. Gough, H. I. Kao, L. F. Nichols and M. M. Labes, Fluctuation-induced conductivity above the superconducting transition temperature in polysulfur nitride, $(SN)_x$, to be published.

6) R. E. Peierls, Quantum theory of solids (Clarendon, Oxford, 1955), pg. 108.

7) J. M. Luttinger, J. Math. Phys. 4, 1154 (1963).

8) D. C. Mattis, E. H. Lieb, J. Math. Phys. 6, 304 (1965).

9) A. Luther, I. Peschel, Phys. Rev. B 9, 2911 (1974).

10) A. Luther, V. J. Emery, Phys. Rev. Lett. 33, 589 (1974).

11) S. Tomonaga, Progr. Theoret. Phys. (Kyoto) 5, 544 (1950).

12) S. Coleman, Lectures International School of Subnuclear Physics, 1975.

13) B. Stolan and L. N Cooper, Phys. Rev. 143, 209 (1966).

14) L. Onsager, Phys. Rev. Lett. 7, 50 (1961).

15) The importance of the coupling of time-reversed states in constructing electron pairs was emphasized by P. W. Anderson; see, for example, P. W. Anderson, J. Phys. Chem. Solids 11, 26 (1959).

16) a) L. C. Hebel, C. P. Slichter, Phys. Rev. 113, 1504 (1959).
    b) A. G. Redfield, A. G. Anderson, Phys. Rev. 116, 583 (1959).

17) a) H. E. Bommel, Phys. Rev. 96, 220 (1954).
    b) R. W. Morse, H. V. Bohm, Phys. Rev. 108, 1094 (1957).

18) M. Fibich, Phys. Rev. Lett. 14, 561 (1965).

19) Sir James George Frazer, The golden bough, in paperback, MacMillan, N.Y. (1960), pg. 419.

POSSIBLE FUTURE STORAGE RINGS AT CERN: p - p and e - p

K. Johnsen

European Organization for Nuclear Research (CERN)
Geneva, Switzerland

1. INTRODUCTION

Colliding proton beams have successfully entered the field of elementary particle physics through the CERN Intersecting Storage Rings (ISR), and it has been natural for CERN to start studying proton storage ring projects for higher energies than the present ISR.

Experience on the ISR, together with studies at several laboratories in the world, have in fact shown the feasibility of building colliding beam p - p devices up to the highest energies of accelerators in existence or under construction and such a facility could be constructed in connection with the SPS at CERN.

A spectrum of possible p - p projects can be envisaged, from a superconducting-magnet ISR conversion giving about 100 GeV in each ring, to a set of accelerating storage rings that might give up to 1000 GeV in each ring. It has been natural to choose for the start of the study of new Large Storage Rings (LSR) an energy equal to the maximum energy of the SPS, as a representative example. This falls somewhat in the middle of the range of possibilities mentioned above, but may nevertheless be in the upper range of realistic possibilities.

Two approaches can be considered, viz.:

(a) 400 GeV rings using normal iron magnets
(b) 400 GeV rings using superconducting magnets

A preliminary study of model (a) has demonstrated that large storage rings with normal magnets and good performance can actually be built. In order to achieve a luminosity of 10^{33} cm^{-2} s^{-1} in a short interaction length of not more than 1 m, as desired by the experimentalists, 7 A of circulating protons will be necessary.

The study has, however, also demonstrated the main drawbacks of such a project: the uncomfortably large circumference needed and the very high power consumption ($\sim$ 120 MW). Adding to this the somewhat higher performance potentiality of a superconducting device and/or the possibility of reduced cost (although little is known about cost yet), we have drawn the conclusion that the study of model (b) should have priority, and the study effort has recently been guided correspondingly.

2. MACHINE LATTICE

The circumference of the machine is made up by two contributions: the normal lattice which occupies the greater part, and the colliding beam insertions which occupy the rest. Consequently, the largest contributions to single beam space charge phenomena come from the normal lattice, while the beam-beam space charge (and high-energy physics) phenomena only occur in the insertions. It has therefore been convenient for the preliminary analysis to consider these two contributions separately.

Experience with the ISR has shown that the betatron tunes (Q-values) have to be controlled with rather high precision if one is to avoid enhanced beam decay rates due to non-linear resonances in the stacked beam. This imposes many tight tolerances on such a machine. In particular, the design of the machine must ensure that the image-dominated incoherent tune-shift is below the acceptable limit, and that the circulating beam is transversely stabilized by the small Q-spread available. These requirements can be met by choosing a sufficiently large aperture over most of the circumference of the machine. This is in fact the determining factor in the choice of aperture of a large-radius storage ring. A large machine aperture further helps to reduce the beam-induced gas desorption vacuum problems.

A formalism which takes these space charge phenomena into account and leads to the physical parameters of a machine was described by Keil at the IXth International Conference on High Energy Accelerators held

… in May 1974. A set of parameters for a superconducting machine arrived at in this manner, is shown in Table I.

Table I

Parameter list for large storage rings

Maximum momentum	400 GeV/c
Maximum bending field	4 T
Circumference	6130 m
Average radius of normal lattice	617 m
Stored current	7 A
Stored energy in beam	57 MJ
Vacuum chamber aperture radius	25 mm
Betatron wave number	36.25
Period length	40.4 m
Quadrupole length	1.6 m
Bending magnet length	3.7 m
Number of periods	96
Half period arrangement	$\frac{1}{2}$FBBB$\frac{1}{2}$D

3. TYPES OF INTERACTION REGIONS

For given energy and stacked current, the maximum design luminosity of an interaction region is limited mainly by two factors, viz. the non-linear electromagnetic beam-beam interaction, and the maximum acceptable values of betatron function in the neighbouring quadrupoles. The first is fundamental but not well quantified, and the second is limited by chromaticity and tolerances. Both factors lead to a situation in which a compromise must be made between luminosity and field-free length around the interaction region.

One is therefore led to consider a machine with several types of interaction regions, each designed to be suitable for a particular class of experiment. So far, three types of interaction regions have been considered at CERN; a high-luminosity low-β region, a general-purpose

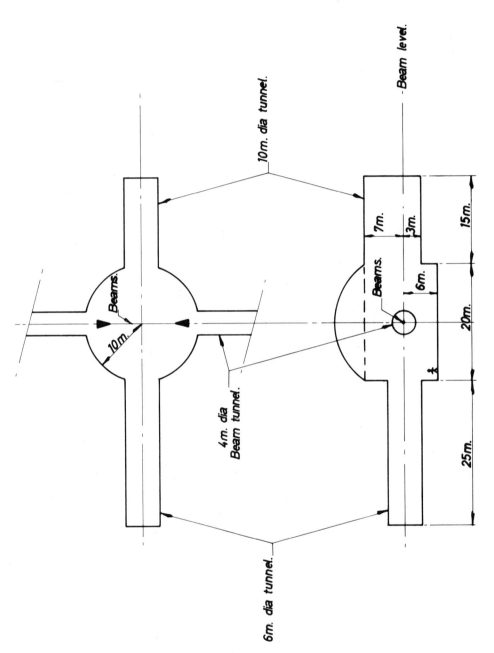

Fig.1: Low-β (High Luminosity) Insertion.

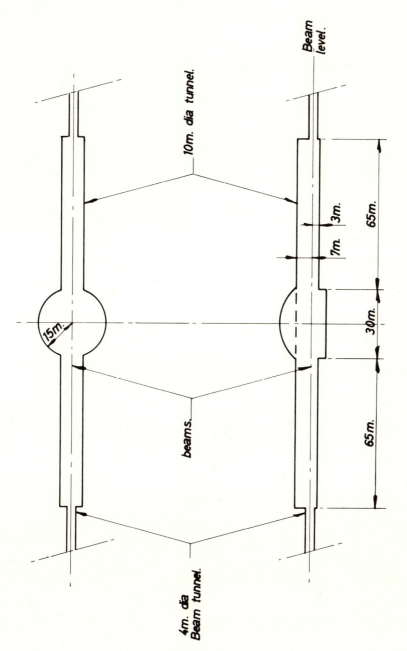

Fig. 2: General Purpose Insertion.

interaction region with plenty of unencumbered space, and a high-β region with special optics for measurements of very small scattering angles. Parameters of examples of such interaction regions are summarized in Table II.

Table II

Performance estimates for three model-insertions

	Low-β	General-Purpose	High-β
Luminosity ($cm^{-2} s^{-1}$)	$1.0 \cdot 10^{33}$	$2 \cdot 10^{31}$	$5 \cdot 10^{30}$
β_v^* (m)	1.0	14	400
β_h^* (m)	6.0	40	300
β_v max, β_h max	550	490	460
Crossing angle (mrad)	2.6	19.4	19.4
Field-free half-length (m)	5	80	18
Total length of insertion (m)	230	320	280

The above data are for:

Stacked current I = 7 A

Normalized emittance $\varepsilon = 30\pi \cdot 10^{-6}$ rad m (both planes)

Energy 400 GeV ($\gamma = 426.3$)

Lay-out and size of experimental halls around these interaction regions have also been studied a little, in particular by an Autumn Study arranged in 1974. Fig. 1 shows a possible arrangement for the low-β region, with a dome around the crossing point and two tunnels to be used for the detection of very high momentum particles at 90°. Fig. 2 shows a possible lay-out for the general-purpose region, again with a dome to accommodate the very large amount of equipment to be used in such a region. There is no tunnel at 90°, but the machine tunnel has been widened to facilitate the analysis of particles coming at small angles. The high-β region would look rather similar, except that the dome is not so high.

4. NUMBER OF INSERTIONS

The number of insertions is determined by the scale and scope of the physics programme which the storage rings are supposed to support. In addition, special insertions will be required for injection and beam dumping. It seems likely that a minimum of six interaction regions will be required for physics experimentation, i.e. two of each of the three different types listed in Table II.

A racetrack configuration with grouped interaction regions has been chosen. Fig. 3 shows a possible lay-out for a superconducting machine assumed to be placed about 60 m underground. Preliminary studies of underground experimental areas are encouraging but further studies are needed. In the model considered at present, the injection and dumping insertions would be in the arcs of the racetrack configuration.

5. POSSIBILITIES OF COLLIDING OTHER PARTICLES

A disadvantage with colliding beam devices is that one can normally only study interactions between particles of one type, p - p, in the kind of project described above. The question therefore arises, what possibilities exist for incorporating other particle options in such a project.

5.1 Anti-protons in one of the rings

The method would be first to fill one of the ISR rings with 400 GeV protons, say to 7 A. These protons are ejected and made to hit an anti-proton-producing target from which 14 GeV anti-protons are guided towards the SPS, injected and accelerated in this machine and then stacked in the other LSR ring. The process is repeated till the available aperture is filled, giving $\bar{p}$ circulating beam of about 2 mA. When this beam collides with the proton beam in the other ring, one might reach a luminosity of 10^{28} cm^{-2} s^{-1} to 10^{29} cm^{-2} s^{-1}. The filling time comes out uncomfortably long, about two days, and operational considerations may therefore restrict the luminosity to, say, an order of

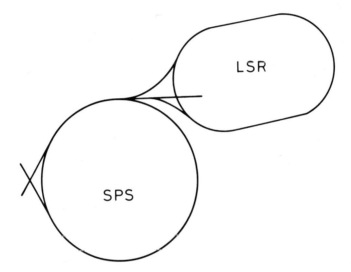

Fig.3: Possible lay-out of 400 GeV storage rings.

magnitude less. The main extra equipment would be ejection/injection devices, a target complex and an additional beam transport channel between the LSR and the SPS.

5.2 e − p option

One question being persistently asked is: "What would it imply to add an electron ring to the 400 GeV proton rings, and what performance could be expected?" Strong wishes have been expressed for 25 GeV electrons against 400 GeV protons. A luminosity of about 10^{32} cm^{-2} s^{-1} is generally considered necessary to achieve adequate event rates.

It may be difficult to satisfy both these requirements simultaneously, however, preliminary evaluations seem to indicate that it <u>is</u> feasible to add an e-ring to give 10^{32} cm^{-2} s^{-1} up to about 20 GeV electron energy, and with about one third of this luminosity at 25 GeV. More understanding is required on proton beam stability and lifetime in the presence of an electron beam, however, before firm conclusions can be drawn.

Using the unbunched proton beam previously described, an e − p collision geometry has been examined with inelastic electron scattering and weak interaction physics in mind. In order to reach the required luminosity of $\sim 10^{32}$ cm^{-2} s^{-1} with a proton beam of 7 A and an RF power limited electron beam of 250 mA at 20 GeV, very small crossing angles, of the order one milliradian, are necessary. In this design, in which the detector is part of the low-β region, use is made of a weak magnetic field to bring the beams into coincidence. The problems associated with the subsequent generation of synchrotron radiation with respect to machine component design and background effects have been studied. Our preliminary studies indicate that an interesting physics programme is possible with such a facility. A tentative list of parameters is shown in Table III.

Table III

Parameters for an e − p option of the LSR

	Electrons	Protons
Beam energy	20 GeV	400 GeV
Centre-of-mass energy	178 GeV	
Circumference	6150 m	6130 m
Bending field	0.13 T	4 T
RF frequency	200 MHz	−
No. of bunches	4100/2050	unbunched
Beam current	0.25 A	7 A
Radiated power	7 MW	−
Total RF power	10 MW	−
Horizontal crossing angle	$1.5 \cdot 10^{-3}$ rad	
Vertical crossing angle	variable	
Distance to first quad.	± 10 m	± 22 m
Max. luminosity	$\sim 0.7 \cdot 10^{32}$ cm^{-2} s^{-1}	

6. CONCLUDING REMARKS

As mentioned already in the introduction, what has been presented in this talk should only be taken as illustration of the possibilities, and other possibilities exist, both less ambitious ones as well as more ambitious ones. Discussions and studies over the next few years will, hopefully, throw more light on the elements that will determine in which direction we should go, and at which speed.

MY LIFE AS A PHYSICIST'S WIFE

L. Fermi

My life as a physicist's wife began at least one year before I got married, the day Fermi and Rasetti took me to see the crocodiles in the old Istituto di Fisica in Via Panisperna, in Rome. It was sometimes in 1927. Enrico Fermi had been called the previous fall to an especially created chair of theoretical physics, and soon afterwards Franco Rasetti had been appointed "aiuto". They were both 25 years old. As a second-year student of general science, I attended courses in the physics building, but had not been in the laboratories.

The two young men had talked so much about their crocodiles, about having to feed them and take care of them, that to this day I am not sure what I expected: I was certainly disappointed when I saw two shabby wooden spectrographs and was told that they were the crocodiles. Spectroscopy was very fashionable in those days, Rasetti was a spectroscopist, and now and then he made Fermi share in an experiment.

I liked young physicists, and some of the older ones. The year before I had taken the "matematichetta", mathematics for chemists and naturalists, which was taught by Fermi's friend Enrico Persico; and in 1927 I was attending Corbino's course of electricity for students of engineering. Senator Orso Mario Corbino was a fascinating man, short, round-headed and round-bodied, with sparkling dots for eyes. I remember him asking in class whether any really worth-while students would like to shift from engineering to physics and be trained by the new faculty members, Fermi and Rasetti. One student only, Edoardo Amaldi, volunteered. We, the girls, teased him for the good opinion he had of himself, but the Roman school of physics was born; a few months later Emilio Segrè and Ettore Majorana joined it.

I liked Persico, Rasetti and Corbino, and I don't need to say that I liked Fermi. But I did not like physics. One day that summer I asked Fermi to quiz me and see if I was well prepared for the approaching exam on the two-year physics course. We were at Ostia, and Fermi was sitting cross-legged on the sand, in his bathing-suit, which came up almost to his neck. As he quizzed me, his usual grin faded and his lips tightened. In the end he said: "I am sorry, Miss Capon, but you don't understand a thing". What an encouragement!

The months went by and we became engaged. Fermi began to weigh on the kitchen scales the silver objects that we were getting as wedding presents. I knew that Fermi was a man of measurement, but as I handed him the objects to be weighed, I felt guilty. In my bourgeois world, the value of a gift was measured only by the giver's intentions. We were married on July 19, 1928, and we went to the Alps for our honeymoon. On a rainy day Fermi said: "I am going to teach you all the physics there is in just two years, and we'll start right now". Physics on my honeymoon did not appeal to me. But what could I do? I reminded him that I didn't understand physics, as he had told me. He replied: "There are no poor students, there are only poor teachers". But I was the exception. After days of coscientious application, we got into an argument about the validity of the Maxwell equations. What has mathematical equality to do with the equality of real phenomena? I wanted to know. And there my training ended.

But physics doesn't always come easy even to great physicists. The next winter was the coldest on record in Rome

and we began talking of storm windows. Fermi pulled out
his slide rule, calculated the effect of drafts on the inside temperature, misplaced the decimal point, and we froze
all winter. Fermi, however, never made the same mistake
twice: he soon taught me always to estimate the order of
magnitude of a result before undertaking to calculate it.

Many years later, when we were already in America, he
was watching me prepare supper, one evening, with the aid
of a pressure cooker. When we had bought it, giving in to
his love of gadgets, he had explained to me the instructions
in physical terms. Now I was holding the weight in my hand,
waiting for the steam to begin to escape. Fermi asked: "Why
don't you put down the weight right away?" "You told me once, I replied, that if I put the weight down while there is
still air in the cooker I would be cooking at the partial
pressure". Fermi said promptly: "If I said so, it must be
true".

=   =   =

I am jumping ahead. We spent ten years in Rome before
going to America. Physics became more pleasant after Fermi
and his group undertook to produce radioactive isotopes by
neutron bombardment, in early 1934. Up to then Fermi had
been mainly a theoretical physicist and had already completed two important papers: his statistics of the monoatomic
gas, and the theory of beta rays.

The theory of beta rays had just caused him some bitterness. He had sent a letter announcing it to the journal
<u>Nature</u>, but the editor had refused to publish it, saying,
in effect, that it was crazy. Fermi was ready to give up
theory for a while and take up experiments. Artificial radioactivity had just been discovered by the Joliot-Curies in
a few light elements that they had bombarded with alpha particles.

Fermi thought that neutrons might be more effective than alphas and decided to try. Geiger counters were not standard laboratory equipment and he had to build his own. He and his friends prepared also the radon-beryllium sources, not an easy operation. I saw the physicists extract the radon from one gram of radium, and try to seal it inside small tubes containing the beryllium. But some times on the flame a tube went "pop" and broke. Segrè's task was procurement: with a shopping bag and a shopping list he made the rounds of all chemist's shops in Rome and even borrowed gold from a jeweller friend.

The group began bombarding elements systematically, in the order of the periodic table. The lightest elements did not react, but from fluorine up most elements became radioactive. I began picking up bits of atomic physics listening to the men talk shop on Sunday hikes. Radioactivity, disintegration, half-life, the head of the manganese and the tail of some other element .... Things began falling into place. In fact, physics was comprehensible, as long as atoms were small planetary systems and discoveries could be made in goldfish ponds .... like the discovery of slow neutrons.

It was the fall of the same year, 1934. Two little physicists, Amaldi and Pontecorvo, had noted some inconsistency in the results of their neutron bombardment. Fermi looked into the matter. One morning he surrounded the neutron source with paraffin, and immediately the activity in the target increased greatly. Over lunch-time Fermi worked out an explanation: in going through paraffin the neutrons collided with hydrogen atoms, were slowed down and then were more easily captured by the target atoms.

Back in the laboratory after their siesta, the group decided to test Fermi's theory using the most abundant hydrogenated substance at hand; and so they plunged neutron source and target in the goldfish pond at the back of the old physics building. Lo and behold! Fermi was right. Water too increased the radiactivity in the target by many times.

Atomic physics was so pleasant at that time that Ginestra Amaldi and I wrote a book about it. I had been pestering Fermi for suggestions of something intellectual that I might do. I had some time on my hands, because a nursemaid was taking care of our little girl, Nella, and our son Giulio had not yet been born. At last Fermi suggested: "Why don't you write a book?" I was taken aback. "A book? What about?" The answer was one word: "Physics." What else is on a physicist's mind? As I protested that I didn't know enough physics, Fermi suggested that I write with Ginestra Amaldi. And so we did. I wrote the classical part, and when electrons began to jump out of their orbits and to claim they were both matter and wave, Ginestra took over . <u>Alchimia del tempo nostro</u> came out in 1936, a bit early for definitive atomic physics, with an introduction signed by Corbino. We sold 2,000 copies, a smash at that time for a book of popular science.

=   =   =

Meanwhile History began to interfere with everyday life. It had interfered in the early times of fascism; then things seemed to quiet down until they took a turn for the worse in the mid-thirties. There was the Ethiopian war and the consequent economic sanctions against Italy, which exasperated Mussolini. Up to then he had been on France's and England's side, trying to check Nazi expansionism. But now he threw himself into Hitler's outstretched arms, let him have his ways, and didn't protest even when he annexed Austria. In the summer of 1938, to emulate his friend, Mussolini promulgated anti-Semitic laws.

We decided to move to the United States, not only because I come from a Jewish family, but also because we felt that Italy was de facto under German rule. Fermi wrote four letters to four American universities, in veiled terms, fearing censorship. But the Americans are smart. They took the hint, and Fermi got five invitations. He accepted the offer of Columbia University in New York.

We left Rome in early December, 1938, with our two children, two and seven years old, and a nursemaid. We stopped in Stockholm to let Fermi pick up his Nobel prize. One day, while strolling in the streets of Stockholm we ran into a mousy little woman with a tense expression. She was Lise Meitner, then a refugee from German persecution. Until the previous July she had been in Berlin where with Hahn and Strassman she had tried to solve the puzzle of element 93. Perhaps I should explain. In 1934, when Fermi and his group were bombarding uranium with neutrons, they detected an activity which they could not attribute to any elements near uranium in the periodic table. On theoretical considerations they thought that they might have created a new element, 93. They couldn't be certain, because in those early days of the art there were no established techniques to separate the extremely small amounts of substances that were produced. Only time would tell.

But Senator Corbino chose to announce the discovery at the royal session of the Accademia dei Lincei, in the presence of the King and Queen of Italy. The announcement created a great commotion inside and outside Italy. The scientific community was by and large skeptical. Fermi was terribly upset, more than I ever saw him before or after. Yet he and his group did not pursue the matter, feeling unsufficiently skilled; and soon they were engrossed in the study of slow neutrons.

Research on the puzzling product of uranium disintegration was picked up by Hahn and Meitner, and later Strassmann.

In the following years they alternately confirmed and denied the existence of element 93. The puzzle had not been solved by the time Meitner left Germany in July 1938.

Fermi and Meitner did not talk about physics that day in the street. To me the significance of that encounter lay in Meitner's tense, almost scared look, a look that I was to see time and time again on the face of other refugees.

On December 24 we boarded the Franconia at Southampton and landed in New York on January 2, 1939, not without having had on the way a lesson in Anglo-Saxon habits. In the boat's elevator we had run into our first Santa Claus — we didn't know who he was; and I became uncomfortably acquainted with New Year's Eve celebrations when a tall female member of the D'Oyle Carte Company, the famous performers of Gilbert and Sullivan operas, bent almost in two to hug and kiss a passive Fermi.

= = =

My adjustment to American life was slow, but Fermi had only a couple of weeks of not knowing what to do with himself, scientifically. At Columbia he had no laboratory and he was preparing to become again a full-time theoretical physicist when news of the discovery of fission broke out. It came to America with Bohr, who had it, as he was leaving Copenhagen, from Lise Meitner and her nephew Otto Frisch. Meitner had received a letter from Hahn informing her that he and Strassmann had identified barium among the products of uranium disintegration. This meant that the uranium atom split in two almost equal parts. Something that had never been known to happen before!

Bohr arrived in New York on January 16, 1939, exactly two weeks after we did; Fermi and I went to meet him at the pier. He did not mention fission then, and from the pier he went to Princeton to be with Albert Einstein. But in a few days he was at Columbia University looking for Fermi, and so full of the news of fission that when he didn't find Fermi in his office, he unburdened his soul with a graduate student.

As a result the student, Herbert Anderson, invited Fermi to work with him and use the cyclotron for which he had built some equipment. Fermi accepted, glad to be in a laboratory again and to resume the kind of work he had done in Rome. By January 26, only ten days after Bohr's arrival, an experiment to verify fission had been completed, and Bohr and Fermi were at the Annual Conference of Theoretical Physics in Washington, where they dropped a bombshell by revealing fission and its implications. Fermi advanced the hypothesis that neutrons might be released in the process, and so the idea of a chain reaction was launched. Soon Fermi and the Hungarian Leo Szilard were exploring the possibility of building an atomic pile.

= = =

Fermi explained fission to me, and its possible role as the "key that might unlock the great stores of energy in the atoms," as people used to say. But soon secrecy fell on the work of atomic scientists. It was self-imposed having been suggested by Szilard, a man of numberless ideas. With the advent of secrecy I went on a long vacation from physics that was to last until the end of the second World War. Then Fermi brought home a mimeographed copy of the so called "Smyth Report", the scientific history of the development of atomic weapons. And slowly, with great difficulty, I caught up with six-years worth of memorable events.

So if I mention something that has to do with atomic physics, you must assume that I pieced it together later on, from the Smyth report and the stories of many friends.

I didn't miss physics at first, busy as I was keeping

house and becoming Americanized. Fermi had made a deal with Anderson and in exchange for teaching him physics he got lessons in Americana. I learned more slowly, absorbing the democratic ways from our children. We had bought a house in Leonia, New Jersey, a friendly suburb of New York that seemed ideal for raising children. But Nella always asked for "more freedom", implying that I was infringing on her rights when I told her to come home after school before going to play with her friends. And four-year-old Giulio once declared "You can't make me wash my hands. This is a free country". My "charming Italian accent", as kindly Americans called it, was a source of steady humiliations. The worst came when I ventured to order groceries on the telephone - I asked for butter and got bird-seed. Life was not always easy.

= = =

In early December, 1941, two important events took place: the American government decided to push as much as possible the effort to develop atomic energy; and a couple of days later the United States entered the second World War. I'd like to take a couple of minutes to tell you how the American government became involved in atomic energy.

In the United States of those days there were no links between government and the universities, such as the ministry of education in other countries; and virtually no channels of comunications were available. So the scientists took the initiative. An early attempt by Fermi to alert the Navy produced little results - although it now seems probable that the first idea of a nuclear submarine dates back to that meeting between Fermi and Navy officiers. A little later, two Hungarians, Eugene Wigner and Leo Szilard, made a more successful attempt. In the typical devious Hungarian way, they agreed with Einstein, the tallest figure in science, that they would write a letter to President Roosevelt, and he, Einstein, would sign it. By the time the letter was ready, Einstein was vacationing in some place that could be reached only by car. So Wigner and Szilard engaged a third Hungarian physicist, Edward Teller, as a chauffeur.

The letter dated August 2, 1939, started: "Some recent work by E. Fermi and L. Szilard .... leads me to expect that the element uranium may be turned into a new and important source of energy in the immediate future ..,". According to Teller, who told me the story, after reading and signing the letter Einstein exclaimed:"For the first time in history, men will use energy that does not come from the sun!" Not trusting the mails, Szilard gave the letter to economist Alexander Sachs, who occasionally talked to the president. Some weeks went by, war broke out in Europe, and finally President Roosevelt received the message. He set up a committee on uranium, but until December, 1941, work remained limited to a couple of universities, and appropriations were exceedingly small. With the decision to push the effort, the uranium project suddenly expanded immensely, and the work on the atomic pile was moved to Chicago.

Fermi began travelling between New York and Chicago, to wound up his experiments at Columbia and start things going in Chicago. But meanwhile we had become enemy aliens, because Italy and the United States were at war. So Fermi was required to obtain a special permit each time he had to travel: his work was so secret that not even the immigration authorities could be told about it. The permit had to come from Trenton, the capital of New Jersey, where we lived.

We had thought we were settled for good in Leonia, but in the summer of 1942 we all moved to Chicago. We expected to be there for the duration of the war; instead, after two years we moved on to Los Alamos, New Mexico; and at each move we came in contact with more European scientists and stricter security measures. Besides Fermi and Szilard, the foreign scientists in Chicago included James Franck, the Nobel Price winning chemist; Eugene Wigner; and Edward Teller.

As for security measures, at Columbia University a couple of rooms were closed to the non-initiated, but I doubt that many people know about them. In Chicago we had our first experience with fake names: the project was called Metallurgical Laboratory, or Met Lab for short, although there wasn't a single metallurgist in it (the only secret about it that I was told). And soon the Met Lab became a part of the protean Manhattan Project, also a code name. Fermi became Eugene Farmer when he was travelling, and like other key scientists he acquired a bodyguard, a big man of Italian descent, John Baudino, who looked much more impressive than Fermi himself. He knew how to bang a fist on a table, which Fermi could not do. We, the wives, were refused access to the physics building by armed guards, and in addition we had to listen to long lectures on the dangers of loose talk.

The way husbands interpreted secrecy varied. At one extreme was Arthur Compton, the director of the Met Lab: in a book that he published in 1958, he revealed that he had obtained clearance for his wife so that he could tell her the secrets - when other wives and I learned this, we were terribly incensed. At the other extreme was Fermi who was completely tight-lipped. An episode will illustrate his attitude: on the evening of December 2, 1942, we had a party for Met Lab people at our house. As they walked in, all men congratulated Fermi, but nobody would tell me why. At last a friend whispered in my ear: "He has sunk a Japanese Admiral with his ship".

I felt as I had many years before about the crocodiles: I couldn't quite believe ... but ... perhaps he had invented death rays ... The friend insisted: "Do you think that anything is impossible to Fermi?"

So the next morning I asked Fermi: "Did you really sink a Japanese Admiral?" But Fermi put on his best poker face: "Did I?" he asked. "So you didn't," I said. "Didn't I?" he replied. There was really no point in trying to extract information from Fermi.

Of course, that day Fermi had led the experiment that achieved the first chain reaction in the first atomic pile. I am told that he gave a great demonstration of showmanship. He was very much in command of the situation, standing on a platform by the pile, self-assured but watchful; moving his gray eyes from the indicators of neutron activity to his slide rule; predicting how the pile would behave at the next step; ordering physicist George Weil on the floor below to pull out the control rod a little more, and then again a little more, and so step by step until the pile chain-reacted. By all accounts Fermi directed the experiment with the precision of a well rehearsed show. But the show had not been rehearsed: construction of the pile had been completed only a few hours earlier, when the night shift under Herbert Anderson laid the last layer and locked all controls in place. At the end of the experiment the scientists made a silent toast with a bottle of Chianti which Eugene Wigner had bought months earlier, with great foresight, before war conditions made Chianti disappear from the Chicago market. The empty bottle, with the signatures of all who were at the experiment, is still making the rounds of museums and exhibits, and you may run into it at some time.

=   =   =

Los Alamos was the climax of my career as a physicst's wife. It was an experience entirely different from any I had before or after, made of elements each with its own striking individuality. There was the contact with the immensely vast wilderness of mesas, canyons, and desert, that has no comparison in Europe. There was the isolation of the town that did not exist for the rest of the world, was not on the map, was not even a part of New Mexico, so that its inhabitants could not even vote. Its code name was Site Y, its address, Post Office Box 1663, Santa Fe (Santa Fe was the nearest city, some 70 chilometers away). The town, on top of a mesa, encircled by a barbed-wire fence, was run by the Army - whatever the Army provided us with, for istance the blankets for our beds, was stamped USED for United States Engineering Detachment.

We were fingerprinted, given passes and badges, and assigned apartments in barrak-style buildings - the size of the apartment depended on the number of children we had, the rent on the salary of the man. Our mail was censored, as Segrè discovered long before the Army officially announced censorship. Once when ha was away from the project on a trip ha placed a strand of hair in a letter to his wife, but when she opened the letter, the strand of hair was gone. There were so many foreign-born. (General Groves, the chief of the Manhattan Project, called them "crackpots") that "it was all a big accent". Emilio Segrè, Bruno Rossi, Edward Teller, Hans Bethe, and Rudi Peierls were among our old friends. Among the new friends we made were Vicky Weisskopf, Hans Staub, mathematician Stan Ulam, and Johnny Von Neumann, who came to Site Y on visit. Also on visit came Uncle Nick -he was Niels Bohr, but his true name was one of the most closely guarded secrets.

The strangeness of Site Y was magnified by the fact that for the first three weeks I was there with my children but without Fermi, who just before we left Chicago had been called to Hanford in the state of Washington. There, three large piles were being built to produce the plutonium needed for atomic bombs. When construction began, only the Chicago pile had ever operated; that was a very small and simple pile, with no shielding and very simple controls. The Hanford piles, as Fermi would have said, were different animals. So Fermi had been asked to be at hand when the first production pile would start operating, just in case something should go wrong. And something did go wrong - and remained secret for the next 14 years: the pile began to chain-react, as it was expected, but soon it shut itself down. It seemed as if the entire Hanford plant might be a failure. So Fermi had been detained to try and solve the problem. With Wigner he found that the pile was being "poisoned" by a product of fission, a Xenon gas, which absorbed large quantities of neutrons.

While Fermi was at Hanford, I had occasion to meet my first and only spy, and so add to my collection of memorable persons. I forgot to tell you about this "collection". There are in it two kings with whom I dined, King Albert of Belgium at a Solvay meeting in Brussels; and King Gustav of Sweden, at the Nobel ceremonies; there is Madame Curie whom I hardly met because she would not take notice of insignificant wives like me; Rutherford, and Einstein, and other sacred objects of the distant past. In Los Alamos I added Klaus Fuchs the spy. The first Sunday I was there a group of friends organized a picnic in a canyon. Our car was needed, but I wouldn't drive in that unknown, wild territory. So the Peierls asked Fuchs, their friend and protegé, to drive my car. He was an attractive young man, German-born, with a quiet look through round eyeglasses, who answered sparingly to my questions. Even as he spoke to me he was leading a double life, that of a competent physicist appreciated by his colleagues, and that of spy. As he was to confess in 1950, he was giving secret information to the Russians on the progress of the atomic bomb. Fermi was to say Fuchs had made it possible for the Russians to make an atomic bomb five to ten years earlier than they would have otherwise.

There is no time to describe the peculiar features of life in Los Alamos. As for the secret work that went on inside the Technical Area, it led to two events of great momentum: the first atomic explosion at Trinity, the code name for Alamogordo in southern New Mexico; and the dropping of the atomic bomb on Hiroshima and Nagasaki.

In the first part of July, 1945, men began to disappear from the mesa to go to Trinity, without telling us why. But soon we learned, somehow, that at Trinity there was going to be a test of some kind.

On the morning of July 16 the news spread that a sleepless patient at the Los Alamos hospital had seen a strange light in the early morning. We thought that the test must have been successful.

Late the same evening Fermi and a few other men came back. He looked dried out and shrunken, baked by the heat of the southern desert. He was dead tired — for the first time in his life he had felt unsafe for him to drive his car, and had let Sam Allison take the wheel. That was all he had to say. But only a few weeks later, when I could ask questions and get answers, I learned from Fermi that he had seen the dazzling light of the explosion (like one thousand suns), but had not heard the sound, a sound that was described as "a strong sustained, awesome roar". His attention had been concentrated on little pieces of paper that he let fall from his hands. The air blast of the explosion dragged them along and they fell at some distance. Fermi, who always liked simple experiments, measured that distance counting his steps, and so he calculated the power of the explosion. His figures were remarkably close to those obtained with precison instruments and complex calculations. Then he explored the site of the explosion: a depressed area 400 yards in radius was glazed with a green, glasslike substance, the sand that had melted and solidified again.

Three weeks after the test at Trinity the bomb was dropped over Hiroshima. In Los Alamos, President Truman's announcement of the bomb was transmitted over the paging system of the Technical Area. There were no telephones in our homes, and the news spread by way of mouth, as soon as the first husband went home to tell his wife. I was informed by Genia Peierls, who lived in the apartment below ours, and rushed upstairs shouting in her thick Russian accent: "Our stuff was dropped over Japan!" She said "our stuff", because not even by the morning after Hiroshima did we fully realize that Los Alamos was making atomic bombs.

Hiroshima and the end of secrecy set off an explosion of words and feelings in Los Alamos. Children became suddenly very proud of their fathers. Wives asked all the questions that had found no answer since the beginning of the uranium project. And the men appeared altogether changed. I had never heard them mention the atomic bomb, and now they talked of nothing else. They had been absorbed in their research in the protective isolation of the Technical Area, and at least at home they had shown no signs of emotion. Now they were troubled and bewildered, and their concern extended to the whole world. They talked of international control of atomic energy, they posed moral questions.

Perhaps they were not emotionally prepared for the absence of that time interval which usually separates scientific discoveries from their applications. Anyhow, driven to action, the men called meetings, exchanged views, formed associations, and made plans to explain atomic energy and its implications to the public. They felt that if everybody understood the issues, atomic power woul not be used again in a war - it would indeed become a deterrent and prevent wars.

Were they right? It is thirty years since the first explosion at Trinity. In about four weeks it will be thirty years since the end of the second World War - between the first and second World War there were only 21 years.

= = =

This is a good point to put an end to my rambling recollections. By comparison with Los Alamos the eight years in Chicago after the war appear grayish, despite the construction of the Chicago synchrocyclotron and other events.

But it is usual to draw a conclusion to a talk. Here is mine. Some physicists' wives believe that their husbands like physics better than they do their wives. And they may have a point. After work in the evening, when a wife is expecting a word of endearment, like "I couldn't live without you", the husband is most likely utterly silent, absorbed in scribbling numbers and symbols on the margins of the evening paper. When she would like to go to the movies, he has a date with an experiment that cannot wait. There are other complaints, some more justified than others. But all in all, life with a physicist is well worth living.

SOME CONCLUSIONS FROM CERN'S HISTORY

L. Kowarski

European Organization for Nuclear Research (CERN)
Geneva, Switzerland

1. THE DOUBLE ORIGIN

CERN's history is a combination of two distinct components : one is the story of the creation and development of a common European effort in high-energy physics and of how this effort contributed to the advance of one branch of frontline scientific research. The other is the story of the relations of this effort with, so to speak, the outside world - with those people for whom high-energy physics is not the most important thing in life. Strange as this may sound, such people do exist. In fact, they constitute the majority of the world's population and it is they who give the money which enables the high-energy physicists to pursue their work and their way of life.

I prefer not to talk here of the first of these two stories; my audience here knows it better than I do - after all I am not a high-energy physicist myself. I shall talk of the other story : why did all these other people give the money ? What did they expect in return ? And how far did they get what they expected ? I can talk about all this because (as Professor Zichichi told you a moment ago) I was involved in the CERN history from its very beginning, or perhaps even before the beginning, through all its stages until the very recent days.

I was encouraged to appear before you with this story by the Round Table which took place here two evenings ago, under Prof. Rabi's chairmanship, and in which all of these questions about the motives and expectations behind the money were actually raised and discussed, not only in relation to CERN, but in a much wider context of fundamental science in general. Answers were given, some of them conflicting. The importance of the CERN story for this debate is that it offers a very clear and complete case history. The Science-Society relations are at present a fashionable subject

of study and CERN's case offers a great deal of interesting evidence for
this kind of study. Let us take a look at this evidence.

Since money is at the centre of the relationship we are interested
in, let us try to define the amounts of money involved, compared with some
other categories of expenses. Recent stastistics (about a year ago),
expressed in prices and exchange levels of mid-1974, indicate that total
expenditure on high-energy physics in Western Europe (this includes not
only CERN, but also all national activities, and comprises the current
building of heavy equipment) - all this expenditure amounts to about three-
tenths of a U.S. cent each day, per head of population in Western Europe.
This is not much compared to some other things people have to pay for. For
example the Americans, when they had to subsidize putting the man on the
moon over about ten years, if you express it all in 1974 dollars, spent
about 15 times as much, nearly five cents a day per head.

Another interesting comparison is what the Europeans nowadays
have to pay extra for oil (petroleum), ever since oil became so much
more expensive. I don't mean all of what they actually pay for oil, I mean
only what they have to pay extra compared to the prices before October 1973.
It turns out that this extra payment is, per head, about 40 cents a day,
that is 130 times what they spend on high-energy physics.

Well, you will say that therefore the cost of high-energy physics is
negligible. Yet to pay out as much as 1/130 part of what we in Europe now
have to hand out to the Arabs (and to those who make the same demands as
the Arabs), is quite a bit of money. In fact, if you multiply by all the
heads in Europe, it will come to about 360 million dollars a year. Which
is not peanuts. Then why spend so much on an activity which, by itself,
does not appeal to those who give the money ? An activity which, at best,
has a purely spiritual value ? Well, history seems to show that outstanding cultural and spiritual achievements often assume an importance
going beyond their own self-contained aims. They may serve as symbols
and triggers preparing the collective mind for mass movements and great
political events.

It is hardly necessary to cite here the great religious movements -

how, for example starting in the VIIth century, Islam led to the establishment of great Mediterranean empires, or - twelve centuries later - the Mormon faith led to the settlement of a great part of the American West. Less sweeping examples, in the same XIXth century, are the role of poetry in the history of Poland or that of the German composers and philosophers in their own country. Also, the contribution of French philosophy to both the American and the French revolutions should not be forgotten in these days so close to the American bicentennial.

This practical prolongation of spiritual values was explicitly understood by the pioneers of the European reconstruction after World War II, on all of its levels - cultural, economic, political. A typical (and by no means unique) manifestation of this connection may be seen in the informal talks which took place in 1947 between a leading physicist (J. R. Oppenheimer) and a French diplomat (F. de Rose). There were other proposals in the course of development at the same time. Some of them did come to my knowledge; no doubt, many others did not. I may cite here the initiative in favour of international laboratories taken by the French physiologist Henri Laugier who, at that time, happened to occupy one of the highest positions at the United Nations. Another example was the activity of H. Kramers, the famous Dutch physicist who was the chairman of a certain commission of experts at the United Nations and took this occasion to spread similar ideas among the physicists who took part in the U.N. discussions (1946-1948) on the international control of nuclear energy. I mention de Rose more specifically because later on he became the <u>one</u> diplomat thus inspired who played a very important role in the foundation and development of CERN.

Not all of the European countries had to face exactly the same reconstruction problems, but many of them could find an obvious advantage in joining their forces and therefore were ready to respond to the cultural claims of advanced science (backed as they were by its practical by-products, and not only the sinister one of Hiroshima). Nuclear, and soon subnuclear physics was the foremost candidate.

As soon as European physicists became aware of the new role which loomed ahead of them, they began to lay down plans of their own. In the year following the end of the war (1946) America showed to the world how new big equipment could powerfully contribute to the acquisition of new physical knowledge. Europe could do the same if willing to accept the increased scale of effort and the necessity of pooling internationally the essential resources. Among these, it seemed quite obvious at the time that no European nation could afford the money on its own. Later on, with increasing prosperity it became clear that at least Britain, France and - very soon - West Germany, would have been quite capable of paying on their own for a machine like CERN's proton synchrotron. But in early 1950's nobody expected that this would happen so soon and for this reason international financing appeared absolutely indispensable. In reality, the need for pooling people was more decisive : even in retrospect it is clear that no single European country could produce full building teams of a human quality comparable to those that CERN showed itself capable of putting together.

The acceptance of an international framework also meant the adoption of such new forms of cooperation as land and machines developed and held in common, multinational teams, or unprecedented channels of hierarchy and recognition.

Not all cultural leaders, and this goes also for physicists, are adventurous enough in spirit to pay such a substantial price for political and financial support. In the years around 1950 the acceptance of the CERN idea was by no means unanimous among the most influential European physicists. Fortunately enough enthusiasm was found on these high levels, including four Nobel prizewinners (at least one of them went through prolonged hesitation before he joined). The politicians and diplomats were delighted! Their chosen cultural mascot for European unification was not only endorsed by prestigious personalitites, but also offered an experiment in new and very concrete forms of international collaboration.

2. FIRST ACTIONS AND EARLY EXPECTATIONS (1950-1953)

The two distinct trends of interest and mutual dependence were now

CONCLUSIONS FROM CERN'S HISTORY

ready to converge. A few encounters between physicists and diplomats took place in late 1949 and early 1950 but the first decisive step was taken at the Unesco assembly of June 1950. A clear proposal to create one or several international research centres was introduced by Prof. Rabi in the name of the American delegation, and a clear commitment was accepted by Prof. Auger in his capacity as the Head of Unesco's Department of Science. Neither Europe, nor physics were explicitly mentioned in this first official exchange, but the underlying motives were, by that time, well understood and underlined by the fact that it took place between two outstanding physicists.

Assured of support, the scientists now could do their best, and the political leaders now could see that their new mascot promised to be of the highest quality. Under Auger's and Amaldi's leadership, a multinational group of physicists produced their first bold plans of action. The aims and expectations they had at that time are well known here and need not be explained. At the same time, however, the statesmen (and this category now included some top-level scientists involved in the shaping of national science policies) began to entertain vast hopes of their own, whose character can best be shown by the following three examples :

1. In Germany, wholehearted participation in CERN was seen as one way out of the cultural estrangement which was still felt in the first post-war years. (This was confirmed to me in a conversation I had two years ago with a very illustrious German physicist .)

2. In France the successful cooperation with Britain was welcome to those politicians who disliked the idea of the restricted Common Market. (I heard this, in 1954, from one of the most prominent among them. He told me quite enthusiastically : if CERN can prove that you *can* work with the British, this will be worth all the money CERN is asking for !)

3. In 1953 Sir John Cockcroft, at an international conference held in Geneva, was able to point out that after only a year or two of work in common, European accelerator designers could already hold their own in the frontline of technical advance, alongside with the Americans.

The convergence of the two original trends was by then complete and was reflected in the ease with which CERN's founders could get all the money they reasonably needed - an exceptional favour which has persisted ever since : the slight hesitations which the money-givers showed just before 1960 and again about ten years later proved to be short-lived.

3. WHAT EUROPE HAS GAINED

In retrospect it seems clear that CERN's success did not fall short of these early expectations. In the last 20 years, step by step, that part of Europe which adds up to CERN's membership has acquired :

a. A flourishing and well-integrated community of scientists engaged in one definite region of advanced research. This asset is of value not only to the specialists which belong to this region (particle physicists, high-energy nuclear physicists and chemists, applied scientists involved in various branches of electronics, electromagnetic technology, computer science, health physics etc.), but also as an example which can be studied and followed in many other domains of European culture.

It is worth noting here one difference between Europe and the United States : in Europe the accelerator laboratory was started before Europe had a high-energy physics community of any size, except in cosmic-ray research which was different in its spirit and techniques. In the U.S. the great universities were well advanced in high-energy physics before the biggest accelerators were provided for their common use. CERN's role in Europe was, therefore, different in scale from that played in the U.S. by Brookhaven or Fermilab. (Note also the intermediate role of Berkeley, neither a self-contained university, nor truly a centre of nation-wide importance).

b. A place of undisputed equality in the worldwide frontline of fundamental science, thus contributing to dispel the fears of Europe becoming a cultural backwater in the post-war world. CERN in fact became a vehicle for re-introduction of some traditionally European qualities on the world stage of advanced physics : a touch of elegance (well shown in the design and execution of the proton synchrotron), a perfectionist care for precision and depth, a readiness to consider a technical problem from an abstract and general point of view (for example in questions of data processing).

c. New standards of industrial excellence. The influence of "CERN quality" on European industries was well demonstrated at a conference which took place at CERN in April 1974. New techniques developed for meeting CERN's demands have been applied by the same manufacturers on other occasions. This new know-how created, for the European countries concerned, a total volume of business vastly exceeding the total financial burden which the construction and running of CERN has represented for the same countries.

d. New concepts in the organization of research (teams converging to meet away from the home bases of their members; "portable physics"; extended sources of procurement, etc.).

e. A wholly new form of international collaboration : the European "Commonwealth" of high-energy physics. CERN is no longer a concept confined to one piece of land and one payroll of resident personnel. Through its commuting physicists, their home-based groups of pupils, the CERN-connected network of conferences and Summer Schools, CERN is somehow present everywhere in Europe - should I cite this School and its eminent Director as one obvious and outstanding example ?

4. CERN'S WORLDWIDE ROLE

Transcending the geographic limits defined by its membership, CERN's attraction and example have acquired a planetary significance. Particle physics is a scientific activity pursued in all parts of the globe; its practitioners feel a need for one universally accepted meeting place and Geneva is now this place as surely as it is the preferred hub of diplomatic activities of global significance (including those of the United Nations in many of their most important aspects).

The spiritual mascot function of CERN found, perhaps, its clearest expression in the pioneering role CERN has played in the long process of re-establishing the East-West relations. In the years of CERN's foundation (early 1950's), the Cold War was at its coldest, and CERN did not escape its corollaries: Soviet sympathizers in Europe branded CERN as a vanguard of American imperialism and attempted to create an adverse trend in public opinion. By 1954 a recognition of the universal validity of a successful cultural achievement began to gain ground, and in 1956 CERN became a mutually accepted centre of East-West collaboration at a time when such collaboration was as yet much less perceptible in other, more down-to-earth, aspects of global relations.

CERN was perhaps, in this way, the first manifestation of that European solidarity which, later on, helped European nations to maintain a position in world affairs - thus going even further than the original hopes which ensured the post-war political support for the CERN idea.

5. CRITICISMS

CERN undoubtedly has been a success, both on its own plane of fundamental research and in its symbolic and triggering role. Taking this success for granted has become a tradition, and the inevitable criticisms, many of them well justified, should be weighed in this perspective.

The multinational nature of CERN's staff, hierarchy and supervising boards has given rise to a new kind of bureaucracy which, perhaps, developed a tendency to play safe and avoid hazardous commitments. In this atmosphere of caution, CERN's research style, especially in the 1960's, laid stress on precision and detailed exploration rather more consistently than it was the case in the American laboratories with their wider opportunities for adventurous anarchy.

European stability of social structures and career patterns also tended to complicate the relationship between "pure physicists" on one hand and applied physicists and engineers on the other. Both sides developed their own standards of excellence, but outstanding novelties produced on the technical side sometimes took a long time to be noticed by their intended users. Here are a few examples taken both from the CERN centre in Geneva and from the surrounding European laboratories :

a. The alternating-gradient principle of acceleration, even after it was adopted in Geneva, still did not appear safe enough to the decision-making physicists in France and Britain. Thus Saturne and Nimrod, respectively, were made fixed-gradient.

b. CERN's storage rings (ISR), first proposed in the early sixties, were fiercely opposed by what appeared to be an influential majority of physicists in Europe. There was, however, a minority in favour of ISR, strongly supported by Weisskopf, then Director-General of CERN, and CERN finally did get this valuable facility. It is interesting to note that on this occasion Europe showed itself more adventurous than America, where the physicists' opposition caused an even longer delay in the acceptance of the storage-rings idea.

c. The principle of the wire chambers was first described in a CERN report by F. Krienen. It was at the time completely neglected. These chambers were re-invented a little later in America and came back to Europe as a newly-developed American device.

d. Data-processing devices transferring whole photographic pictures, or whole portions of pictures into a computer, and then dealing with them by software, were first developed at CERN. Their acceptance suffered considerable delays in Europe, almost until the development in America of a second generation of devices, in which the "tricky software" principle was successfully combined with other ideas.

The same kind of separatist inertia may have been responsible for the fact that CERN's example found relatively few imitators in other fields of science and technology. The inertia was shown by specialists in those other fields, of which I shall mention a few examples in a moment : "why should we learn from those remote, arrogant high-energy physicists?" But curiously enough, CERN people themselves exhibited their own kind of separatism : "we are the thing, we are the best and foremost, why should we spend our attention and effort on teaching those other fellows how to work successfully?"

The closet analogy occurred in the foundation of ESRO (in which both

Amaldi and Auger were personally involved); the imitation was however
not complete and neither was the success. The less ambitious and, as yet,
not fully deployed ventures, on molecular biology (EMBO) and astronomy
(ESO) were started explicitly in contact with CERN. The absence, so far,
of a substantial joint European effort in computer science and in plasma
physics is, however, all the more conspicuous since in these two cases
CERN would be able to provide not only the organization pattern, but even
an initial nucleus drawn from its own experience in informatics and magnetodynamics.

Even in those fields where CERN's experience did in fact contribute
to the success of other international projects, the influence was implicit
and almost unnoticeable. This applies, for example to some of the European
initiatives in nuclear reactor technology . I may cite here the initiatives
of the Nuclear Agency of the Organization for Economic Co-operation and
Development (O.E.C.D., Paris) which led to such successful developments as
the high-temperature reactor "Dragon", located in England, or the high-
flux research reactor at Grenoble (now run jointly by France, West Germany
and Britain). It would be interesting to study how CERN's influence contributed to these achievements of O.E.C.D., while the much more costly efforts
of Euratom - in the same field, but without any connection with CERN's organizational methods - turned out to be much less effective. We may conclude,
on the whole, that CERN's promise as an example to follow in other domains
of scientific and technological cooperation was never fully realized.

6. SUMMING UP

The balance of successes and shortcomings remains, however, overwhelmingly in CERN's favour, and this could not have been possible if
the founders, the builders and the researchers had not applied their
full endeavour to make the best of the exceptional opportunities which
were made available to them from the very beginning.

The final lesson to be drawn from this story seems to be this : a
craftsman, an artist, a thinker who follows wholeheartedly his inclinations
and inspirations for their own sake need not feel frivolous or unfeeling
when made conscious of wider and more down-to-earth problems of his time.

If his chosen field has any humanly perceivable relevance at all, his achievement will produce decisively beneficial effects in seemingly remote, and sometimes even unpredictable ways. It can almost be said that spiritual values are, in the last resort, the most practical ones.

CLOSING CEREMONY

The Closing Ceremony took place on Thursday, 31 July 1975. The Director of the School presented the prizes and scholarships to the winners as specified below.

PRIZES AND SCHOLARSHIPS

Prize for *Best Student* - awarded to Mr Michael S. PESKIN, Cornell University, Ithaca, N.Y., USA.

The following Students received *honorary mentions* for their contributions to the activity of the School:

Michael E. PESKIN, Cornell University, Ithaca, N.Y., USA.
David J. MILLER, University College, London, UK.
H. David POLITZER, Harvard University, Cambridge, MA, USA.
Thomas E. DE GRAND, Massachusetts Institute of Technology, Cambridge, MA, USA.
Michael J. CREUTZ, Brookhaven National Laboratory, Upton, L.I., N.Y., USA.
Laurence JACOBS, Massachusetts Institute of Technology, Cambridge, MA, USA.
Ivan ANDRIC, Universität Bielefeld, Bielefeld, Germany.
Friedrich DYDAK, CERN, Geneva, Switzerland.
Daniel G. CALDI, Rockefeller University, New York, N.Y., USA.
Martin ULEHLA, Massachusetts Institute of Technology, Cambridge, MA, USA.
Reinhold RÜCKL, Max-Planck-Institute, München, Germany.
Gerald P. CANNING, Københavns Universitet, København, Denmark.
Eldad GAL-EZER, Tel-Aviv University, Tel-Aviv, Israel.
Heinz J. ROTHE, Universität Heidelberg, Heidelberg, Germany.
Murat GUNAYDIN, Scuola Normale Superiore, Pisa, Italy.
Girish C. JOSHI, University of Melbourne, Melbourne, Australia.
Witold KOZANECKI, Harvard University, Cambridge, MA, USA.
Roberto PETRONZIO, University of Rome, Rome, Italy.
Kensuke YOSHIDA, Università di Salerno, Salerno, Italy.
William P. BAURESS, Lincoln College, Oxford, UK.
Miguel CALVO, Universidad Simon Bolivar, Caracas, Venezuela.
Steven GOTTLIEB, Princeton University, Princeton, N.J., USA.
Frixos TRIANTIS, CERN, Geneva, Switzerland.
Douglas A. JENSEN, Princeton University, Princeton, N.J., USA.

Pietro MENOTTI, Scuola Normale Superiore, Pisa, Italy.
Iftkhar MOHAMMAD, University of Manchester, Manchester, UK.
Kurt SUNDERMEYER, Gesamhochschule Wuppertal, Wuppertal, Germany.
R. Keith ELLIS, University of Rome, Rome, Italy.
Koji YOSHIKAWA, Scuola Normale Superiore, Pisa, Italy.
Kwan Je KIM, Max-Planck-Institute, München, Germany.
David L. KREINICK, DESY, Hamburg, Germany.
Meral SERDAROGLU, Middle East Technical University, Ankara, Turkey.
Jan FINJORD, NORDITA, København, Denmark.
Niels HEDEGARD-JENSEN, NORDITA, København, Denmark.
M. Hossein PARTOVI, Arya-Mehr University of Technology, Tehran, Iran.
Ronald C. SHELLARD, University of California, Los Angeles, Calif. USA.
Paul J. STEINHARDT, Harvard University, Cambridge, MA, USA.
Geert-Gonne WINTER, DESY, Hamburg, Germany.

Thirteen scholarships were open for competition among the participants. They were awarded as follows:

Patrick M.S. Blackett Scholarship - awarded to Mr Michael E. PESKIN, Cornell University, Ithaca, N.Y., USA.

James Chadwick Scholarship - awarded to Dr H. David POLITZER, Harvard University, Cambridge, MA, USA.

Amos De-Shalit Scholarship - awarded to Dr Thomas E. DE GRAND, Massachusetts Institute of Technology, Cambridge, MA, USA.

Gunnar Källen Scholarship - awarded to Dr Laurence JACOBS, Massachusetts Institute of Technology, Cambridge, MA, USA.

André Lagarrigue Scholarship - awarded to Dr **Dan**iel G. CALDI, Rockefeller University, New York, N.Y., USA.

Giulio Racah Scholarship - awarded to Mr Martin ULEHLA, Massachusetts Institute of Technology, Cambridge, MA, USA.

Giorgio Ghigo Scholarship - awarded to Dr Eldad GAL-EZER, Tel-Aviv University, Tel-Aviv, Israel.

Enrico Persico Scholarship - awarded to Dr Murat GUNAYDIN, Scuola Normale Superiore, Pisa, Italy.

Peter Preiswerk Scholarship - awarded to Dr Witold KOZANECKI, Harvard University, Cambridge, MA, USA.

Gianni Quareni Scholarship - awarded to Dr Roberto PETRONZIO, University of Rome, Rome, Italy.

Antonio Stanghellini Scholarship - awarded to Dr Miguel CALVO, Universidad Simon Bolivar, Caracas, Venezuela.

CLOSING CEREMONY

Alberto Tomasini Scholarship - awarded to Dr Steven GOTTLIEB, Princeton University, Princeton, N.J., USA.

Ettore Majorana Scholarship - awarded to Dr R. Keith ELLIS, University of Rome, Rome, Italy.

Prize for *Best Scientific Secretaries* - awarded to Dr Roberto PETRONZIO, University of Rome, Rome, Italy and Dr R. Keith ELLIS, University of Rome, Rome, Italy.

The following participants gave their collaboration in the scientific secretarial work:

Ivan ANDRIC	Paolo MARCOLUNGO
Daniel G. CALDI	William A. McNEELY
Thomas A. DE GRAND	David J. MILLER
John A. DIXON	Iftkhar MOHAMMAD
Friedrich DYDAK	Terence A. MULERA
R. Keith ELLIS	Michael E. PESKIN
Babak ETEMADI	Roberto PETRONZIO
Jan FINJORD	H. David POLITZER
Steven GOTTLIEB	Reinhold RÜCKL
Murat GUNAYDIN	Ronald C. SHELLARD
Laurence JACOBS	Paul J. STEINHARDT
Denis P. JOHNSON	Jozko STRAUSS
Kwan Je KIM	Martin ULEHLA
Harold KIRK	Pierre VALIN
Witold KOZANECKI	L. VANRYCKEGHEM
Masaaki KURODA	Koji YOSHIKAWA

PARTICIPANTS

Guido ALTARELLI	Istituto di Fisica dell'Università Piazzale delle Scienze, 5 00185 ROMA, Italy
Ivan ANDRIC	Universität Bielefeld Fakultät für Physik Abteilung Theoretische Physik Herforder Strasse 28 48 BIELEFELD, Germany
Marcello BALDO	Istituto di Fisica Teorica dell'Università Corso Italia, 57 95129 CATANIA, Italy
Giuseppe BANDELLONI	Istituto di Scienze Fisiche Viale Benedetto XV, 5 16132 GENOVA, Italy
Guido BARBIELLINI	Istituto di Fisica dell'Università Piazzale delle Scienze, 5 00185 ROMA, Italy
Barry C. BARISH	California Institute of Technology Lauritsen Laboratory of Physics PASADENA, CA 91109, USA
William P. BAURESS	Lincoln College OXFORD, OX1, UK
Martin BREIDENBACH	Stanford Linear Accelerator Center P.O. Box 4349 STANFORD, CA 94305, USA
Gennaro BROSCO	INFN - Sezione di Pisa Via Livornese SAN PIERO A GRADO (Pisa), Italy
Paolo BUTERA	Istituto di Scienze Fisiche "Aldo Pontremoli" via Celoria, 16 20133 MILANO, Italy
Nicola CABIBBO	Istituto di Fisica dell'Università Piazzale delle Scienze, 5 00185 ROMA, Italy
Daniel G. CALDI	Rockefeller University NEW YORK, NY 10021, USA

Miguel CALVO	Universidad Simon Bolivar Departamento de Fisica Apartado Postal 5354 CARACAS 108, Venezuela
Gerald P. CANNING	Københavns Universitet Niels Bohr Institutet Blegsdamvej 17 2100 KØBENHAVN Ø, Denmark
Giovanni CHIEFARI	INFN - Sezione di Napoli via Tari, 3 80138 NAPOLI, Italy
Sidney R. COLEMAN	Harvard University Department of Physics CAMBRIDGE, MA 02138, USA
Maurizio CONSOLI	Istituto di Fisica Teorica dell'Università Corso Italia, 57 95129 CATANIA, Italy
Leon N. COOPER	Brown University Department of Physics PROVIDENCE, RI 02912, USA
Michael J. CREUTZ	Brookhaven National Laboratory Associated Universities, Inc. UPTON, LI, NY 11973, USA
Gerardo A. CRISTOFANO	Istituto di Fisica Teorica dell'Università Mostra d'Oltremare - Pad. 19 80125 NAPOLI, Italy
James W. CRONIN	The University of Chicago The Enrico Fermi Institute 5630 Ellis Avenue CHICAGO, IL 60637, USA
Richard H. DALITZ	The University of Oxford Department of Theoretical Physics 12 Parks Road OXFORD, OX1 3PQ, UK
Thomas A. DE GRAND	Massachusetts Institute of Technology Department of Physics CAMBRIDGE, MA 02139, USA
Albert N. DIDDENS	CERN - NP Division 1211 GENEVE 23, Switzerland

PARTICIPANTS

John A. DIXON	University of Oxford Department of Theoretical Physics 12 Parks Road OXFORD, OX1 3PQ, UK
Sidney D. DRELL	Stanford Linear Accelerator Center P.O. Box 4349 STANFORD, CA 94305, USA
Friedrich DYDAK	CERN - NP Division 1211 GENEVE 23, Switzerland
R. Keith ELLIS	Istituto di Fisica dell'Università Piazzale delle Scienze, 5 00185 ROMA, Italy
Babak ETEMADI	Arya-Mehr University of Technology Department of Physics Eisenhower Avenue P.O. Box 3406 TEHRAN, Iran
William M. FAIRBANK	Stanford University Department of Physics STANFORD, CA 94305, USA
Graeme T. FAIRLEY	University of Durham Department of Mathematics Science Laboratories South Road DURHAM DH1, UK
Shlomo FEINBERG	Tel-Aviv University Department of Physics and Astronomy RAMAT-AVIV, TEL-AVIV, Israel
Jørgen FENHANN	Københavns Universitet Niels Bohr Institutet Blegsdamsvej 17 2100 KØBENHAVN Ø, Denmark
Thomas FERBEL	The University of Rochester River Campus Station ROCHESTER, NY 14627, USA
Laura FERMI	5532 S. Shore Drive CHICAGO, IL 60615, USA
Jan FINJORD	NORDITA Blegsdamsvej 17 2100 KØBENHAVN Ø, Denmark
Mauro FIORI	Scuola Normale Superiore Piazza dei Cavalieri 56100 PISA, Italy

P.H. FRAMPTON	Syracuse University Department of Physics 201 Physics Building SYRACUSE, NY 13210, USA
W.R. FRANKLIN	Imperial College of Science and Technology Department of Physics Prince Consort Road LONDON, SW7 2BZ, UK
Steven FRAUTSCHI	California Institute of Technology Lauritsen Laboratory of Physics PASADENA, CA 91190, USA
Eldad GAL-EZER	Tel-Aviv University Department of Physics and Astronomy RAMAT-AVIV, TEL-AVIV, Israel
Giuseppe GIANSIRACUSA	Istituto di Fisica Teorica dell'Università Corso Italia, 57 95129 CATANIA, Italy
Ryszard GOKIELI	CERN - TC Division 1211 GENEVE 23, Switzerland
Steven GOTTLIEB	Princeton University Department of Physics Joseph Henry Laboratories Jadwin Hall P.O. Box 708 PRINCETON, NJ 08540, USA
Michel GOURDIN	Université Paris VI Laboratoire de Physique Théorique et Hautes Energies Tour 16 - 1er Etage 4, Place Jussieu 75230 PARIS CEDEX 05, France
David GROSS	Princeton University Department of Physics Jadwin Hall P.O. Box 708 PRINCETON, NJ 08540, USA
Murat GUNAYDIN	Scuola Normale Superiore Classe di Scienze Piazza dei Cavalieri 56100 PISA, Italy
Donald L. HECKATHORN	University of Chicago Enrico Fermi Institute 5630 Ellis Avenue CHICAGO, IL 60637, USA

PARTICIPANTS

Niels HEDEGARD-JENSEN
: NORDITA
Blegsdamsvej 17
2100 KØBENHAVN Ø, Denmark

Andreas HEINZ
: Universität Hamburg
II. Institut für Theoretische Physik
Luruper Chaussee 149
2 HAMBURG 50, Germany

Clemens A. HEUSCH
: University of California, Santa Cruz
Division of Natural Sciences
SANTA CRUZ, CA 95060, USA

A. Barbosa HENRIQUES
: University of Glasgow
Department of Natural Phylosophy
GLASGOW G12 8QQ, UK

Gabriele INGROSSO
: Istituto di Fisica
dell'Università
Via Arnesano
73100 LECCE

Laurence JACOBS
: Massachusetts Institute of Technology
Department of Physics
CAMBRIDGE, MA 02139, USA

Douglas A. JENSEN
: Princeton University
Joseph Henry Laboratories
Jadwin Hall - P.O. Box 708
PRINCETON, NJ 08540, USA

Hans JOHANSSON
: University of Stokholm
Department of Physics
Vanadisvägen 9
113 46 STOCKHOLM, Sweden

Kjell JOHNSEN
: CERN
1211 GENEVE 23, Switzerland

Denis P. JOHNSON
: Université Libre de Bruxelles
Faculté de Science
Service de Physique
des Particules Elémentaires
Avenue F.D. Roosevelt 50
1050 BRUXELLES, Belgium

Girish C. JOSHI
: University of Melbourne
Department of Physics
MELBOURNE, Victoria, Australia

H. Jeffrey KATZ	California Institute of Technology Mosher-Jorgensen 317 PASADENA, CA 91126, USA
Kwan Je KIM	Max-Planck-Institut für Physik und Astrophysik Institut für Physik Föhringer Ring, 6 8 MÜNCHEN 40, Germany
Harold KIRK	III. Physikalisches Institut Charlottenstrasse 14 51 AACHEN, Germany
Ian KIRKBRIDE	University of Oxford Department of Nuclear Physics Nuclear Physics Laboratory Keble Road OXFORD OX1 3RH, UK
Lew KOWARSKI	40 Avenue William Favre GENEVE, Switzerland
Witold KOZANECKI	Harvard University Department of Physics High Energy Physics Laboratory 42 Oxford Street CAMBRIDGE, MA 02138, USA
David L. KREINICK	DESY Notkestieg 1 2 HAMBURG 52, Germany
Masaaki KURODA	University of Tokyo Department of Physics 3-1 Hongo 7-Chome Bunkyo-Ku 113 TOKYO, Japan
Herbert LOTSCH	Technische Hochschule Wien Institut für Theoretische Physik Karlsplatz 13 1040 WIEN, Austria
Luciano MAIANI	Istituto Superiore di Sanità Laboratorio di Fisica Viale Regina Elena, 299 00161 ROMA, Italy
Paolo MARCOLUNGO	Istituto di Fisica "G. Galilei" dell'Università Via Marzolo, 8 35100 PADOVA, Italy

PARTICIPANTS

André MARTIN	CERN - Th Divsion 1211 GENEVE 23, Switzerland
Albert MAS-PARAREDA	Universität Graz Institut für Theoretische Physik UniversitätPlatz 8010 GRAZ, Austria
C. Brian A. McCUSKER	University of Sidney School of Physics SIDNEY, N.S.W. 2006, Australia
William McNEELY	DESY Notkestieg 1 2 HAMBURG 52, Germany
Pietro MENOTTI	Scuola Normale Superiore Classe di Scienze Piazza dei Cavalieri 56100 PISA, Italy
David J. MILLER	University College Department of Physics Gower Street LONDON, WC1E 6BT, UK
Leah MIZRAHI	Tel-Aviv University Department of Physics and Astronomy RAMAT-AVIV, TEL-AVIV, Israel
Iftkhar MOHAMMAD	University of Manchester Department of Physics MANCHESTER M13 9PI, UK
Giacomo MORPURGO	Istituto di Scienze Fisiche dell'Università Viale Benedetto XV, 5 16132 GENOVA, Italy
Terence A. MULERA	Argonne National Laboratory HEP 3625ZGS User's Office 9700 South Cass Avenue ARGONNE, IL 60439, USA
Nobuyuki MURAI	Westfield College Department of Physics Kidderpore Avenue LONDON NW3 7ST, UK

M. Hoseein PARTOVI	Arya-Mehr University of Technology Department of Physics Eisenhower Avenue P.O. Box 3406 TEHRAN, Iran
Michael E. PESKIN	Cornell University Laboratory of Nuclear Studies ITHACA, NY 14853, USA
José PETERNELJ	University of Alberta Department of Physics EDMONTON, Alberta, Canada
Roberto PETRONZIO	Istituto di Fisica dell'Università Piazzale delle Scienze, 5 00185 ROMA, Italy
G.B. PIANO MORTARI	Istituto di Fisica dell'Università Piazzale delle Scienze, 5 00185 ROMA, Italy
Andrew PICKERING	Daresbury Laboratory Science Research Council DARESBURY, Warrington WA4 4AD, UK
H. David POLITZER	Harvard University Department of Physics CAMBRIDGE, MA 02138, USA
Francesco POSA	Istituto di Fisica dell'Università Via Amendola, 173 70126 BARI, Italy
Giuliano PREPARATA	CERN - Th Division 1211 GENEVE 23, Switzerland
Isidor I. RABI	Columbia University Department of Physics P.O. Box 135 NEW YORK, NY 10027, USA
Jesus RAMOS	Universidad de Madrid Facultad de Ciencias Departamento de Fisica Teorica MADRID-3, Spain
Claudio REBBI	Massachusetts Institute of Technology Center for Theoretical Physics CAMBRIDGE, MA 02139, USA

PARTICIPANTS

Heinz G. ROTHE	Universität Hiedelberg Institut für Theoretische Physik Philosophenweg 16 69 HIEDELBERG, Germany
Carlo RUBBIA	CERN - NP Division 1211 GENEVE 23, Switzerland
Reinhold RÜCKL	Max-Planck-Institut für Physik und Astrophysik Institut für Physik Föhringer Ring 6 8 MÜNCHEN, Germany
Roberto SACCO	Istituto di Fisica Superiore dell'Università Corso Massimo d'Azeglio 10125 TORINO, Italy
Abraham SEIDEN	CERN - NP Divsion 1211 GENEVE 23, Switzerland
Meral SERDAROGLU	Middle East Technical Univerity Department of Physics ANKARA, Turkey
Ronald C. SHELLARD	University of California Department of Physics 405 Hilgard Avenue LOS ANGELES, CA 90024, USA
Paul J. STEINHARDT	Harvard University Department of Physics CAMBRIDGE, MA 02138, USA
Jozko STRAUSS	CERN 1211 GENEVE 23, Switzerland
Kurt SUNDERMEYER	Gesamhochschule Wuppertal Faculty of Physics Hofkamp 82-84, Postfach 13 02 22 56 WUPPERTAL 1, Germany
Kyriakos TAMVAKIS	Brown University Department of Physics PROVIDENCE, RI 02912, USA
Gerard 't HOOFT	Instituut voor Theoretische Fysica Rijksuniversitèit Utrecht Princetonplein 5 UTRECHT 2506, The Netherlands

Samuel C.C. TING	Massachusetts Institute of Technology Department of Physics CAMBRIDGE, MA 02139, USA
Luciano TRASATTI	Laboratori Nazionali di Frascati CNEN Casella Postale 56 00044 FRASCATI, Italy
Frixos TRIANTIS	CERN - TC Division 1211 GENEVE 23, Switzerland
Martin ULEHLA	Massachusetts Institute of Technology Department of PHysics CAMBRIDGE, MA 02139, USA
Pierre VALIN	Harvard University Department of Physics CAMBRIDGE, MA 02138, USA
L. VANRYCKEGHEM	University of Liverpool Department of Theoretical Physics P.O. Box 147 LIVERPOOL L69 3BX, UK
Claude-Michel VIALLET	Laboratoire de Physique Théorique et Hautes Energies Université de Paris VII 2 Place Jussieu 75221 PARIS CEDEX 05, France
Kadayam S. VISWANATHAN	Institut des Hautes Etudes Scientifiques 91440 BURES-SUR-YVETT, France
Victor F. WEISSKOPF	Massachusetts Institute of Technology Department of Physics CAMBRIDGE, MA 02139, USA
Milton G. WHITE	Princeton University Department of Physics Jadwin Hall P.O. Box 708 PRINCETON, NJ 08540, USA
Björn H. WIIK	DESY Notkestieg 1 2000 HAMBURG 52, Germany
Kenneth G. WILSON	Cornell University Laboratory of Nuclear Studies ITHACA, NY 14850, USA

PARTICIPANTS

Richard WILSON
: Harvard University
Department of Physics
Lyman Laboratory of Physics
CAMBRIDGE, MA 02138, USA

Geert-Gonne WINTER
: DESY, F 39
Notkestieg, 1
2 HAMBURG 52, Germany

Kurt WITTMAN
: Universität Wien
Institut für Theoretische Physics
Boltzmanngasse 5
1090 WIEN, Austria

Kensuke YOSHIDA
: Istituto di Fisica
dell'Università
Via Vernieri, 42
84100 SALERNO

Koji YOSHIKAWA
: Scuola Normale Superiore
Classe di Scienze
Piazza dei Cavalieri, 6
56100 PISA, Italy

INDEX

Adler anomaly, 20
Alpha particles, 27, 247
Amplitudes, scattering
  See Scattering amplitudes
Amplitude analysis
  as one energy, 436
  using various energies and angles, 439
Angular momentum states, decoupling, 208
Antibaryon annhilation, 785
Antikinks, 309, 338
Antineutrinos, 869
  dimuon events from, 924
  elastic scattering, 1021
  neutral currents from, 967
  scattering, 916, 925, 982
  Y distributions, 92
Antipartons, 890
Antiprotons, 785
Antiquarks, 43
  colour, 29
  movement, 154
  spectrum, 891
Antisolitons, 309, 372
Argonne-Wisconsin experiment, 939
Asymptotic freedom, 17, 18, 71, 78, 221, 261, 291, 295, 409, 485, 489, 493, 861
Asymptotic states, 144

Bags, 145
  See also Bubbles
  Bjorken scaling, 556
  boundary motion, 535, 555

Bags (cont'd)
  boundary variables, 541, 546, 548, 549, 554
  coupling in, 174, 176, 177, 178, 181, 182, 190, 213
  deformed, 159, 553
  degrees of freedom, 555
  experimental confirmation of structure, 185
  Fermi fields in, 259
  formation of, 147
  gluon fields in, 186
  implications of, 260
  instability, 187
  interactions, 186, 557
    with quarks, 145
  massless quarks in, 171
  MIT, 188, 209, 241-260, 531
    axial currents, 255
    collapse, 259
    comparison with reality, 251
    coupling, 190
    description, 241
    free quark model, 243
    gluon field in, 247, 258
    hadron mass, 251
    in one space dimension, 554
    introduction of gluon field, 242
    magnetic moments, 253, 257
    massive quark model and, 234
    mean square fluctuation, 556
    numerical predictions, 170
    quark movement in, 170
    quarks in, 242
    relation to SLAC model, 169, 170, 185

Bags (cont'd)
  MIT (cont'd)
    small oscillations, 553
    spectrum of, 191
    SU(6) symmetry, 242
    with more than three
      quarks, 255
    with more than three
      nonstrange quarks, 246
    zero point energy, 248
  quark propagation in, 201
  radii, 254
  Regge states and, 223
  shape of, 163, 165
  six quarks in, 256
  SLAC, 148
    Bjorken scaling in, 189
    boson coherent states, 152
    definition of Hamiltonian, 152
    exotic states, 156
    Fermion states and
      Bogoliubov
      transformation, 153
    Fock space, 151
    magnetic moments, 156
    relation to MIT model,
      169, 170, 185
    surface area, 158
  small oscillations, 533–538
    Hamiltonian for, 539,
      544, 547
  spectrum of, 191
  stabilization, 186
    surface of, 188, 189
  time and space, 201, 207,
    208
  total field energy, 164
  with single charged scalar
    field, 534
Baryons, 40, 241, 1055
  charmed, 50, 65
    mass, 52, 64
  configuration, 32, 513
  coupling constants in
    decay, 245
  20-dimensional
    supermultiplet, 42
  excited, 128
  Fermi statistics for, 840

Baryons (cont'd)
  formation, 290
  mass, 63, 64, 92
  multiplets, 42, 74
  propagators, 92
  SU(6) multiplets, 74
Baryon-antibaryon bound state,
  527
Baryon-antibaryon systems, 1055
Baryon number, black holes and,
  739
Beg-Zee model, 1006
Bell-Jackiw anomalies, 284
Bethe-salpeter approach, 202,
  223, 235
Big bang theory, 733
Bjorken scaling, 106, 468
  in lepton-hadron scattering,
    762
  neutral currents and, 976, 980
  tests for, 763
Black holes
  collision with earth, 738
  decay of, 733
  elementary particle model,
    733, 735
  emission from, 734, 737
  energy obtained from, 739
  formation of, 733
  lifetime of, 736
  mini-, 731, 732
    detection of, 736
    practical use of, 738
  spectrum of, 731, 732
  statistical bootstrap model,
    733, 735
Bogoliubov transformation, 153
Bohr's correspondence principle,
  374
Bohr-Sommerfeld quantization,
  376
Bosons, 419
  free massive, 387
  Goldstone, pion as, 157, 187
  relation to fermions, 518
Bubbles
  shape of, 165, 168
  static equations, 167
  time-dependent, 168
  two-dimensional, 166

INDEX

Cabibbo theory, 34, 61, 67
Caltech-Fermilab neutrino
    experiment, 897-927
Cancer radiotherapy, 704,
    708, 711
CERN
  history of, 1201
    early, 1204
    intersecting storage rings,
      1173, 1209
      anti-protons in one ring,
        1179
      e-p option, 1181
      insertions, 1179
      interaction regions, 1175
      machine lattice, 1174
      performance estimates, 1178
    spiritual value of, 1203
    wire chambers, 1209
    worldwide role of, 1207
Charged current interactions,
    913
Charm, 628, 747, 859
  discovery of, 660
  evidence for, 62
  hidden, 812, 823
  motivation for, 809
  new states in, 810
  scale of, 36
Charm changing transitions,
    481
Charm spectroscopy, 33-67
  charmonium states, 53
  group theoretical methods,
    46
  $1P_1$ state, 57
  $3P_j$ state, 54
    decays, 56
  P wave states, 53
  SU(4) supermultiplets, 36
Charmed particles, 27
  decay, 61, 63, 66
    non-leptonic, 489
  mass, 43
Charmonium model, 40, 53,
    656, 660
Chicago-Harvard,Pennsylvania-
    Wisconsin neutrino
    experiment, 939

Chicago-Princeton experiment on
    nucleon-nucleon
    collisions, 938, 948
Chiral selection, 477
Chiral symmetry, 137, 283, 294
Clusters, spin of, 793
Coherence effects, 1159
Colour, 628, 747, 762, 804, 837
  dynamical significance, 132
  electromagnetic interactions,
    842
  interactions, 147, 842
  invisibility of, 23
  local invariance, 24
  motivations for, 838
  possible schemes, 844
  tests for, 847
  string theory, 78, 526
  three-triplet models, 841
Colour confinement, 264
Colour currents, 137
Colour gauge groups, 261
Colour gauge theory, 69, 79,
    103, 143
  lattices and, 73, 130
    discrete, 70
  renomalization of, 70
Colour singlet states, 21, 23, 39
Columbia-Fermilab-Illinois-
    Cornell Hawaii group, 949
Columbia-FNAL experiment, 939,
    948
Cosmic radiation, 1135-1150
  development of large showers,
    1149
  distribution of P_T, 1138
  hadrons in, 1140
    Centauro, 1142
    Tien Shan experiment, 1141
  high transverse momenta, 1135
  interaction at $>10^{17}$ éV, 1149
  long-lived particles in, 1144
    Kolar gold field experiment,
      1145
    Niu's X particle, 1144
    time of flight experiments,
      1147
    Yunnan event, 1144
  particles delayed behind air
    shower front, 1148

Cosmic radiation (cont'd)
  showers
    P_T for, 1137
    electron distribution in, 1135
    multicoloured events, 1136
    single and multi, 1136
    transverse momenta, 1140
Cosmic rays, 710
Coulomb interference, 459
Currents
  See Neutral currents, etc.

DASP, 635
  description of, 637
DHN formula, 377
Dalitz decays, 934, 963
Deck model, 685, 690
$\Delta T = \frac{1}{2}$ rule, 467, 475, 490
Derrick's theorem, 310
Diffractive dissociation, 787
Diffraction peaks, 444
  geometric scaling, 449, 460
Dimuons
  equal sign, 886
  mass, 923
  production, 756, 877
    energy dependence, 924
Dimuon events, 866, 894, 896, 897
  beam configurations, 868
  branching ratios, 870
  decay and, 895
  distribution of, 923
  experimental details, 866
  from antineutrinos, 924
  from neutrinos, 905
  from nucleon-nucleon collisions, 957
  main features of, 870
  origin of, 872, 875, 912
  quadruple focusing beam, 893
  selection and background, 868
  source of, 911
Dirac monopoles, 346, 391
Dirac string, 401, 411
Dissipation, 297, 298
DORIS, 635
  description of, 635

DORIS (cont'd)
  parameters of, 637
  results with, 640
Doublet solution, 373
Drell-Yan process, 954
Drift chambers, 796
Dual resonance models, 494
  relation to gauge field theory, 493-532
    quantum mechanical strong coupling example, 515
    Regge slope expansion, 519
    spontaneous breakdown, 520
Dual string model, 508

Elastic cross sections
  Pomeranchuk-like theorem for, 441, 463
Electromagnetic interactions in colour, 852
Electrons
  definitions, 653
  from nucleon nucleon collisions
    calculation of yield, 954
    Cern-Columbia-Rockefeller-Saclay experiment, 942
    Columbia-FNAL experiment, 942
    measurement, 933, 939
    neutral pion lifetime experiment, 943
    Pennsylvania-Stony Brook experiment, 943
  movement of, 25
Electron pair transitions, 1153
Electron scattering, scaling in, 597
Electroproduction, 766, 991
Elementary particle model, 733, 735
ETA particle, 283, 286
Ettore Majorana Centre, 7, 11

Fermi, Enrico, 1183
Fermilab, experiments with muons, 591
Fermilab energies, neutron dissociation at, 663-702

INDEX 1233

Fermions, 312, 419, 500
  as coherent state, 386
  quarks becoming, 236
  relation to bosons, 518
Fermion systems, 153, 1151-1171
  one dimensional, 1151
  three dimensional, 1158
  two dimensional, 1156
Feynman diagram for composite meson propagator, 111
Feynman rules
  free scalar particles, 82, 83
  gauge theory, 264
  lattice theory, 133
  three quarks on lattice, 88
Feynman scaling, 780
Field theories, 298
  quark confinement schemes in, 143-191
    coupling, 174, 176, 177, 178, 181, 182, 190
    MIT bag, 169
    properties, 182
    quantum corrections, 172, 191
    SLAC bags, 148
Firesausages, 208
  decay of, 228
  dimensions of, 239
Fock space, 151
Free particle transitions, 82

Gauge theory, 17, 25, 316, 863
  association of element of G with space time path, 319
  charm changing transitions, 481
  Feynman rules for, 264
  for strong interactions, 261
    comparison with σ model, 281
    η-3π decay, 286
    $\Delta T = \frac{1}{2}$ rule, 467, 475, 490
    GIM model, 470, 480, 481, 483, 484
    infrared divergence and, 292
    in lattices, 526, 529
      quark loops in, 506

Gauge theory (cont'd)
  for strong interactions (cont'd)
    in one space one time dimension, 271
    large N limit, 266
    long distance behaviour, 501
    mass of bound state, 277, 294
    mass of ETA particle, 283
    Reggeization of, 495
    relation with dual resonance models, 493-532
      quantum mechanical strong coupling, 515
      Regge slope expansion, 519
    Salam-Weinberg, 1036
    short distance singularities, 467, 473, 474, 475, 483, 485, 489
    singular and nonsingular, 398
    spontaneously broken, 314
    string models and, 16, 26
    surfaces, 270
    symmetry, 395
      breaking, 470
    theorem, 393
    theories of, 311
    two dimensional, 284, 298
    two dimensions, 271
    weak nonleptonic amplitudes in, 465
Glasgow-Illopoulos-Maiani model, 470, 480, 481, 483, 484
Gluons, 70
  understanding, 131
Gluon field in MIT bag, 242, 247, 258
Gluon flux with no quarks, 409
Goldstone boson, 293
Goldstone mode, 420
Goldstone phenomenon, 290
Gravitons, in black holes, 734
Green's functions, 196, 206, 212
  at high energy, 220
  hadronic amplitudes in terms of, 197
Gross-Llewellyn-Smith sum rule, 1012

Hadrodynamics with quarks, 193-240

Hadrons
　as lumps, 391
　azimuthal distribution, 752, 753
　break up of shell, 159
　charmed, 62
　collisions, final states, 219, 223
　constituents of, 262
　current vertex, 216
　decay, 621, 850
　　muon yield from, 930
　decoupling of higher angular momentum states, 208
　elementary scattering amplitudes, 197
　energies, 46
　energy distributions, 907
　excited, 31, 158
　exotic states, 1061
　geometrical description, 211
　G_t contribution and Regge states, 222
　-hadron scattering, 223
　in mini black holes, 732
　interactions, 211, 778, 958
　　correlations, 1132
　　　large transverse momentum particle, 1122
　　　one particle in fragmentation region, 1116
　　diffraction dissociation, 1134
　　elastic scattering, 1134
　　high energy, 1067-1134
　　　charge exchange scattering, 1098
　　　correlation studies, 1115
　　　diffraction dissociation, 1107
　　　elastic scattering, 1078
　　　inelastic cross section, 1074
　　　low multiplicity exclusive channels, 1112
　　　many body final states, 1107

Hadrons (cont'd)
　interactions (cont'd)
　　high energy (cont'd)
　　　many particle production, 1115
　　　polarization in elastic scattering, 1093
　　　total cross sections, 1067
　　　two body, 645
　　　two body final states, 1067
　　irreducible vertex functions, 198, 213
　　lepton scattering by, 593
　　linear decay chains, 224
　　mass, 251
　　non-charmed, spectroscopy, 1053-1066
　　normalization, 203
　　particles, 194
　　(η π) phases, 793
　　polarizations, 1130, 1131
　　possible states, 37
　　production, 611
　　scattering, 241
　　　at high energies, 226
　　　length, 793
　　slope-mass correlations, 783
　　space time description of, 199
　　spin-like behaviour, 752
　　spin of clusters, 793
　　spin-spin interactions, 250
　　structure of, 37
　　SU(6) multiplets, 1057
　　SU(6) symmetry, 1053
　　thin shell model, 150
　　three-bag vertex, 217
　　wave functions, 197, 212
　　weak and e m currents, 198, 215
Han-Nambu triplet scheme, 839
Harari-Freund duality, 237
Harvard-Pennsylvania-Wisconsin experiments, 957
Heavy ions
　acceleration to relativistic energies, 703-730
　AGS parameters, 718
　applications of, 705, 708
　Bevalac parameters, 722
　biological applications, 711
　in nucleon structure research, 709

INDEX

Heavy ions (cont'd)
  acceleration to relativistic
      energies (cont'd)
    in radiotherapy, 708
    multi-nucleon character
      of, 706
    properties of, 705
    requirements of
      accelerator, 712
Hedgehogs, 200, 330
  See also Monopoles
Higg's scalar particles, 17
High energy Scales, 19
Hilbert space, 262
Homomorphism, 341
Homotropic, definition of, 323
Homotropy classes, 321, 335, 338
Homotropy groups, 330, 412
  elements of, 335, 338
  mathematical definitions, 333
Hydrogen, neutron diffractive
      production on, 680

Infrared divergencies, 470
  gauge theory and, 292
Interactions,
  gauge theory for, 261
    comparison with σ model, 281
    large N limit, 266
    masses of bound states, 277, 294
    mass of ETA particle, 283
    nonleptonic amplitudes
      in, 465
    two dimensions, 271
Ions, heavy
  See Heavy ions

J, 229
  definition of, 123
J particles
  See under Particles
J plane, 237
Jets, 747, 752
  observed at ISR, 789
  theories of, 754

Kaons, 905, 930
Kinks, 305, 309, 338
Kink-antikink solutions, 188
Klein paradox, 734
Kolar gold field experiment, 1145

Lattices
  basic theory of, 25, 69, 526
  causality, 131
  colour gauge theory and, 130
  Euclidean time, 82
  excited meson states, 74
  Feynman rules, 133
  free particle behaviour, 104, 105
  free particle movement, 82
  in relation to gauge theory
      to dual resonances, 501
  movement of mesons in, 80
  propagator, 90, 135
  quarkless states, 76
  quark lines on, 92
  quarks and strings on, 69-142
    angular momentum, 75
    Bjorken scaling, 106
    colour gauge theory, 79
    continuum limit, 85, 90, 103, 113, 127, 130, 132, 135, 136, 138, 141
    continuum models, 75
    dual resonance model, 130
    dynamics, 80
    exotic states, 76
    Feynman rules, 82, 83, 88, 91
    free quark behaviour, 105
    gauge transformation, 108
    higher order gauge field
      contractions, 118
    loops, 109, 110
    numbers present, 106
    orbital angular momenta, 129
    propagators, 89
    quantum mechanics, 73
    quantum operators, 78
    renormalization, 110, 135, 138, 141
    scalar particle spectrum, 86
    scattering by exchange, 82
    single particle states, 74

Lattices (cont'd)
  quarks and strings on (cont'd)
    static theory, 71
    transition operators, 81
  self-coupled neutral scalar boson field on, 174
  spacing, 71
  string particles in, 31
Leptons, 35, 61
  hadron scattering, 593
  heavy, 747
    production and decay, 876
  known sources of, 948
  production, 759, 835, 912
    in nucleon-nucleon collisions, 929-965
    in pairs, 958, 965
    yield of, 963, 964
  scattering, 602
  yields, 193
Lumps, 297-421
  Abelian and non-Abelian magnetic monopoles, 339
  and their quantum descendents, 297
  as particles, 307
  concept of, 299
  construction of, 351
  energy of, 305, 359, 360
  hadrons as, 391
  in sine-Gordon theory, 305
  mesons interacting with, 357
  nature of classical limit, 350
  patching together distant, 330, 337
  periodic, 371
  quantitative and qualitative information, 388
  small oscillations and stability, 304
  solitons as, 389
  spherically symmetric, 360
  stability angles, 377, 378
  time dependent
    coherent-state variational method, 361
    in one space dimension, 300
    power series expansion, 352

Lumps (cont'd)
  topological conservation laws, 312
  ultraviolet divergences, 363
  variational method, 414

Magnetic flux, definition of, 342
Magnetic moments
  of MIT bag, 253, 257
  of proton and neutron, 257
Magnetic monopoles, 391, 513
Massive quark model, 185, 194
  criterion and spectrum, 206
  degeneracy of lower lying resonances, 238
  Feynman scaling, 239
  MIT bag and, 234
  momentum space analysis, 204
  time oscillations, 239
Mathur-Okubo-Kim model, 1006
Mesons, 14, 23, 241, 375, 1054
  charm, 811
    decays of, 61, 852, 860
    detection of, 834
    mass, 49
    production of, 836
  decay, 63, 127
  degenerate, 157
  equations of motion, 200
  excited states, 74
  from nucleon-nucleon collisions, 949
  group theoretical analysis, 48
  hidden charm, 813
  in sine-Gordon theory, 387, 388
  interactions, 292, 357
    with lumps, 357
  in Thirring model, 387
  $J^P = 4^+$, 776
  mass, 92, 111
  movement on lattice, 80
  production, 951
  propagators, 92, 95, 97, 99, 104, 105, 111, 139
  pseudoscalar, 820, 822, 859, 860, 1054
  qq wave function, 210
  quarkless, 31
  spectroscopy, 776
  spectrum, 134

INDEX 1237

Mesons (cont'd)
  vector, 520, 1054
    Reggeization, 496
  wave functions, 197, 199
Meson-meson scattering, 528
Meson supermultiplets, 39
K mesons, 27
μ Mesons, 27
ρ Meson
  instability, 81
  production of, 605, 951
  propagators, 100
  search for, 603
π Mesons
  instability, 81
  mass, 111
  propagator, 100
φ Mesons, production, 951
ω Mesons, production, 951
MIT bag, 188, 209, 241-260, 531
  axial currents, 255
  collapse, 250
  comparison with reality, 251
  coupling in, 190
  description, 241
  energy, 248
  free quark model, 243
  gluon field in, 242, 247
  hadron mass, 251
  in one space dimension, 554
  magnetic moments, 253, 257
  massive quark model, 234
  mean-square fluctuation, 556
  numerical predictions, 170
  quarks in, 242
    movement, 170
  relation to SLAC model, 169, 170, 185
  small oscillations, 533
  spectrum of, 191
  SU(6) symmetry, 242
  with more than three quarks, 255
MIT-BNL experiments, 566
Molecules
  diatomic, 353, 354
  polyatomic, 353, 356
Momentum space, wave functions in, 204

Momentum space analysis, 204
Monopoles, 300
  Dirac, 346, 391
  magnetic, 330, 391
    Abelian and non-Abelian, 339
  relation to strings, 347
Motion, equations of, 298, 299, 302
Multilepton final states
  See Dimuon events
Muons
  as probes of nucleon structure, 589
  Chicago-Harvard-Pennsylvania-Wisconsin experiment, 939
    Chicago-Princeton experiment, 938
    Columbia-FNAL experiment, 939
    measurement, 930, 959
    polarization, 938
    Serpukov experiment, 936
    Yale-Brookhaven experiment, 941
  from nucleon-nucleon collisions, 939
    Argonne-Wisconsin experiment, 939
  from pion decay, 930
  higher precision of, 771
  identification of, 638
  pair production of, 949
  polarization of, 938
  production, 965
    from hadron decay, 930
  radiation, 592
  scattering at 150 GeV, 589-608
  transverse momentum of, 872
Muon beam, design of, 591
Muon-proton scattering, 597
Muproduction, 766

Nambu model, 146, 512, 530
Neutral currents, 755-776, 897, 927
  analysis of inclusive reactions, 992
  conservation properties, 1014
  coupling and, 902

1238   INDEX

Neutral currents (cont'd)
  crossing properties, 982
    (V, A) case, 989
    positivity domains,
      983, 986
  elastic neutral form
    factors, 1042
  elastic reactions, 1017, 1052
    cross sections, 1020
    electron target, 1026
    kinematics, 1017, 1040
    models, 1032
    nucleon target, 1019
    (V, A) case, 1021, 1028
  evidence for, 1051
  hadronic, 995
  inclusive reactions, 1037
  internal symmetry, 995, 1011
    Beg-Zee model, 1006
    Mathur-Okubo-Kim model,
      1006
    model I, 1001
    model II, 1004
    model III, 1005
    model IV, 1006
    model V, 1007
    model VI, 1008
  investigations, 897
  isoscalar and isovector
    components, 996
  isospin structure, 996
  isotopic spin, 995
  isovector assumptions,
    988, 1003
    inclusive, 999
  leptonic, 775
  Lorentz structure of, 969
    differential cross
      section, 970
  polarization, 975
  scaling assumptions,
    976, 1049, 1050
    averaged hadronic energy,
      977
    positivity domain, 977
    pure (S P) case, 979
    pure (V, A) case, 980
  second class, 1049
  symmetry, 967
  theoretical models, 776
  theories of, 967-1052

Neutral pion lifetime experiment,
  943
Neutral semileptonic currents,
  775
Neutrinos
  angular distribution of, 902
  Caltech-Fermilab experiment,
    897-927
  charged current interactions,
    913
  coupling, 902
  deep inelastic scattering,
    878, 910, 1021
  dimuon production, 756, 905
  Gargamelle energy, 893
  in black holes, 734
  missing energy, 900
  neutral currents induced by,
    967
  new particle production by, 865
    simple quark-parton model,
      879
  y anomaly, 881
  reactions
    events of ($\Delta S = \Delta Q$), 759
    new effects in, 754
  relation to antineutrino, 903
  scaling, 924
Neutrino-antineutrino
  annihilation channel,
    1038
Neutrons
  elastic scattering, 677
  magnetic moments, 257
Neutron dissociation, 663-702
  angular distributions, 684
  Coulomb excitation, 664, 666,
    671, 673
  (p π) system, 690, 700
  Deck mechanism, 685
  diffractive excitation, 664,
    668, 671
  from hydrogen, 680
  helicity amplitudes at low
    mass, 698
  using nuclear targets, 667
Neutron-nuclear total cross
    sections, 676
Nielsen-Olesen confinement,
  138, 391
Niu event, 836

INDEX

Nuclei, de-excitation of, 1051
Nucleons, 14, 426
  formation of, 23
  structure, 709
  muons as probes of, 589
Nucleon-nucleon collisions, 929-965
  direct lepton production in, 929
  measurement, 930
  electrons from, 939
    calculation of yields, 954
    CERN-Columbia-Rockefeller-Saclay experiment, 942
    Columbia-FNAL experiment, 942
    measurement, 933
    neutral pion lifetime experiment, 943
  J resonance from, 952, 956
  mesons from, 949
  muons from, 936
    Argonne-Wisconsin experiment, 939
    Chicago-Harvard-Pennsylvania-Wisconsin experiment, 939
    Chicago-Princeton experiment, 938
    Chicago-FNAL experiment, 939
    measurement, 930, 959
    polarization, 938
    Serpukov experiment, 936
    Yale-Brookhaven experiment, 941
  ρ and ω production, 949
  ϕ production, 951

Octet enhancement, 491
Okubo-Zweig rule, 804, 822, 846, 859, 860
  evidence for, 812
  hidden charm and, 812
  trajectories and, 861
Omitron, 704
One space one time dimensional models, 13, 17, 172

P states, 53, 826
$1p_1$ states, 57
$3p_j$ states, 54
  decays, 56
Particles
  identification of, 638
  in a potential, 376
  interactions between, 308
  lumps as, 307
  new, 27
    charm scheme, 810
    couplings, 808
    decays, 805, 825
    experimental facts, 804
    mixing and gas problem, 817, 820
    P states, 826
    production of, 865, 889
    review of, 741
    search for, 747
    theoretical models, 803-863
    weak vector boson W^0, 806
  production yields, 587
  ψ J, 559, 741, 803
    as handron, 741
    decay, 830, 851
    decay rates and widths, 743
    discovery of, 559
    origin of, 570
    position of the η_c, 823
    production of, 587
    search for, 560
    strong decay, 831, 832
    unstable states, 81
    zero width of, 741
  ψ', 851
  Y, 886, 894, 916
    kinematics and properties, 884
Pauli-Luders theorem, 30
Peierls transition, 1152, 1153
Pennsylvania-Stony Brook experiment, 943, 948, 963
Phase shift analysis, 423, 455
Photons, 285
  collinear, 647
  decays, 846, 862
  definition, 653

Photons (cont'd)
  exchange, 668
  in black holes, 734
Photoproduction, 766, 808
Pions, 187, 600, 905, 1115
  decay of, 590, 661
    muons from, 930
  role of, 157
  yield from nucleon-nucleon collisions, 938
Pionization of matter, 706, 711
Pomeranchuk-like theorem, 441
Pomeranchuk exchange, 668
Pomerons, 228, 528, 1131
  primeval, 222
Pomeron-like excitations, 127
Pomeron poles, 795
Positrons, 361
Potential, particle in, 376
Primakoff effect, 668
Protons, 785
  in black holes, 739
  magnetic moments, 257
Proton constituents
  fractional charge structure, 769
Proton-neutron quark mass difference, 281

QED checks, 771, 772
Quantum chromodynamics, 264
Quantum field theory, 13
  renormalization of, 18
  two dimensional, 293
Quantum lumps, 350
  construction of, 351
Quarks, 13-32, 1061
  See also Antiquarks, Massive quark model, etc.
  allowed orbits, 200
  angular momentum, 158, 166
  anticolour and, 31
  as carriers of space time and quantum number characteristics, 195
  basic assumptions, 21
  basic types, 36
  becoming fermions, 236
  binding, 14, 505
    See also Strings

Quarks (cont'd)
  binding forces, 50
  Bjorken scaling, 235
  changing concepts of, 20
  charmed, 73
  coloured, 21, 143
    See also Colour
  colour gauge theory, 69, 79
  combined with strings, 23
  confinement, 194, 234, 258, 470, 494, 503, 795
    colour interaction and, 147
    mechanism, 508
    permanent and approximate, 144
    theories, 144
  confinement schemes in field theory, 143-191
    coupling, 174, 176, 177, 178, 181, 182, 190
    deformed bags, 159
    MIT bag, 169
    properties, 182
    quantum corrections, 172, 191
    SLAC bag, 148
  connected on a vertex, 136
  constituents, 528
  continuum models, 75
  current and consistent, 530
  decoupling from weak interactions, 34
  dynamics, 50, 66, 80, 137, 193
  elusiveness of, 234
  evidence for, 35, 143
  excitations, 31, 128, 259
  exotic states, 156
  on lattice, 76
  Feynman rules, 82, 83, 91
  four model, 36
  free
    behaviour, 104, 105
    in MIT bag, 243
  gluo-electromagnetic interactions, 248
  gluo-magnetic moment, 253
  Green's functions, 196
  hadrodynamics with, 193-240
  in GIM model, 484
  in MIT bag, 242

INDEX

Quarks (cont'd)
  interactions, 241, 247, 254
    energy between, 249
    with bags, 145
  invisibility of, 23, 144
  lines, 92, 143, 214
  loops, 140, 214
    behaviour of, 109, 110
    in lattice gauge theory, 506
  mass, 63, 128, 129, 189, 258
  mass relations, 51
  movement of, 150, 154, 170
  multi-bound states, 154
  Okubo-Zweig rule and, 861
  on lattice, 69-142
    angular momentum, 75, 129
    Bjorken scaling, 106
    exotic states, 76
    gauge transformations, 108
    higher order gauge field contractions, 118
    propagators, 89
    quantum operators, 78
    renormalization, 135
    scalar particle spectrum, 86
    scattering by exchange, 82
    single particle states, 74
    static theory, 71
  original hypothesis, 20
  potential between antiquarks and, 263, 818
  potential energy of system, 147
  propagate, 201
  quantum numbers, 38, 193, 196
  Regge behaviour, 196, 206, 209
  relative time problems, 235
  renormalization, 110, 138
  rescattering, 557
  scattering by exchange, 82
  shell model, 1056, 1058
  space-time description of, 199
  species of, 21
  strange, interactions, 254
  string theory, 96
  wave function, 276

Radiotherapy, 704, 708, 711
Rayleigh-Ritz method, 367
Regge
  triple failure, 791
Reggeization of gauge theory, 495, 525
Regge states
  G_t contribution and, 222
  of quarks, 196, 206, 209
Relative scaling, 791
Renormalization, 18, 103, 135, 141
Resonances
  J (ψ), 609-634, 641, 646, 659
    coupling, 627
    decays, 617, 621, 625, 634
    isospins, 620
    measured cross sections and decay widths, 611
    production of, 952, 956
    properties of, 614
    spin and parity, 614
  ψ, 642, 659
  colour models for, 840
  decay, 1133
  narrow, 609, 747, 803, 849
    existance of, 570
    search for, 572, 652
  P_c, 635
    cascade display, 650
    decay modes, 650, 652, 653
    $\gamma\gamma$ final states, 647
    hadronic decay, 645
    identification, 659
    inclusive spectra, 640
    particle ratios, 640
  Roper, 1064
Roper Resonance, 1064
Rotational invariance, 32

Salan-Weinberg gauge theory, 1036
Scalar fields
  free massless, 383
  in one space one time dimension, 301
  theory, 141, 311
    time independent solutions, 310
    three dimensional, 392
Scalar particles, 17, 133
  spectrum, 86

Scaling
  See also Bjorken and
      Feynman scaling
  breakdown of, 780
  relative, 791
Scattering
  deep inelastic, 910
    by neutrinos, 878
  lepton, 602
  muon-proton, 597
  of leptons by hadrons, 593
Scattering amplitudes
  analysis as one energy, 436
  analyticity, 432, 445
  angular momentum, 459
  diffraction peaks, 444
    geometric scaling, 449, 460
  elastic, 425
  elastic unitarity, 427
  inelastic unitarity, 428
  phase and modulus of, 423
  Pomeranchuk-like theorem,
      441, 463
    restraints on, 462
  using various energies and
      angles, 439
Scientific research, 4
Serpukov experiment, 936
Shock waves, 710, 711
Short distance singularities,
      467, 473, 474, 475,
      483, 485, 489
Sine Gordon theory, 419, 518
  as Thirring model, 416, 419
  equation, 302, 309, 315, 350,
      359, 367, 370, 372,
      385, 413, 515
  lumps, 305
  qualitative and quantitative
      information from, 388
  soliton/meson mass ratio, 389
  topological charges in, 410
SLAC bag, 148, 391
  Bjorken scaling, 189
  boson coherent states, 152
  definition of Hamiltonian,
      152
  exotic states in, 156
  Fermion states and Bogoliubov
      transformations, 153

SLAC bag (cont'd)
  Fock space, 151
  magnetic moments, 156
  relation to MIT model, 169,
      170, 185
  surface area, 158
SLAC-LBL magnetic detector, 609
Solid State physics, 710
Solitons, 299, 309, 372, 409, 415
  See also Lumps
  annihilation operators, 517
  as lump, 389
  concept of, 415
  creation operator for, 418, 517
  finite length, 512
  introduction of, 525
  oscillations, 373, 375
  scattering, 525
  secret of, 385
  stability of, 525
  strong coupling classical
      vortex, 509
  vortex, 521
    energy-density, 511
  vortex line, 417
Soliton-antisoliton doublet, 373,
      375, 388, 419
Soliton problem, 414
Space, 4
SPEAR, 631
Statistical bootstrap model,
      733, 735
Strangeness and charm conserving,
      482
Strings, 14, 347
  breaking of, 30
  coloured, 22, 29, 78
    continuity of, 30
    mass of, 24, 26
  colour gauge theory, 79
  combined with quarks, 23
  dual model, 16, 142, 236
  dynamics, 80
  Feynman rules, 82, 83, 91
  gauge theory and, 16, 26, 79
  mass, 73, 110, 128
  multiple, 127
  on lattice, 69-142
    angular momentum, 75, 129
    colour gauge theory, 79

INDEX

Strings (cont'd)
  on lattice (cont'd)
    continuum limit, 85, 91,
        103, 113, 127, 130,
        132, 135, 136, 138,
        141
    gauge transformation, 108
    higher order gauge field
        contractions, 118
    number present, 106
      quantum mechanics, 73
    quantum operators, 78
    quarkless states, 76
    renormalization, 135
    single particle states, 74
    static theory, 71
    transition operators, 81
  path integral, 96
  relation to monopoles, 347
  representations, 106
  spin of, 132
  transition operators, 81
  types of, 29
SU(3), mass formulae, 67
SU(3) properties, 39
SU(4) multiplets
  structure of, 37
  supermultiplets, 36
SU(6) symmetry
  in MIT bag, 242
  quark masses, 73
Superconductors, 326, 412,
        1151, 1156, 1158,
        1164, 1165
Symmetry
  breakdown, 302, 318, 529
Synchrotron
  AGS parameters, 718
  Bevalac parameters, 722
  heavy ion, 712

Tachyons, 520
Thirring model, 381, 385, 416,
        419, 515, 518
  massless, 418
  mesons in, 387
Time, cosmic, 4

Topological conservation laws,
        312, 348, 412
  basic principles, 312
  finding, 321
  in sine-Gordon theory, 410
  in three dimensions, 322, 329,
        331, 336
  in two dimensions, 321, 325,
        331
  non-dissipative solutions and,
        315
  structure of, 342
Topological stability, 411
Two body bound state, 661

Ultraviolet divergences, 363

Veneziano model, 209

WKB formula, 373
Weak currents, 969
Wick's theorem, 365, 384
Winding numbers, 326, 411, 1156
Wormholes, 270, 289

Yale-Brookhaven experiment, 941,
        956, 964
Yang-Mills theory, 135
Yunnan event, 1144

Zweig's rule, 65, 136, 158, 159,
        238, 289, 859